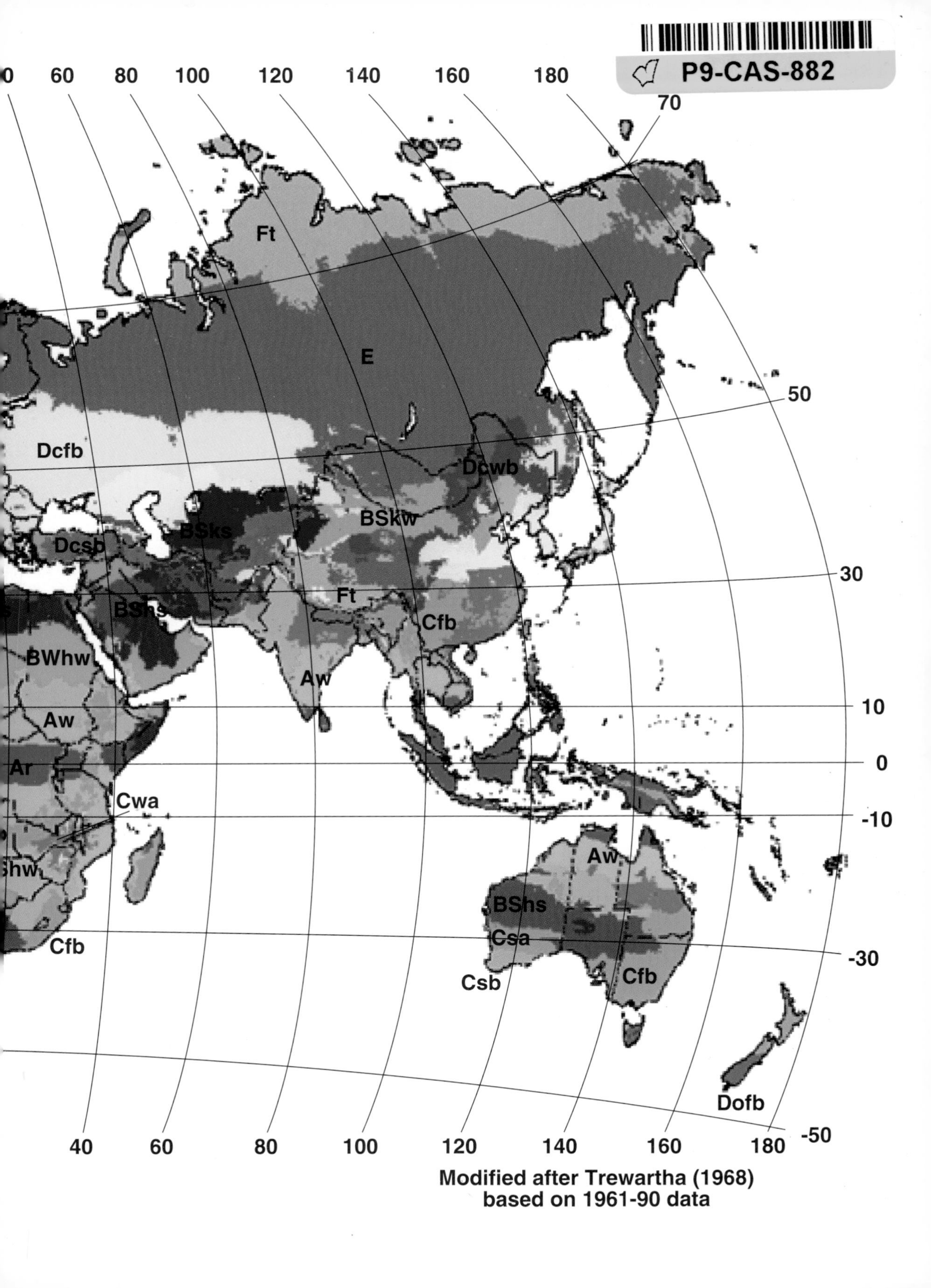
P9-CAS-882
Ft
E
Dcfb
Dcwb
BSkw
BSks
Dcsb
Ft
BShs
Cfb
BWhw
Aw
Aw
Ar
Cwa
Aw
BShs
Csa
Cfb
Csb
Cfb
Dofb
Modified after Trewartha (1968)
based on 1961-90 data

TURFGRASS MANAGEMENT

Fifth Edition

A. J. Turgeon
Professor of Agronomy
Pennsylvania State University

PRENTICE HALL
Upper Saddle River, New Jersey 07458

Library of Congress Cataloging-in-Publication Data

Turgeon, A. J. (Alfred J.), 1943–
Turfgrass management / A.J. Turgeon. -- 5th ed.
p. cm.
Includes bibliographical references (p. 356) and index.
ISBN 0–13–628348–9
1. Turf management. 2. Turfgrasses. I. Title.
SB433.T84 1999
635.9'642--dc21 99-34063
CIP

Dedication

To Jean, my wife and best friend
Carolyn and Catherine, our children
Arthur and Beatrice, my parents
The numerous mentors I've been fortunate to have who shaped my life
And my students who have taught me so much

Publisher: *Charles Stewart*
Editorial Assistant: *Jennifer Stagman*
Production Editor: *Lori Harvey, Carlisle Publishers Services*
Production Liaison: *Eileen M. O'Sullivan*
Director of Production & Manufacturing: *Bruce Johnson*
Managing Editor: *Mary Carnis*
Production Manager: *Marc Bove*
Composition/Page make-up: *Carlisle Communications, Ltd.*
Cover Art Director: *Jayne Conte*
Cover Designer: *Miguel Ortiz*
Printer/Binder: *R.R. Donnelley Harrisonburg*

Printed in the United States of America

10 9 8 7 6 5 4 3 2 1

ISBN 0-13-628348-9

Prentice-Hall International (UK) Limited, *London*
Prentice-Hall of Australia Pty. Limited, *Sydney*
Prentice-Hall Canada Inc., *Toronto*
Prentice-Hall Hispanoamericana, S.A., *Mexico*
Prentice-Hall of India Private Limited, *New Delhi*
Prentice-Hall of Japan, Inc., *Tokyo*
Simon & Schuster Asia Pte. Ltd., *Singapore*
Editora Prentice-Hall do Brasil, Ltd., *Rio de Janeiro*

Contents

Preface

This book has been designed as a basic text for beginning students of turfgrass science and management. In covering the important features of turfgrass systems, interactions between and among system components, and principles of turfgrass management, it attempts to unlock some of the mysteries of turf and establish the role of cultural interventions for achieving specific objectives. Illustrations are used generously throughout the text to help students grasp concepts, processes, and relationships of importance in turfgrass systems. Each chapter concludes with a series of questions to test the reader's comprehension of the material.

The fifth edition of *Turfgrass Management* employs the same organization as earlier editions. The first chapter includes an introduction to turf quality, and it characterizes turfgrass management as the means by which turf quality can be sustained. The second chapter focuses on the turfgrass plant and how it grows and develops into a sustainable turfgrass community. An expanded treatment of metabolism has been included in this edition. The third chapter provides detailed information, including botanic descriptions, environmental adaptations, cultural requirements, and uses of turfgrass species, along with a climatic classification system helpful in determining where turfgrass species are adapted and a taxonomic scheme useful in determining where specific turfgrasses fit in relation to other members of the grass family. The fourth chapter deals with the components of the environment—atmospheric, edaphic and biotic—in which turfgrasses must grow, compete, and survive. An expanded treatment of traffic effects and management has been included in this edition. Chapters five and six cover the broad array of primary and supplementary cultural practices, respectively, for sustaining turf at desired levels of quality. The seventh chapter covers important aspects of turfgrass pest management, including those involved in the management of weeds, diseases, nematodes, insects, and large-animal pests. While the role of pesticides is emphasized in this chapter, the entire text is concerned with pest management to the extent that pest problems can be reduced or, in some cases, essentially eliminated by providing conditions that favor healthy turfgrass growth. Turfgrass propagation is covered in the eighth chapter. As many problems encountered in the management of existing turfs are directly attributable to improper establishment, the previous chapters set the stage for an enlightened discussion of propagation and its importance throughout the life of a turf. Finally, the ninth chapter attempts to bring it all together into integrated cultural systems for sustaining specific types of turf.

As with earlier editions, the fifth edition contains updated information and specific improvements based, in part, on feedback from many users of the text who were kind enough to share their thoughts and constructive criticisms with the author. I am especially indebted to those individuals who have made contributions to the first and subsequent editions. These include Joe Russo (ZDEX Corporation, Boalsberg, PA) for assisting in the development of the revised climatic map inside the front cover; Floyd Giles for his painstaking drawing and redrawing of the many illustrations in the text; April Pahl, Trudy Zohn, and Jennifer Cooney who also contributed illustrations; Judy Verbeke, who provided some of the leaf cross-section pictures from which illustrations in Figure 3.2 were drawn; and the reviewers of portions of the manuscript, including B.J. Augustin, R. Bacon, J.B. Beard, R. Boufford, P. Busey, A.E. Dudeck, R.E. Engel, T.A. Gaskin, V.A. Gibeault, R.L. Goss, G. Hamilton, M. Hendricks, D. Henley, D. Huff, K. Killian, R.B. Malek, C. Mancino, L. Marty, A. McNitt, W.A. Meyer, H.G. Myers, B. Nelson, R. Randell, B. Rehberg, P.E. Rieke, D. Rodrigues, M.C. Shurtleff, J.M. Vargas, D.V. Waddington, T.L. Watschke, and J.R. Watson. Finally, the assistance of my wife, Jean, in editing the manuscript is gratefully acknowledged.

A. J. Turgeon

CHAPTER 1

Introduction

Turfgrasses are plants that form a more or less contiguous ground cover that persists under regular mowing and traffic. An interconnecting community of turfgrasses and the soil adhering to their roots and other belowground organs form a turf. The terms turf and turfgrass are thus different in that one refers only to the plant community (turfgrass), while the other represents a higher level of ecological organization (turf) by including a portion of the medium in which the turfgrasses are growing. When the surface layer of a turf is harvested for transplanting it is called sod.

Turfgrasses are used for a variety of purposes (Figure 1.1). A utility turf exists primarily for soil stabilization. The binding effect of an interconnecting system of fibrous roots prevents erosion from wind and water. The protective cover of aerial shoots further stabilizes the soil while providing a cooling effect during warm weather. Along roadsides, turfgrasses absorb many toxic emissions from vehicles and, thus, have a cleansing effect on the air. Along airport runways, turfgrasses reduce dust to prolong engine life; at small airports, the runways themselves are often turfed.

Lawn turfs serve a decorative function: their uniform, green appearance enhances the beauty of a landscape. Lawns also provide inviting arenas for recreational activities and relaxation and offer relief from heat-absorbing roadways, buildings, and other structures.

Sports turfs provide enjoyment for participants and observers alike. Football, baseball, and soccer are just a few of the many sports played, for the most part, on turf. The cultural requirements of an athletic field turf are unique because of the intense traffic that compacts the soil and causes considerable wear of the turfgrass.

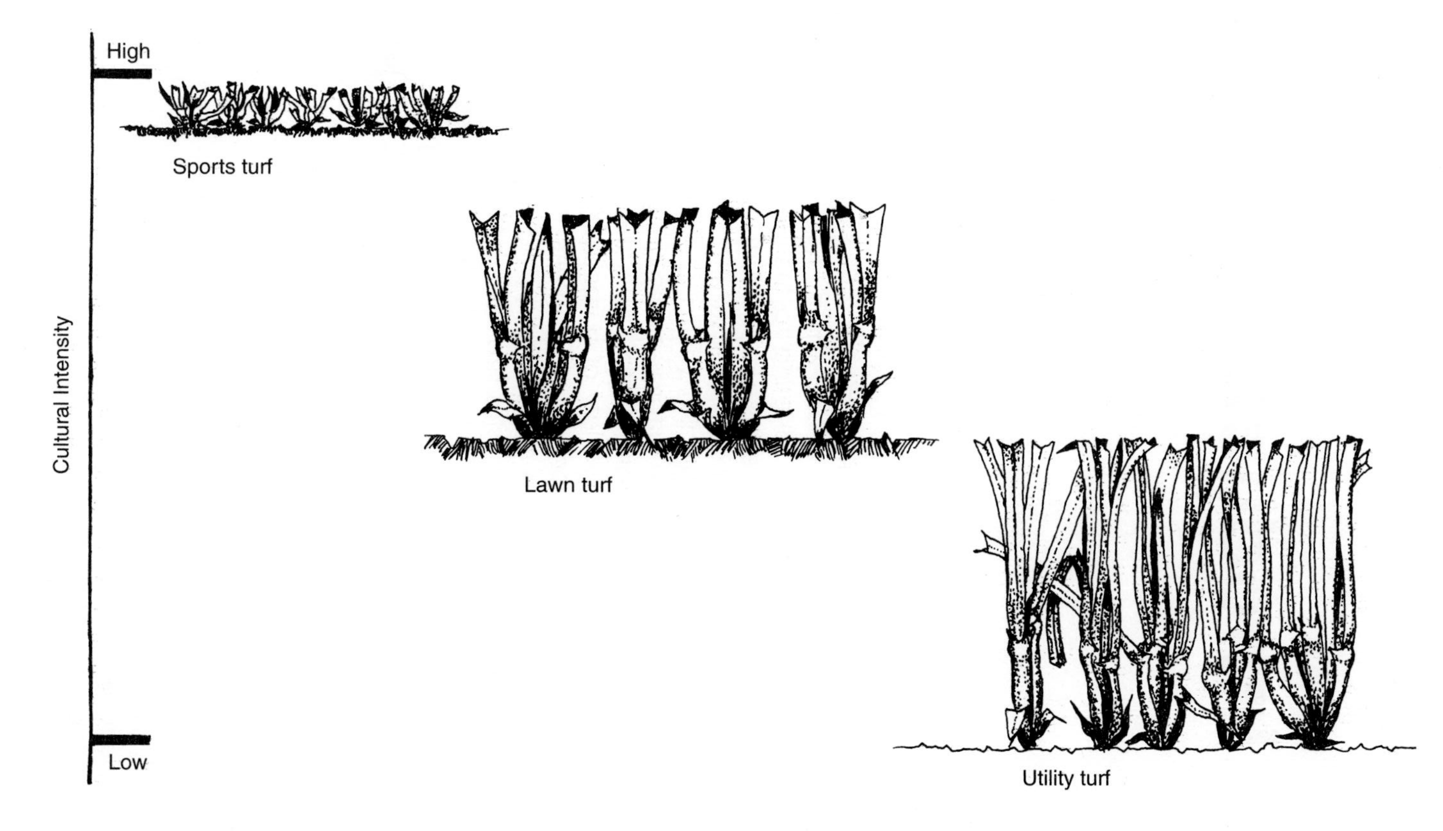

Figure 1.1. Turf types: sports turfs are cultured intensively to provide playable surfaces; lawn turfs serve primarily a decorative function; and utility turfs are established primarily for soil stabilization.

Among sports, golf enjoys the longest and closest association with professional turfgrass managers. The evolution of turfgrass science and technology is largely the result of efforts by golf course superintendents to improve techniques and solve problems. Local, state, and national organizations of golf course superintendents have, for decades, sponsored turfgrass research and encouraged academic institutions to develop turfgrass programs. Golf greens represent the highest intensity of turfgrass culture; they also present the greatest challenge.

CAREERS IN TURF

At one time nearly all formally trained persons entering the turfgrass professions became golf course superintendents. A relative few entered academic, business, or professional-service careers. Today the industry is composed differently (Table 1.1). A person with formal training in turfgrass management is likely to be found managing all types of turfgrass facilities, including athletic fields, parks, institutional grounds, and, of course, golf courses. The burgeoning lawncare industry has absorbed many trained personnel, and many producers and distributors of products for sale within the turfgrass industry are employing turf-trained personnel. The development and expansion of turfgrass programs

Table 1.1. Career Opportunities in the Turfgrass Industry

Grounds Superintendent	
Golf course	Government facility
Athletic facility	Transportation facility
Recreational grounds	Business facility
Institutional grounds	Residential complex
Manufacturing/Sales Representative	
Equipment	Seed
Fertilizers	Sod, stolons
Pesticides	Miscellaneous
Professional-service Contractor	
Consulting	Lawncare
Testing	Landscaping
Design	Construction
Technical Writer	
Trade magazines	Newsletters
Technical journals	Commercial papers
Scientist/Educator	
Academic institutions	
Industrial-research operations	

at academic institutions have created an increased demand for personnel with graduate degrees in turfgrass science.

The turfgrass industry has undergone rapid growth in its attempt to meet the public's increasing demands for products and services. Furthermore, professionally trained personnel are being sought for occupations that previously did not require formal training. This reflects the increasing technical sophistication of turfgrass management. A salesperson must not only know a product but must have a thorough understanding of the cultural system in which the product will be used. The lawn-service operator must understand the complexities of lawn culture, have diagnostic skills sufficient to recognize existing or potential problems, and be able to implement appropriate corrective measures. Sod and seed growers must be knowledgeable, not only in production techniques, but in how new turfgrass cultivars influence turfgrass quality and cultural requirements. The turfgrass facility manager must also be able to incorporate new technologies into cultural operations to sustain an acceptable level of turfgrass quality at the lowest cost. At the same time, he or she must be diligent in efforts to avoid any adverse environmental consequences.

People who would be successful in turfgrass careers must have appropriate technical credentials as well as the capacity to grow in response to technological, legal, and business developments.

TURF QUALITY

The quality of a turf is a function of its utility, appearance, and, in the case of sports turf, its playability during the growing season. A utility turf must be sufficiently rooted and persistent for soil stabilization. An ornamental (lawn) turf should be dense, uniform, and of pleasing color. A sports turf should provide the playability characteristics desired in a particular sport. Football fields should have firm footing, resiliency to cushion impact, wear resistance, and strong recuperative growth following injury. Golf fairways should afford suitable lies so that the ball is held atop the turf with no obstructions. Golf greens should provide sufficient ball-holding capacity for property directed approach shots and true putting to the hole from any position on the green.

The characteristics of each of these turfs vary widely; therefore, quality is related to function and subjective requirements. At fifty-five miles per hour, a uniform stand of tall fescue may appear quite attractive, but in a lawn turf this species may be considered inferior to Kentucky bluegrass because of its coarse texture and low shoot density. Kentucky bluegrass can be an excellent turfgrass for lawns, but it is unsuitable for greens. The extremely high density of a creeping bentgrass turf provides an excellent playing surface on greens, but in a lawn it tends to become puffy and unattractive unless it is sustained under intensive culture.

Visual Quality

Many factors influence turf quality. The most visible determinants of quality include density, texture, uniformity, color, growth habit, and smoothness (Figure 1.2).

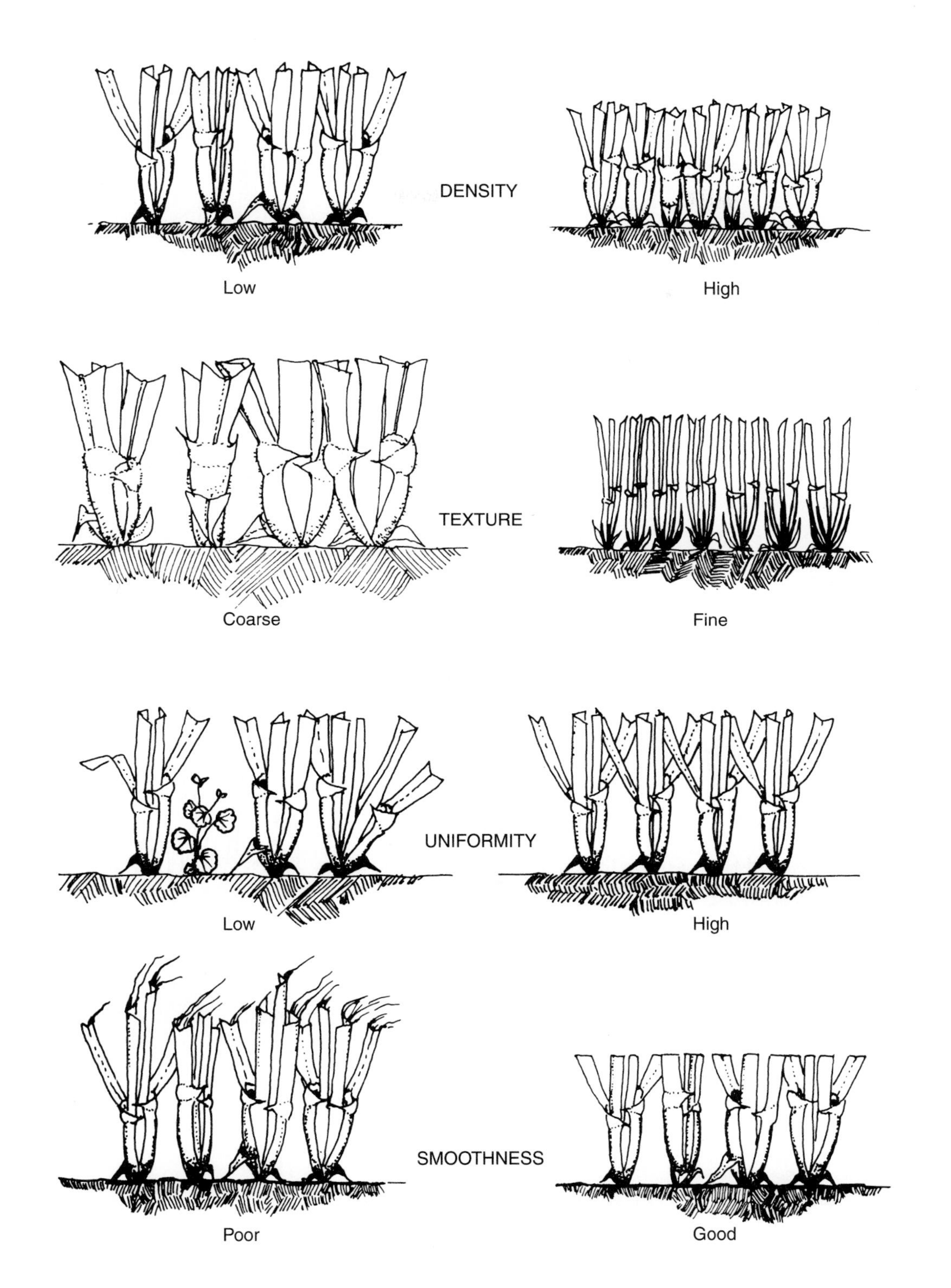

Figure 1.2. Visual characteristics of turf quality.

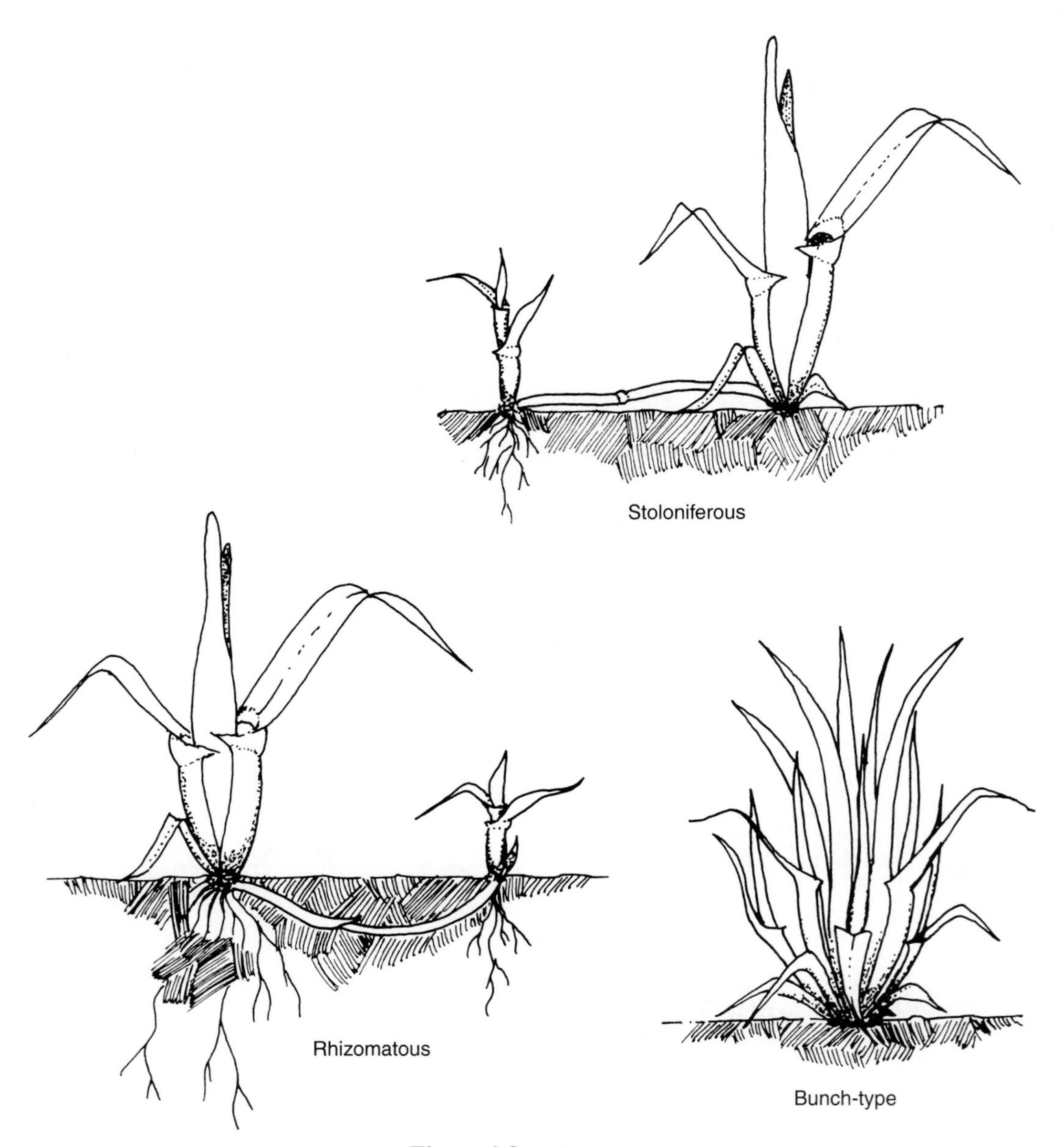

Figure 1.2. Continued

Density is a measure of the number of aerial shoots per unit area. It can vary with genotypic, natural environmental, and cultural factors. Highest turfgrass densities are obtainable with some bentgrasses and bermudagrasses, especially where they are closely mowed, receive abundant supplies of fertilizer and water, and are protected from disease-causing organisms, insects, and other pests. Within a turfgrass species, cultivars can differ widely in density; Pennlinks creeping bentgrass forms a denser turf than does Penncross, and a Midnight Kentucky bluegrass turf is typically denser than Kenblue when all are growing favorably under the same conditions.

Texture is a measure of the width of the leaf blades. Fine-textured turfgrasses, such as red fescue and rough bluegrass, have narrow leaves. At the opposite extreme are the coarse-textured tall fescue and St. Augustinegrass. Texture influences the compatibility of turfgrasses in mixtures. Fine- and coarse-textured turfgrasses are usually not planted together, as this would result in a nonuniform-appearing turf. Density and texture are often related features of a turfgrass in that, as density is increased, texture becomes finer.

Uniformity is an estimate of the even appearance of a turf. There are two dimensions to uniformity. One is compositional, dealing with the mass of aerial shoots; the other is a surface characteristic that deals with the evenness of a turf surface. Unlike texture and density, uniformity cannot be measured accurately; it is influenced by many features of the turf. Differences in texture, density, species composition, color, mowing height, and other features determine uniformity and, therefore, visual quality and playability of a turf.

Color is a measure of the light reflected by turfgrass. Different species and cultivars vary in color from light to very dark green. These differences are usually more apparent in the early and late portions of the growing season. Annual bluegrass and Kentucky bluegrass are often difficult to separate in summer, but are readily distinguishable in early spring when their color differences are most apparent. Old stands of Kentucky bluegrass may have a variegated appearance due to the presence of many different-colored genotypes of this species that have formed distinct patches.

Color is a useful indicator of the general condition of the plants. A yellow or chlorotic appearance may indicate nutritional deficiencies, disease, or some unfavorable factor influencing growth. Unusually dark color may be evidence of excessive fertilization, wilting, or the early stages of some disease. Mowing quality can also influence the color of a turf. Improperly mowed turfgrass with ragged leaf ends may appear gray to brown at the surface. Use of a sharp, properly adjusted mower can easily correct this problem.

Growth habit describes the type of shoot growth evident in a particular turfgrass. The three basic types are *bunch-type, rhizomatous, and stoloniferous.* Bunch-type turfgrasses spread primarily or exclusively by tillering. Where seeded at sufficient rates, they can form a uniform turf. However, at low seeding rates or where growth proceeds from isolated individual plants, small clumps develop, resulting in a nonuniform surface. Clumpiness in a turf is characteristic of perennial ryegrass, tall fescue, annual bluegrass, and other bunch-type turfgrasses.

Rhizomatous turfgrasses spread by belowground shoots called *rhizomes.* Because of the emergency of rhizome terminals at positions away from the mother plant, strongly rhizomatous turfgrasses tend to form uniform turfs with aerial shoots oriented in a more or less upright position. Differences in stem elongation and leaf orientation among cultivars influence turfgrass quality and close-mowing tolerance.

Stoloniferous turfgrasses spread by aboveground, lateral shoots, called *stolons.* Turfs formed from stoloniferous turfgrass may appear to have most aerial shoots oriented in a predominantly decumbent position. St. Augustinegrass turfs often appear this way, and creeping bentgrass sustained at relatively high mowing heights ($>$ 3/4 inch) forms a distinctly grainy turf with most shoots growing horizontally. Graininess varies among turfgrass cultivars; Arlington creeping bentgrass is especially prone to form a grainy turf, even under greens culture in which mowing heights are 1/4 inch or lower. Grain influences not only visual quality, but putting quality as well.

Smoothness is a surface feature of a turf that affects visual quality and playability. With improper mowing (i.e., dull blades), the leaf ends may appear ragged and discolored. In golf turf, putting quality is reduced where leaf ends are not smooth. The velocity and duration of ballroll are reduced where the turf's surface is not smooth and uniform.

Functional Quality

The functional quality of a turf is determined not only by some of the visual characteristics already discussed, but by other characteristics as well, including rigidity, elasticity, resiliency, yield, verdure, rooting, and recuperative capacity (Figure 1.3).

Rigidity is the resistance of the turfgrass leaves to compression and is related to the wear resistance of a turf (see Chapter 3). It is influenced by the chemical composition of the plant tissue, water content, temperature, plant size, and density. Zoysiagrasses and bermudagrass form very rigid turfs of excellent wear resistance. Kentucky bluegrasses and perennial ryegrass form less rigid and less wear-resistant turfs. Creeping bentgrass and annual bluegrass rank somewhat lower; rough bluegrass ranks very low. Softness is the opposite of rigidity. Given sufficient wear resistance, softness may be a desirable feature of some turfs depending on the intensity and type of use to which they are subjected.

Elasticity is the tendency of the turfgrass leaves to spring back once a compressing force is removed. It is an essential property of any turf since some traffic is inevitable with mowing and other activities. The elasticity of a turf is dramatically reduced when the turf is frozen. For this reason traffic should be withheld on frosted turfs during the growing season. Frost disappears and elasticity increases naturally as diurnal temperatures increase; however, the process can often be accelerated with an early-morning syringing.

Resiliency is the capacity of a turf to absorb shock without altering its surface characteristics. Some resiliency is contributed by the turfgrass leaves and lateral shoots. However, resiliency is largely a feature of the medium in which the turfgrasses are growing. Layers of thatch and thatchlike derivatives add substantially to the resiliency of a turf; soil type and structure are also important contributing factors. Golf greens should be sufficiently resilient to hold a properly directed approach shot. On football fields, resiliency reduces the potential for player injuries.

Ball roll is the average distance a ball travels upon being released to a turf surface. Mechanical devices may be used to release a ball at a consistent speed to obtain reliable measurements. An example is the "stimpmeter" used to measure ball roll on golf greens as an estimate of putting speed.

Yield is a measure of clippings removed with mowing. It is an indication of turfgrass growth as influenced by fertilization, irrigation, and other cultural as well as natural environmental factors. In experimental plots clippings are first dried, then weighed, to provide yield data. On greens, superintendents often measure yield in number of baskets

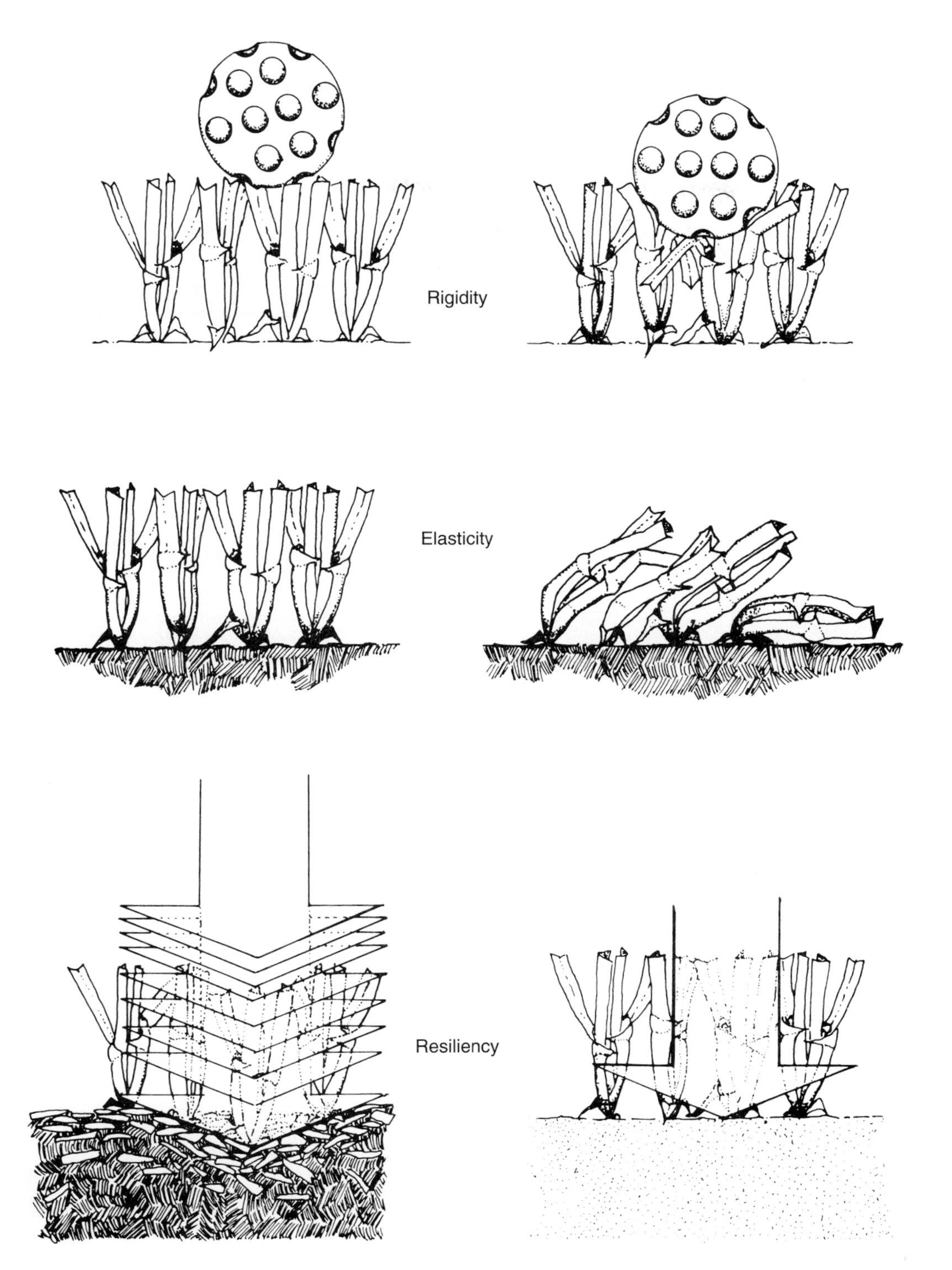

Figure 1.3. Functional characteristics of turf quality.

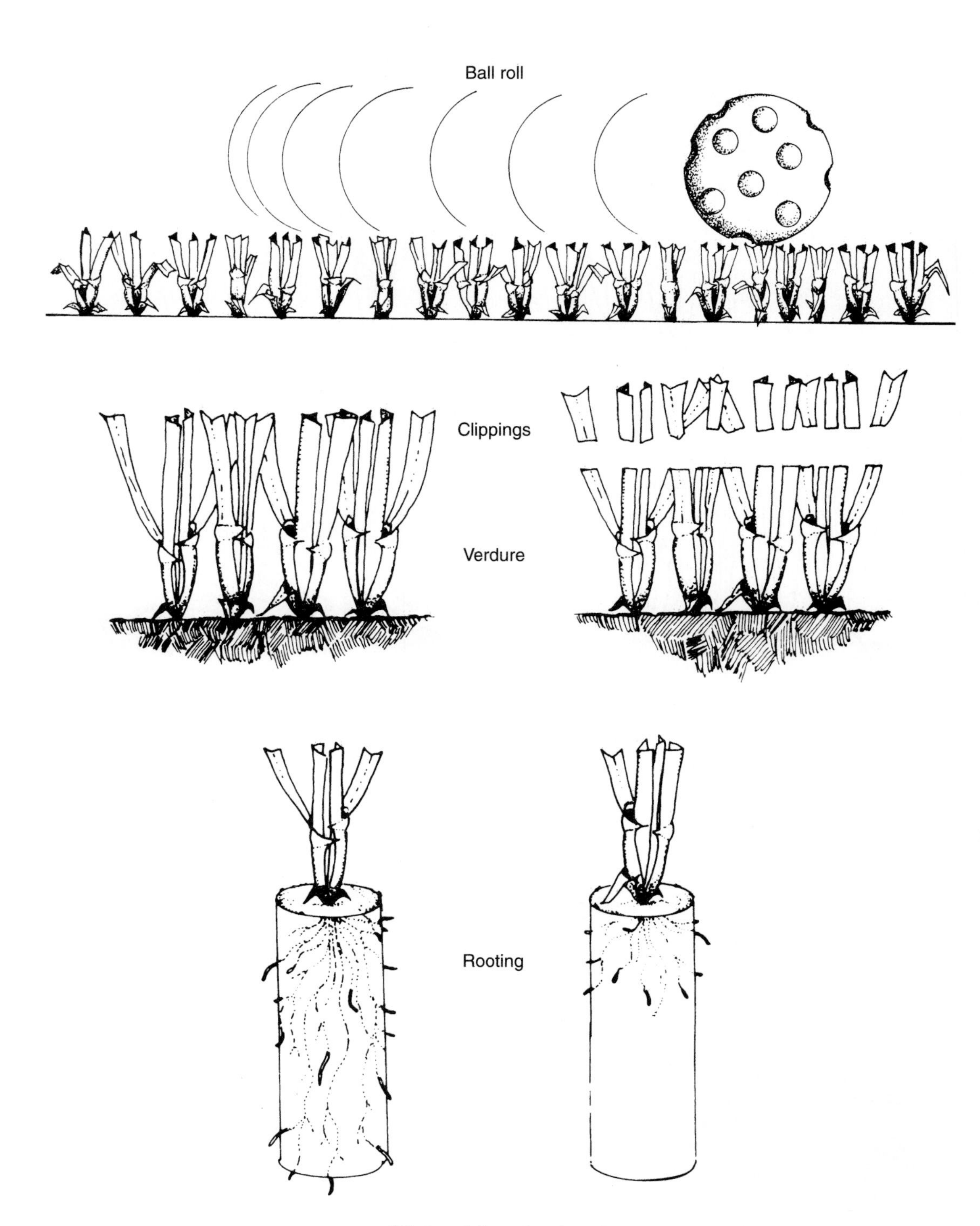

Figure 1.3. Continued

of clippings removed. The latter procedure is less accurate, but provides a quick estimate of vertical shoot growth. Since the objective in turfgrass culture is to sustain a satisfactory turf, and not to produce high yields of clippings, yield measurements do not provide a comprehensive assessment of turf quality. However, considered along with other criteria such as density, color, and rooting, they do provide an indication of the turf's response to culture and natural environmental conditions. Excessive use of fertilizers, especially nitrogen, can lead to excessively high yields accompanied by shallow rooting, reduced stress tolerance, and increased disease incidence and severity.

Verdure is a measure of the aerial shoots remaining after mowing. Within a particular turfgrass genotype increasing verdure is correlated with increasing resiliency and rigidity. At higher mowing heights the same genotype will often have more verdure and generally better wear resistance. With the same genotype and mowing height, verdue is directly proportional to density. Comparisons of the verdure of different genotypes under the same growing conditions indicate the relative competitive ability among genotypes that would exist in a mixed turfgrass community. In effect, the amount and appearance of plant material comprising the verdure over time strongly influence both the visual and functional qualities of a turf.

Rooting is the amount of root growth evident at any one time during the growing season. It can be estimated visually by extracting a turf core with a soil probe or knife and carefully working the soil free with the fingers to expose the roots. Numerous white roots extending to a depth of several inches indicate favorable rooting in a turf. Where rooting is shallow or largely confined to a thatch layer, problems can be anticipated, especially during stress periods. Important objectives in turfgrass culture, at least with cool-season turfgrasses, are to develop a strong root system during favorable conditions in the spring, to maintain as much of the root system as possible during the summer, and to generate new root growth in the fall. Cultural and natural environmental factors that influence root growth will be discussed in subsequent chapters.

Recuperative capacity is the capacity of turfgrasses to recover from damage caused by disease organisms, insects, traffic, and the like. Recuperative capacity varies with different turfgrass genotypes and is strongly influenced by cultural and natural environmental conditions. Factors that reduce turfgrass recuperative capacity include severely compacted soils, inadequate or excessive fertility and moisture, unfavorable temperatures, insufficient light, toxic soil residues, and disease. Generally, conditions that favor the growth of a turfgrass also favor its capacity to recover from injury.

TURFGRASS MANAGEMENT

Turfgrass management is defined as the range of activities, including cultural practices, for establishing and sustaining turf at a desired level of quality. Where turfgrass quality is below an acceptable level, it is usually due to mismanagement. Proper turfgrass management involves the following:

1. Selection of well-adapted turfgrasses
2. Acceptable establishment procedures

3. Proper mowing, fertilization, and irrigation practices
4. Proper cultivation and associated practices
5. Proper pesticide selection and use

Turfgrass culture is largely a matter of selecting plant genotypes compatible with a natural environment and then modifying the environment through cultural practices to promote the survival and desired growth of the plants (Figure 1.4). No amount of cultural expertise can guarantee the survival and turfgrass quality of bermudagrass in subarctic climates or Kentucky bluegrass in the Tropics, but a reasonably adapted turfgrass can provide a turf of acceptable quality with proper culture.

Inevitably, some conflicts arise between the culture of turf and its use. The often-heard remark by homeowners that "my neighbor's lawn looks better than mine, and he hardly does anything to it" is evidence of the interdependence of all components of a turfgrass cultural program. Turfgrasses grow best in well-aerated soils with adequate moisture and nutrients. However, many turfs are subjected to intensive traffic, which compacts the soil and causes extensive wear of the turfgrass plants. The demand for low-growing, dense turfgrass of rapid recuperative growth ultimately results in small plants that are less

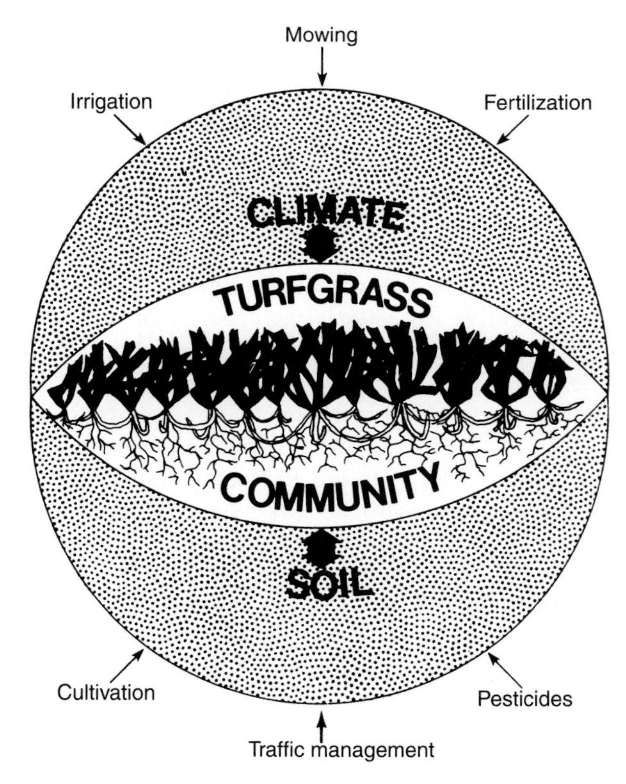

Figure 1.4. The essence of turfgrass culture is the modification of a natural environment in such a way that the turf is sustained at a desirable level of quality.

tolerant of stress and more susceptible to disease. Reducing mowing height to achieve higher shoot density requires increased use of fertilizers, irrigation water, pesticides, and cultivation methods. With each additional operation the potential for error increases.

Turfgrass management, then, involves determining the specific level of turfgrass quality desired and developing a comprehensive program of culture that will achieve and maintain that level of quality. As cultural intensity increases so do required levels of technical expertise and operational capability. Except where cultural intensity and expectations are relatively low, it is hardly a game for amateurs. The professional turfgrass manager blends technical and mechanical skills from practical experience, scientific knowledge from study and observation, and, in large operations, his or her ability as a personnel manager to achieve the desired objectives.

QUESTIONS

1. Differentiate between **turf** and **turfgrass.**
2. List and characterize three **turf types.**
3. List and explain six **visual** determinants of turf quality.
4. List and explain seven **functional** determinants of turf quality.
5. List five factors contributing to **turf mismanagement.**

CHAPTER 2

Growth and Development

Grass plants are composed of a complex array of leaves, stems, and roots that arise from seed and various vegetative propagules. The activity of these organs and the manner in which they form are subjects that merit intensive investigation by students of turfgrass management. The capacity to sustain a dynamic and complex turfgrass community depends, in part, on a thorough understanding of how turfgrasses grow and develop.

The turfgrass plant, a low-growing monocot, differs substantially in structure and growth pattern from typical dicotyledonous species. Tolerance to frequent defoliation by mowing and to traffic are unique features of turfgrasses. These tolerances exist because of the position of the growing point atop an unelongated stem, called a crown, located at or near the surface of the ground. Leaves continually arise from the growing point to provide a contiguous cover of green shoots. Because older leaves eventually fall to the surface of the ground and are replaced by newly emerging leaves, a relatively constant number of leaves per shoot is maintained. Under some conditions the growing point may undergo changes that result in the emergence of a flowering culm. This is eventually followed by the death of the shoot, since the growing point has changed morphologically and can no longer give rise to new leaf primordia. Axillary buds located at nodes along the crown develop into new tillers that emerge from within enclosing leaf sheaths. Alternatively, the axillary buds may give rise to horizontally growing shoots, called rhizomes and stolons, that burst through the enclosing leaf sheaths (if still attached) and grow outward from the parent shoot. Rhizomes and stolons can, in turn, give rise to new shoots at their terminals or nodes. Roots develop adventitiously from nodes along the crown and grow into the underlying soil or growth medium.

Therefore, an understanding of the turfgrass shoot, an integration of a shoot and its associated leaves, nodes, axillary buds, and flowers, is essential to learning how turfgrasses grow and develop. The components of a turfgrass plant and the processes involved in their life cycles will be discussed in detail in the following sections.

THE GRASS PLANT

The most obvious components of grass plants are the leaves that occur alternately on opposite sides of each shoot. The lower portion of the leaf, called the *sheath,* is tightly rolled or folded around the main axis of the shoot, while the upper portion, the leaf blade or *lamina,* is relatively flat and extends outward at an angle from the sheath (Figure 2.1). Thus, mature leaf blades appear as separate structures, while many emerging leaves are not entirely visible because they are enclosed within other leaf sheaths. At the junction of the blade and the sheath on the inner side of the leaf is the *ligule,* a membranous or hairy structure varying in size and shape. Opposite the ligule, on the outer side of the leaf, is a light green or whitish band called the *collar,* which varies among grass species. In some grasses the base of the leaf blade extends into two clawlike appendages called *auricles.* The ligule, collar, and auricles are important features in distinguishing different turfgrass species.

At the base of the leaves and partially hidden within the enclosing leaf sheaths is the *crown.* In the vegetative stage of growth, the crown is a highly compressed stem with a succession of nodes separated by very short internodes. Elongation of the internodes occurs during flowering, which signals a transition from vegetative to floral growth and development. A flowering culm emerges from within the enclosing leaf sheaths and terminates in an inflorescence.

Roots of a grass plant are of two types: *seminal* and *adventitious.* The seminal (also called *primary*) roots develop during seed germination. These survive for a relatively short period. Adventitious (also called *secondary* or *nodal*) roots arise from nodes along a stem, and in a mature turfgrass community these usually constitute the entire root system.

In addition to the crown and flowering culm, other stems of the grass plant include those associated with rhizomes and stolons. Rhizomes grow below the surface of the ground and give rise to new shoots at their terminals and nodes. Adventitious roots may also develop from the nodes. Likewise, stolons produce new shoots and adventitious roots; however, they differ from rhizomes in that they grow along the surface of the ground. In a new turfgrass plant from seed, a rhizomelike structure called the *mesocotyl* may be present. Where seeds germinate from a position below the surface of the ground, the growing point rises to a position closer to the surface via the mesocotyl. Where elongation of the mesocotyl occurs, the primary and adventitious root systems may be entirely separated from each other.

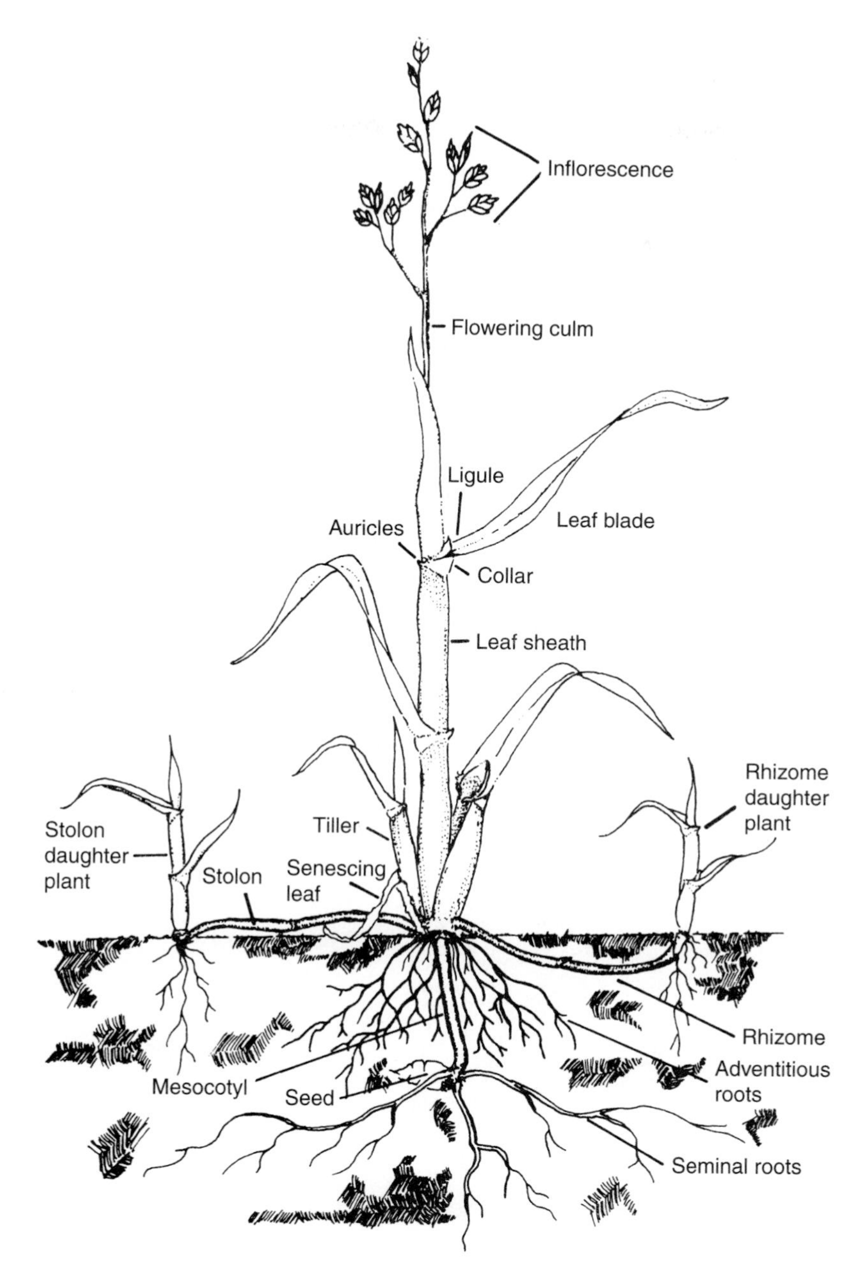

Figure 2.1. Diagram of the grass plant.

GERMINATION AND SEEDLING DEVELOPMENT

Mature florets harvested from the inflorescence of flowering grass plants constitute what is commonly referred to as grass seed. The floret is composed of a caryopsis sandwiched between two floral bracts called the *lemma* (outer bract) and *palea* (inner bract) (Figure 2.2). At the base of the palea is the rachilla, a short stemlike structure (Figure 2.3). The caryopsis, or dried fruit, contains the true seed surrounded by a pericarp (ovary wall). Just inside the seed coat is the aleurone layer, a thin proteinaceous material that plays an important role in germination. Also contained within the seed are the embryo, which is a miniature plant, and the endosperm, the ood supply for sustaining the plant during germination until it is capable of producing its own food through photosynthesis.

The germination process begins when water is absorbed (imbibed) by the seed. This initiates several biochemical and morphological events that ultimately result in the development of a seedling plant. Hydrolytic enzymes are produced that function in breaking down the starch in the endosperm to simpler carbohydrates for nourishing the embryo. These enzymes are produced in the aleurone layer in response to hormones

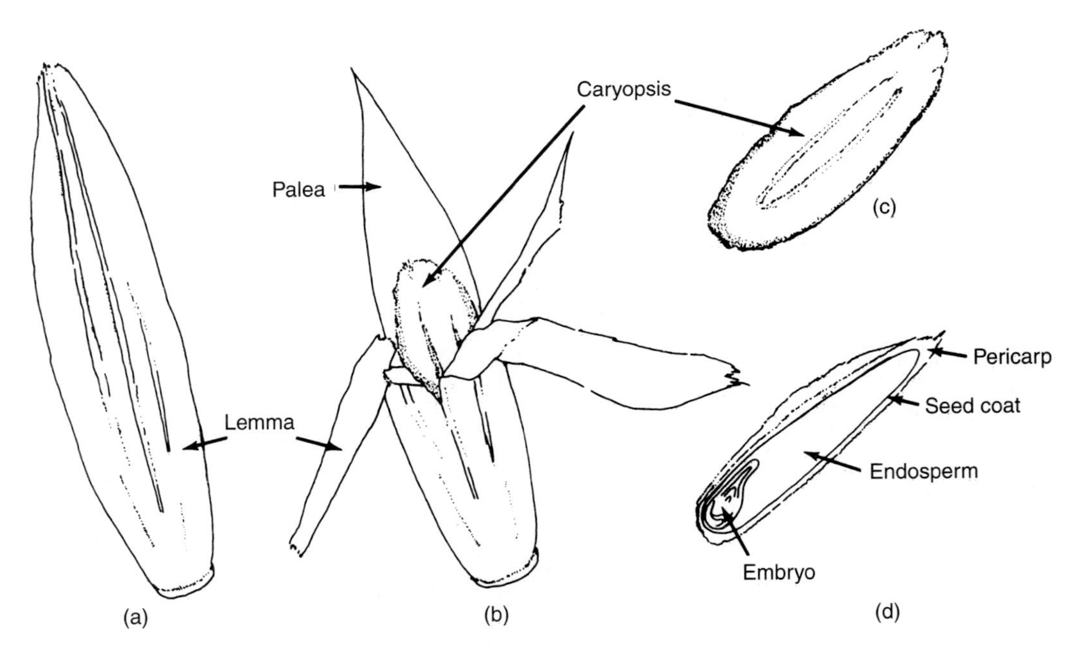

Figure 2.2. Components of the grass floret: (a) abaxial side showing the lemma, (b) lemma partially removed exposing the palea and caryopsis, (c) caryopsis, and (d) the true seed consisting of a seed coat, endosperm, and the embryo.

(gibberellins) produced in the scutellum. Carbohydrates from the endosperm are absorbed by the scutellum and transmitted to other parts of the embryo. All structural components of the seedling grass plant arise from the embryo.

The structure of the embryo is shown in Figure 2.4. The first morphological development evident during germination is the enlargement of the coleorhiza via cell elongation, accompanied by the emergence of root-hairlike structures from the coleorhiza that anchor the embryo to the soil and, presumably, function in absorbing water (Figure 2.5). Then the primary root (radicle) pushes through the side of the coleorhiza and penetrates downward through the soil. At about the same time, the coleoptile, a sheath of translucent tissue surrounding the growing point, emerges above the soil surface. In some grass species upward growth of the coleoptile may be associated with elongation of the mesocotyl, an internode located between the scutellar node and the coleoptile. The extent of mesocotyl elongation depends on the depth of the seed in the soil. While growth of the coleoptile is promoted by light, mesocotyl growth occurs in the dark and is inhibited by light. Thus, elongation of the mesocotyl initiates at the embryo and terminates at or near the soil surface, regardless of planting depth. In grass species lacking a mesocotyl, emergence of the coleoptile above the soil is entirely

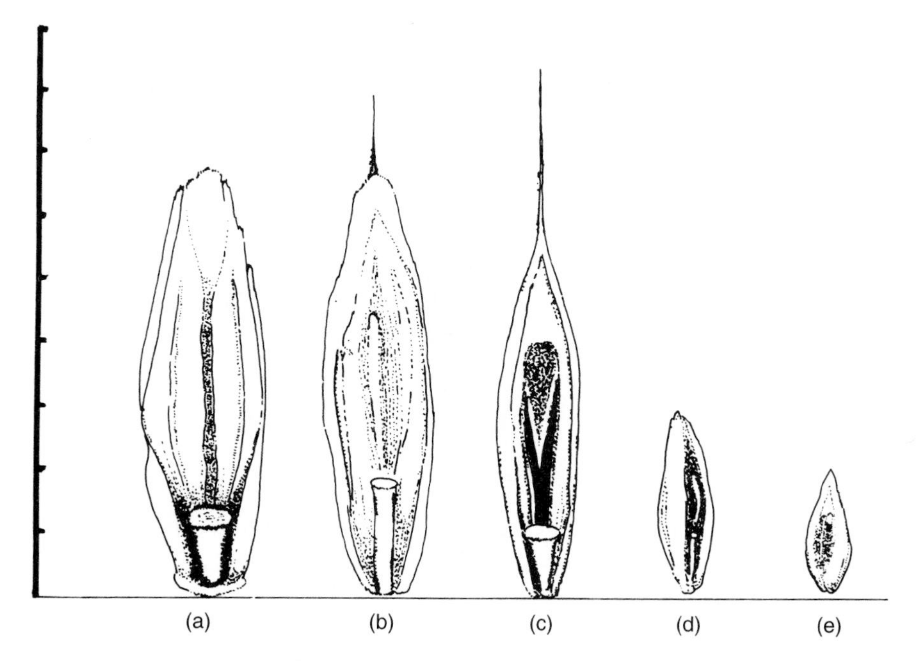

Figure 2.3. Comparison of the adaxial side of grass florets showing the palea and rachilla of (a) perennial ryegrass, (b) tall fescue, (c) red fescue, (d) Kentucky bluegrass, and (e) creeping bentgrass.

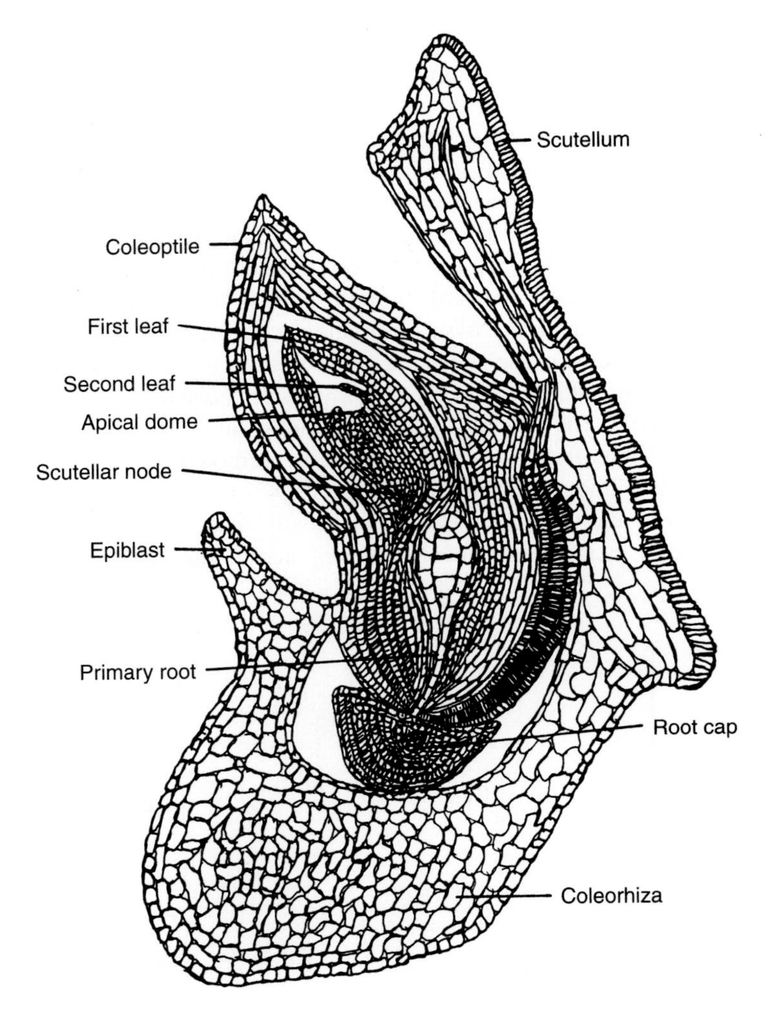

Figure 2.4. Typical grass embryo.

dependent on growth of the coleoptile itself, and these species may require a shallower planting depth for successful seedling establishment.

Within the emerging coleoptile the first leaf elongates and pushes out through a pore at the tip of the coleoptile. Photosynthetic activity begins, and soon the seedling becomes entirely independent of the endosperm for its food supply. At this point the seedling is said to be *autotrophic*. (A *heterotrophic* condition exists when the developing plant is entirely dependent on the endosperm for its food.) If seeds are buried too deep in the soil, food reserves in the endosperm may become depleted before the seedling is

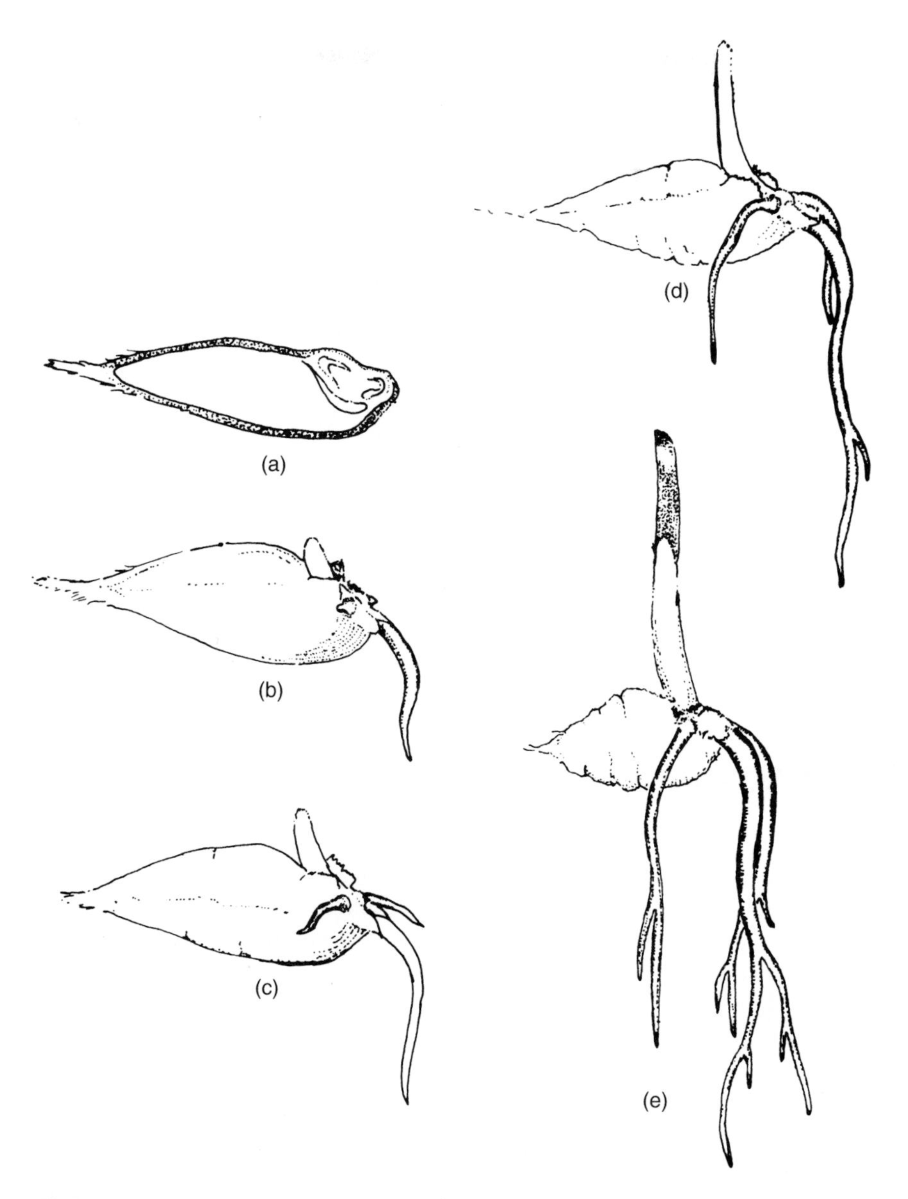

Figure 2.5. Process of germination, showing (a) caryopsis with embryo positioned at one end, (b) emergence of a primary root and coleoptile, (c) emergence of additonal seminal roots, (d) root branching, and (e) emergence of the first leaf through the top of the coleoptile.

capable of manufacturing all of its complex organic compounds through photosynthesis and associated biochemical processes. In this case the seedling dies.

The growing point of the seedling is enclosed within the coleoptile; consequently, the second leaf also grows through the coleoptile and within the fold (or roll) of the first emerging leaf. Eventually the coleoptile withers away, and only leaves are evident above the soil surface. Each succeeding leaf develops from the growing point and upward within the older enclosing leaves.

The next seedling structures to form are the adventitious roots, which develop from nodes at the base of the new shoot. Thus, two types of roots may be present in a newly planted turf. Eventually the primary root system dies, so in a mature turf the entire root system is adventitious.

Successful seedling development and subsequent survival are dependent on planting depth, available moisture, temperature, sufficient light, and the amount of food contained within the endosperm. Emerging seedlings are highly prone to desiccation since their capacity to secure moisture from the soil is limited by a relatively undeveloped root system and rapid water loss by evaporation from the surface soil. In heavily shaded environments, the seedlings may not receive sufficient sunlight for photosynthetic production of food in quantities necessary to sustain growth. Finally, where the endogenous food supply is insufficient to sustain the seedlings until they reach the autotrophic state, death may occur. This condition is frequently associated with deep planting of seed or with the use of older seed in which viability has been reduced.

LEAF FORMATION

Turfgrasses are well adapted to frequent mowing because leaf formation continues after each defoliation. As long as the plants continue vegetative growth virtually all meristematic tissues (those that contain cells capable of dividing to produce new cells) remain near the surface of the ground and below the mower blades. The growing point, located at the top of an unelongated stem (crown), continually forms leaf primordia, which eventually develop into fully expanded leaves. These appear initially as small protuberances just below the apical meristem (Figure 2.6). The number of leaf primordia visible at any time varies from a few to as many as twenty or more depending on species, plant age, and environmental conditions. Most turfgrasses have from five to ten leaf primordia present in various stages of development. The entire length of the growing point is usually less than 1 millimeter.

Leaf primordia arise due to cell division below the apical meristem. Rapid division of cells at the midpoint of each leaf primordium results in the formation of the leaf tip. Subsequent meristematic activity is restricted to the basal portion of the leaf primordium, establishing the intercalary meristem. Thus, two types of meristems are present in the growing point: the apical meristem that produces new cells to continue stem development at the top of the crown, and the intercalary meristem that produces leaves just below the apical meristem.

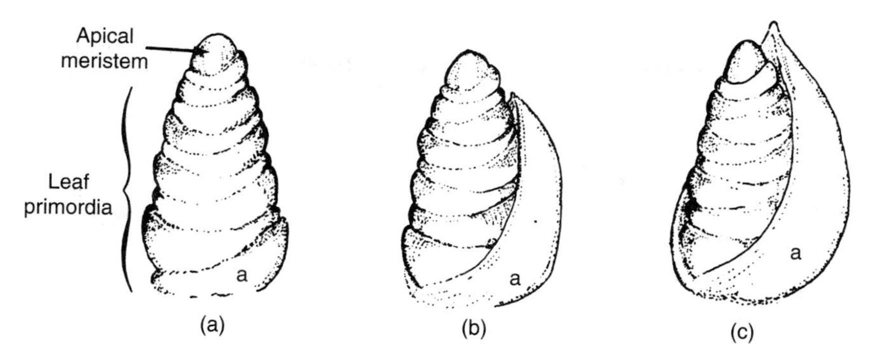

Figure 2.6. Leaf emergence from the growing point. The basal leaf primordium (a) grows around upper leaf primordia (a → c) in the process of forming a newly emerging leaf.

As the leaf primordium continues to develop, its intercalary meristem divides into two distinct meristems: an upper intercalary meristem that produces cells for growth of the leaf blade, and a lower intercalary meristem that remains at the base of the leaf to continue development of the leaf sheath. Cell division at the upper intercalary meristem usually ceases by the time the leaf tip emerges from the enclosing leaf sheaths. Further expansion of the leaf blade is due to cell elongation, primarily at its base. Meristematic activity at the base of the leaf sheath usually proceeds for some time after the leaf blade has been fully formed. Therefore, the oldest portion of a leaf is the tip, while the youngest is the base of the sheath. Leaf expansion may continue after a portion of the leaf blade has been removed by mowing. Following the emergence of a new leaf above the enclosing leaf sheaths, the new blade and sheath assume different shapes. The blade unfolds (or unrolls) to form a relatively flat structure, while the sheath remains in a folded or rolled configuration surrounded by older leaf sheaths (Figure 2.7). As newer leaves originate from higher positions along the crown, each succeeding leaf sheath occurs at a higher position than the next older leaf sheath.

Eventually a turfgrass leaf undergoes senescence (the plant growth phase from maturity to death), beginning at the tip and extending downward, and falls away from the shoot. As the number of leaves per shoot generally remains constant under a specific set of environmental conditions, the rate of new leaf emergence is approximately the same as the rate at which older leaves die. Measurements of the photosynthetic activity of grass leaves have shown that newly emerging leaves may use all the food they manufacture plus some additional photoassimilates from other leaves. Young, fully expanded leaves have the highest photosynthetic rate and contribute photoassimilates to various growing parts of the plant plus some for storage (carbohydrate reserves), principally in the crowns. Older leaves contribute little to the rest of the plant since their photosynthetic activity declines as they approach senescence. Prior to the initiation of photosynthetic activity, emerging leaves are totally dependent on carbohydrate reserves in storage organs and

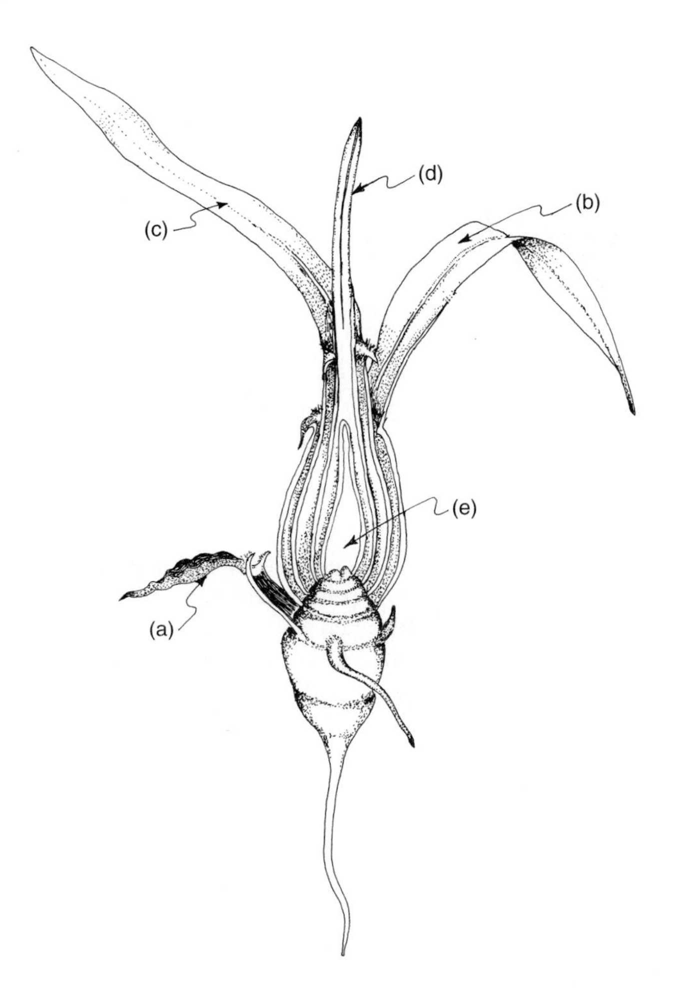

Figure 2.7. Organization of leaves within the grass shoot: (a) senescing leaf; (b) mature leaf; (c) fully expanded leaf; (d) emerging leaf; and (e) immature leaf enclosed within older leaves.

from other leaves. Hence, excessive defoliation from mowing may severely reduce turfgrass vigor. Leaf growth rate varies with age; the youngest leaves grow most rapidly, while the oldest leaves cease growing (Figure 2.8).

Vertical development of a leaf is generally coordinated with that of the next leaf in succession (Figure 2.9). As the tip of a leaf begins its upward movement within the shoot the next older leaf initiates sheath elongation. Subsequent expansion of the sheath occurs at approximately the same rate as that of the enclosed leaf blade. Thus, the growth of different morphological units of each leaf in a pair is synchronized so that little or no friction is generated until elongation of the enclosing sheath ceases.

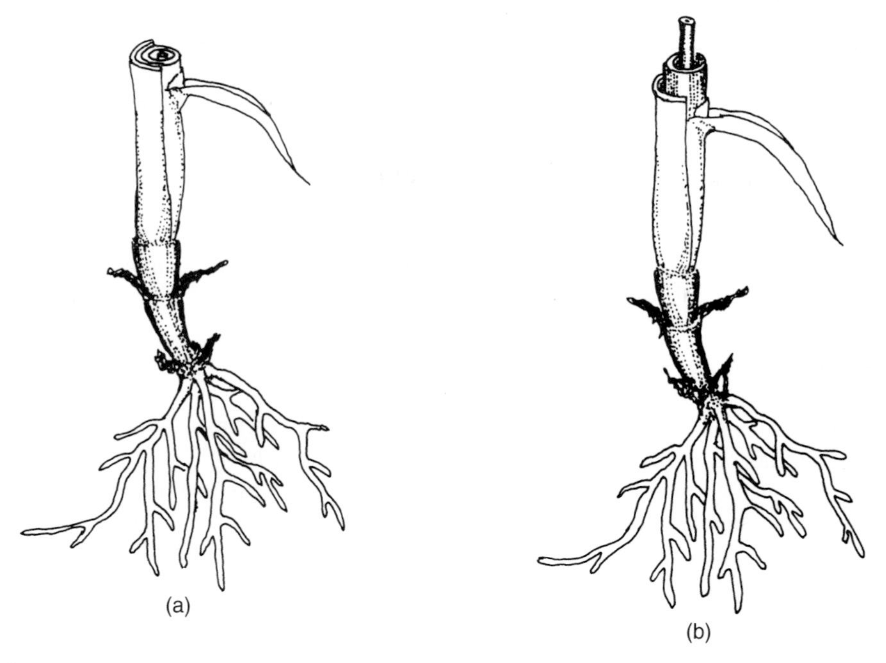

Figure 2.8. Grass shoots (a) immediately after mowing down to sheath height and (b) several days after mowing. Subsequent elongation of sheaths is age dependent; the youngest sheaths grow most rapidly, while older sheaths grow more slowly or not at all.

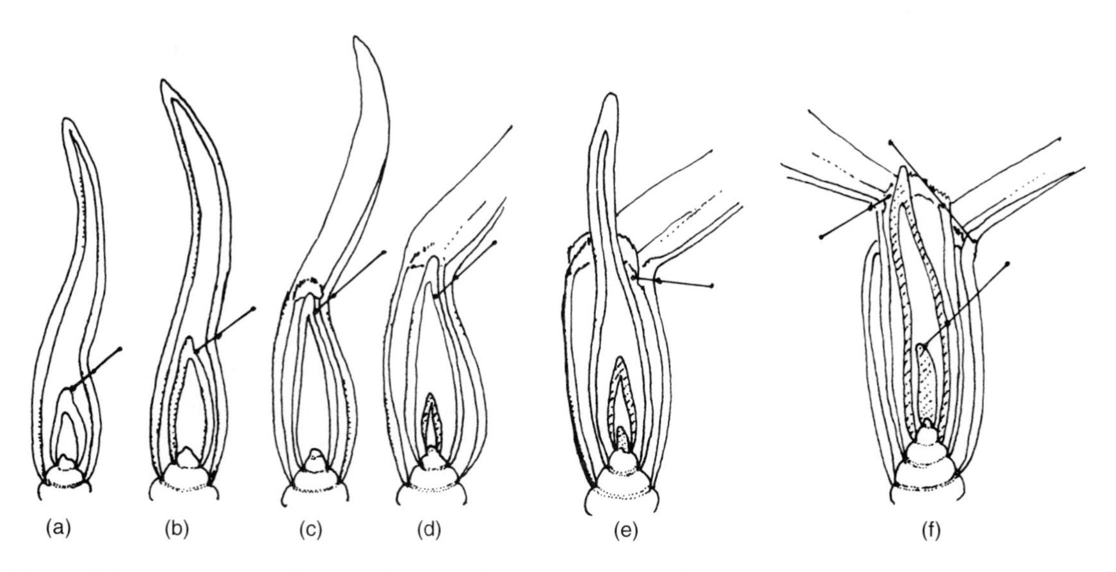

Figure 2.9. Elongation of enclosed leaf blades occurs at approximately the same rate as that of the enclosing leaf sheaths (a–d) until sheath elongation ceases (e). The enclosed leaf continues its upward growth through blade, then sheath, elongation. This process is repeated with subsequently formed leaves (f).

The rate at which new leaves appear varies among different turfgrass species and, presumably, different cultivars. Climatic conditions and fertilization practices also affect appearance rate. The time interval between the appearance of successive leaves is called a *plastochron* and is usually measured in days. The shortest plastochrons occur under optimum temperatures, high light intensities, high levels of nitrogen fertilization, and optimum soil moisture conditions.

TURFGRASS STEMS

Three principal types of stems occur in turfgrasses: the crown, flowering culm, and the lateral stems (associated with rhizomes and stolons). All but the crown have elongated internodes and are easily recognizable as stems. In contrast, the crown is a highly contracted stem with its nodes appearing to be stacked one on top of the other (Figure 2.10). Given its position at or below the surface of the ground and the fact that its upper portions (including the growing point) are entirely enclosed within the bases of several leaf sheaths, the crown is an elusive organ that is difficult to visualize or comprehend. Yet, the crown is a key organ giving rise to leaves, roots, tillers, and elongated stems of the turfgrass plant. Crowns also serve as storage organs for carbohydrate reserves to support the growth of new plant organs.

Crowns form wherever new shoots develop: from the embryo of germinating seed, from axillary buds and terminals of rhizomes and stolons, and from axillary buds on the crown that develop into new tillers. The highly contracted nature of crowns is what establishes the mowing tolerance of the several dozen grass species used for turf. Other stems of importance in turfgrasses are those associated with the horizontally growing rhizomes and stolons. These are elongated stems that arise from axillary buds on the crown. A newly developing lateral stem breaks through the enclosing leaf sheaths (if still present). This process is called *extravaginal branching.* During early internode elongation the entire stem segment between nodes may be meristematically active. As the internode continues growth, cell division becomes restricted to regions directly ahead of each node, forming the stem intercalary meristems.

Stolons grow along the surface of the ground and form roots and new shoots at the nodes. A new aerial shoot may also arise from the stolon terminal if the apex turns upward. Branching may occur at the nodes, forming a complex network of lateral stems. Stoloniferous turfgrasses include creeping bentgrass, rough bluegrass, and zoysiagrass.

Rhizomes grow beneath the surface of the ground and may be of two types: determinate and indeterminate. Determinate rhizomes are usually short and turn upward to form a new aerial shoot (rhizome daughter plant). Growth of these rhizomes occurs in three distinct phases: downward (plagiotropic) growth from the parent shoot, horizontal (diageotropic) growth, which accounts

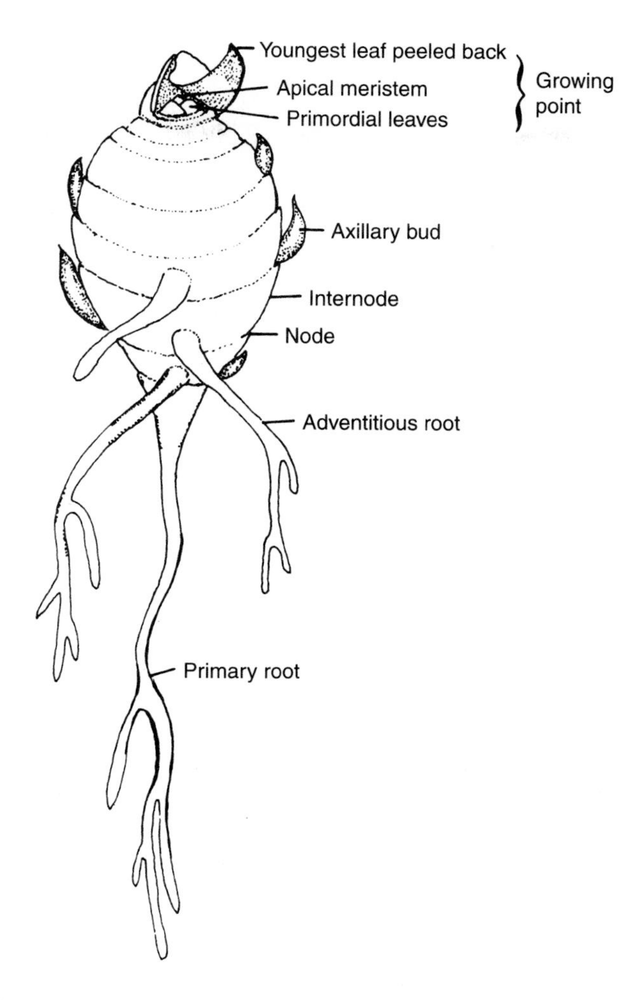

Figure 2.10. Grass crown and its associated leaf primordia, axillary buds, and roots.

for most of the elongation of the rhizome, and upward (negatively orthogeotropic) growth to a position near the surface, where light interception results in a cessation of internode elongation and the formation of a new aerial shoot. Turfgrasses having determinate rhizomes include Kentucky bluegrass, creeping red fescue, and redtop. Of these, Kentucky bluegrass is the most vigorous rhizome former. Its rhizomes grow with a boring-type (circumnutational) motion that aids penetration of compacted soils.

Indeterminate rhizomes are long and tend to branch at the nodes. Aerial shoots arise from axillary buds along these submerged stems. Bermudagrasses have indeterminate rhizomes and may also be stoloniferous, depending on the relative position of the lateral stems with respect to the surface of the ground. The extent of rhizome growth varies from

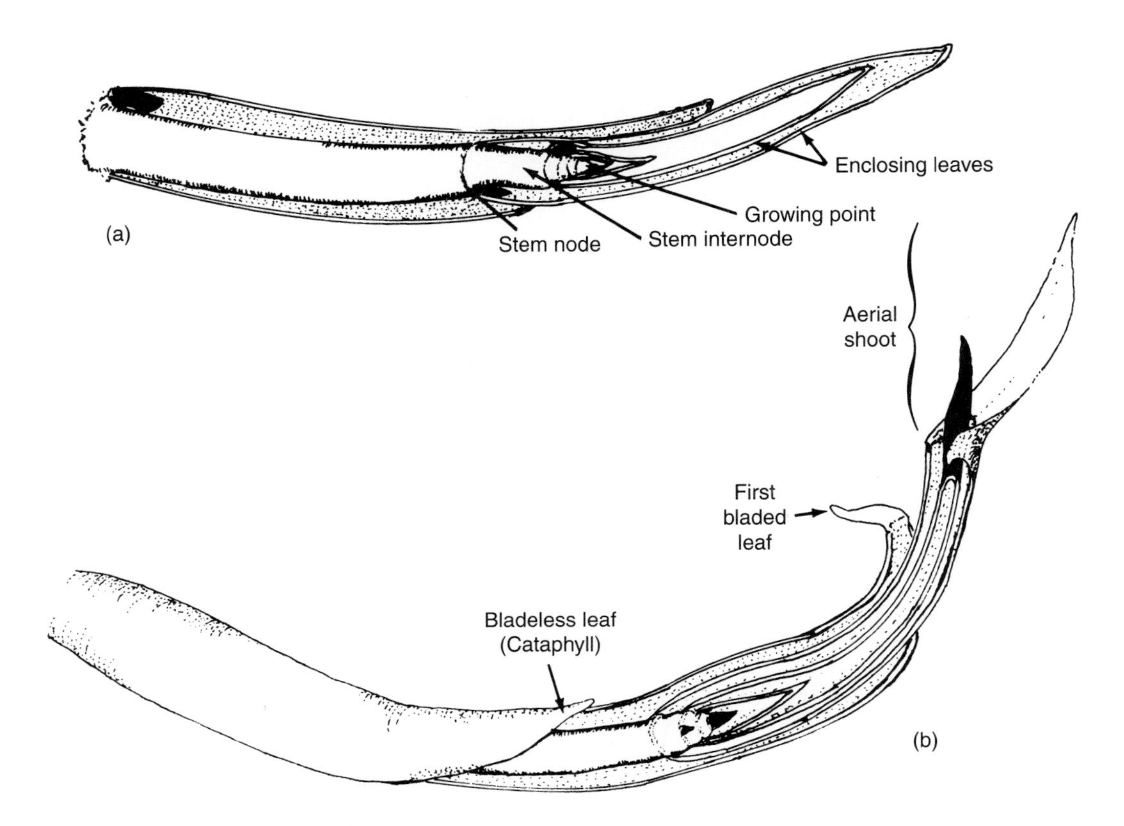

Figure 2.11. Rhizome growth is by leaf and internode elongation. The growing point is enclosed within several leaves in a fashion analogous to aerial shoots (a). The appearance of a bladed leaf signals the turning up of a rhizome to form an aerial shoot (b).

almost none (rhizome tip turns up almost immediately) to several inches or longer. Unlike roots, the rhizome does not merely add cells at the tip but, rather, grows in somewhat the same fashion as an aerial shoot. It is composed of alternately arranged leaves, a growing point, nodes, internodes, and axillary buds. The rhizome differs from an aerial shoot in that the internodes elongate and its leaves (cataphylls) are usually bladeless and scalelike in appearance (Figure 2.11). The rhizome tip is a small, conical leaf that encloses the growing point. On older parts of the rhizome these leaves hang loosely at each node and partially conceal the next internode. In the axil of each leaf is an axillary bud that can give rise to a branch rhizome or an aerial shoot. Adventitious roots may develop near the axillary buds. The rhizome tip is pushed through the soil by cataphyll elongation initially and then by internode elongation. Some cataphylls have very short blades at the tip. Blade development usually signals an upward turning of the rhizome and subsequent formation of an aerial shoot. When the bladed leaf reaches the light, elongation of the internode beneath it ceases. With successive leaf formation from the growing point, a new crown develops that is similar to that found in young seedlings.

TILLERING

The process by which new aerial shoots emerge intravaginally from axillary buds is called *tillering*. In contrast to rhizome and stolon emergence, tillers grow upward (apogeotropically) and within the sheaths of enclosing leaves (Figure 2.12). The result is a dramatic increase in the number of new shoots occurring immediately adjacent to the parent shoot. Considering the impact of tillering in conjunction with lateral growth of stolons and/or rhizomes, which also produce tillering shoots, we can visualize how an entire turfgrass community can eventually develop from a single seedling. Although turfgrass establishment from a single seed is never recommended, it is not necessary to seed so heavily that a contiguous seedling cover is achieved. In fact, a satisfactory turf can be developed from a relatively sparse stand of seedlings where rhizomatous or stoloniferous grasses have been planted. In an existing turfgrass community individual shoots eventually die and must be replaced by new shoots to maintain a desired density level. Turfgrasses are perennials not because individual shoots survive indefinitely, but because the plant community is dynamic, with dying members continually being replaced by new shoots and roots. The life of an individual shoot is usually not more than one year, and frequently it is less. Tillers formed in the fall are important to winter survival and spring regrowth of the turf but may die during summer. Tillers formed in the spring may, in turn, be important for summer survival. Under conditions of environmental stress the newly formed tillers are usually the first to die. Those tillers initiating inflorescences in spring usually die before the end of summer.

A young tiller depends on the parent shoot for photoassimilates until it has developed several leaves and an adequate root system. Although a mature tiller may appear to function as an independent entity, some relationship apparently exists between tillers interconnected by a common vascular system. Thus, a grass plant appears to be a highly organized system rather than a collection of competing tiller entities.

TURFGRASS ROOTS

The root system of grasses may include two types: the primary roots that develop from the embryo during seed germination, and the adventitious roots that emerge from nodes of the crown and lateral stems. Primary (seminal) roots usually do not live beyond the first year following planting. Adventitious (secondary, nodal) roots begin forming soon after the first leaf emerges from within the coleoptile following seed germination, and subsequent formation occurs from lower nodes of rhizomes and stolons. Although root initiation usually takes place at or below the surface of the ground, some root formation may occur above the surface (prop roots) in dense turfs where a favorable microclimate exists.

Figure 2.12. Intravaginal (tillering) (b–d) and extravaginal (rhizome growth) (e–g) branching from a parent shoot (a).

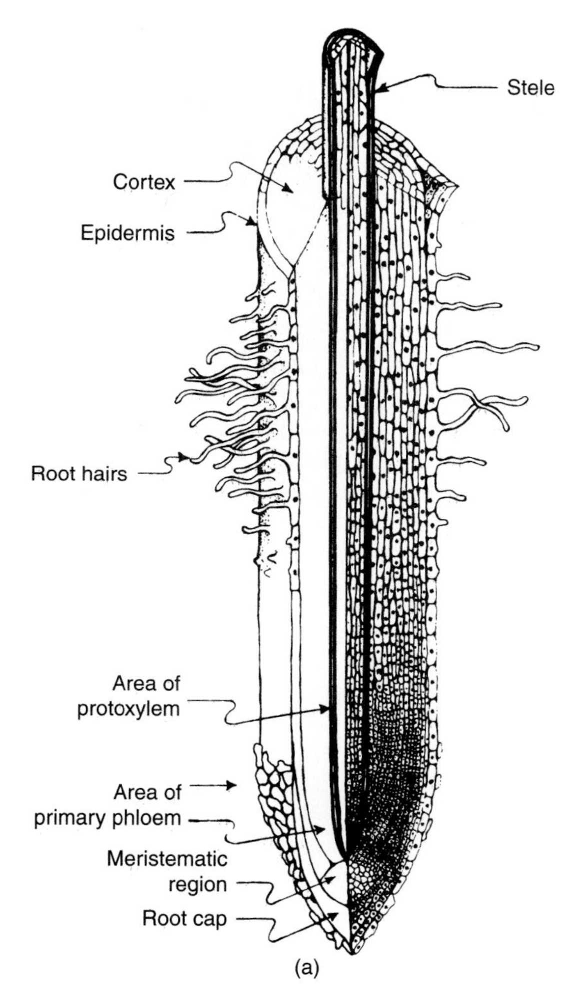

Figure 2.13. The grass root: (a) longitudinal section, (b) cross-section, and (c) tip.

The life span of adventitious roots may be as long as that of the shoot they support; however, climatic stresses and unfavorable soil conditions may cause death of roots, while their associated shoots survive. This is most likely to occur in cool-season turfgrasses during midsummer stress periods. Most root initiation and growth of cool-season grasses occur in spring and, to a lesser extent, during cool weather in the fall. Root growth of warm-season grasses is most active during the summer months. Turfgrasses differ in the extent to which their roots are replaced each year. Kentucky bluegrass retains a major portion of its roots for more than one year and is referred to as a perennial rooting grass. Some bentgrasses, as well as perennial ryegrass, bermudagrass and rough bluegrass, replace most of their root systems each year and are considered annual rooting types.

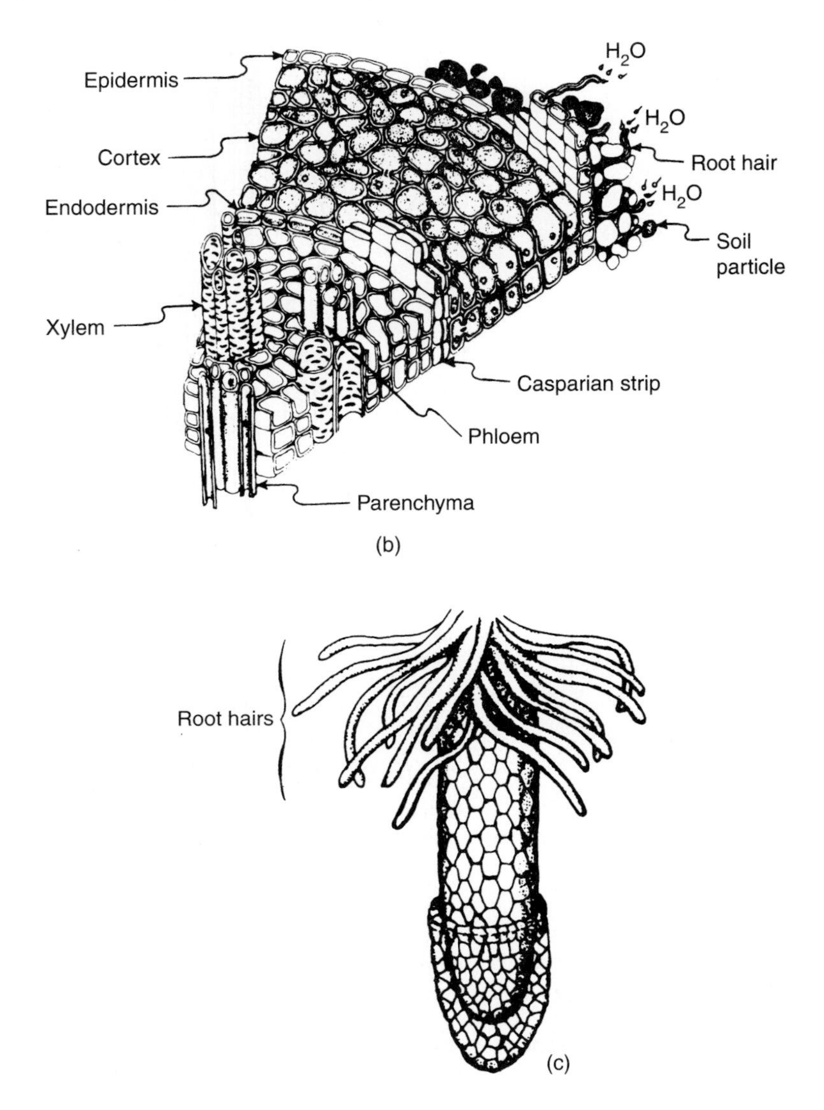

Figure 2.13. Continued

The root is made up of an organized arrangement of cells produced by the division of meristematic cells located just behind the root cap (Figure 2.13). The root cap protects the root meristem from abrasive effects of soil particles as growth proceeds through the soil. Meristematic cells of the root apical meristem replenish the root cap and provide for tip growth of the root itself. Following division, the new cells elongate and push the root cap through the soil. Maturation and differentiation of elongated cells result in the development of specialized tissues for absorbing and transporting water and nutrients to other parts of the plant. This upward movement of substances absorbed from the soil is called *acropetal transport.* Downward movement of photoassimilates from leaves to the roots is called *basipetal transport* and is essential for sustaining root growth and respiration. Within the zone of differentiation and maturation, root hairs arise from epidermal cells. In some grass species only specialized epidermal cells, called *tricoblasts,* can produce root hairs. These delicate extensions of the root epidermis greatly increase the surface of the root for absorbing water and nutrients. Acropetal transport of these materials is through the xylem tubes located within the stele. The stele also includes phloem tubes for basipetal transport of photoassimilates. Movement of materials between epidermal cells and the stele is by diffusion through the living cortex cells (symplastic movement) or in pores within the cell walls (apoplastic movement).

Separating the stele from the cortex is a layer of specialized cells called the *endodermis.* The inner surface of the transverse and radial primary walls of the endodermal cells exhibits a band of suberin, known as the Casparian strip, which impedes the apoplastic movement of water and dissolved materials into the stele. Water is diverted into the endodermal cells following an osmotic gradient through the pericycle and into the conducting cells of the xylem. Movement of dissolved materials also must take place through the membranes of the living endodermal cells and, therefore, depends on energy from respiration. Where turfgrasses are growing in severely compacted or waterlogged soils oxygen for root respiration may be so deficient that transport of materials within the roots is restricted. This condition can result in *wet wilt* and other adverse effects commonly observed in turfgrasses.

Recently initiated roots appear thick and white. With age, roots become thin and darker in color. Root decay begins in the cortex and eventually spreads to the stele. The cortex may slough off (decortication) in older portions of a root; however, the bare stele may still be capable of transporting water and nutrients from the region of absorption to aerial parts of the plant.

INFLORESCENCE

The flowering portion of the grass shoot is called the *inflorescence.* Although leaf sheaths enclose and provide support for a portion of the flowering culm, the inflorescence itself is devoid of leaves except for the floral bracts (modified leaves) within the spikelets. The spikelet contains one or more flowers, each enclosed within two floral bracts called the *lemma* and *palea* to form florets and is delimited at its base by two empty floral bracts called the *glumes.* The glumes and florets are alternately arranged on the rachilla. Spikelets may be attached directly (sessile) to the main inflorescence axis or borne on short stalks (pediceled).

The arrangement of spikelets within an inflorescence is of three primary types: raceme, panicle, and spike (Figure 2.14). In a spike all spikelets are sessile on the main axis; bermudagrasses, ryegrasses, and wheatgrasses have spike-type inflorescences. A raceme has spikelets borne on individual flower stalks, called *pedicels,* that are attached directly to the main axis. Carpetgrass, St. Augustinegrass, and bahiagrass have raceme-type inflorescences. In a panicle spikelets are not sessile or individually pediceled on the main axis but occur along various flowering branches. Bluegrasses, bentgrasses, and many other turfgrass genera have panicle-type inflorescences.

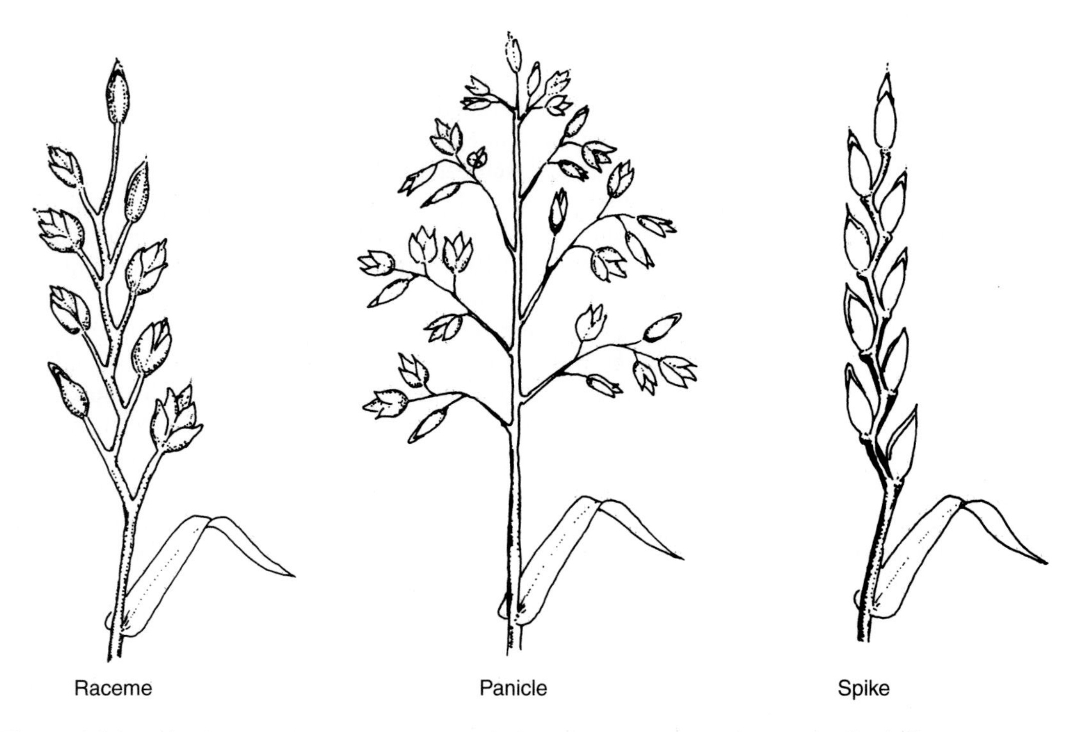

Figure 2.14. The three primary types of grass inflorescence. In the raceme and spike inflorescence types, the main flowering axis is called the rachis, while in a panicle, the term *rachis* is applied to the lateral branches.

The number of florets per spikelet varies among grasses. Panicoid grasses have two-flowered spikelets with a perfect (contains both stamens and pistil) upper floret and a staminate (contains only stamens), or sterile, lower floret. Festucoid and eragrostoid grasses have either single- or multiple-flowered spikelets. Disarticulation (separation at the nodes at maturity) of the spikelet occurs below the glumes in panicoid grasses or above the glumes and frequently between florets in festucoid and eragrostoid grasses.

The grass flower develops on a rachilla and within the axil of a lemma. Immediately below the flower is the palea, which, together with the lemma, enfolds the flower (Figure 2.15). The lowermost organs of the grass flower are the two lodicules, which function in opening the floret to expose the flower at anthesis. The lodicules become turgid at a particular stage of floral maturation and force the lemma and palea to separate for the exsertion of stigmas and anthers. The male portion of the flower is composed of three stamens that consist of pollen-bearing anthers and their filaments. The female portion is the pistil, which is composed of a single ovary and usually two feathery stigmas and styles. At anthesis the slender filaments elongate rapidly and then shrivel up after releasing their pollen. Subsequent transfer of pollen to the stigmas is termed *pollination.* Both cross- and self-pollination occur within the grass family. Pollination within the

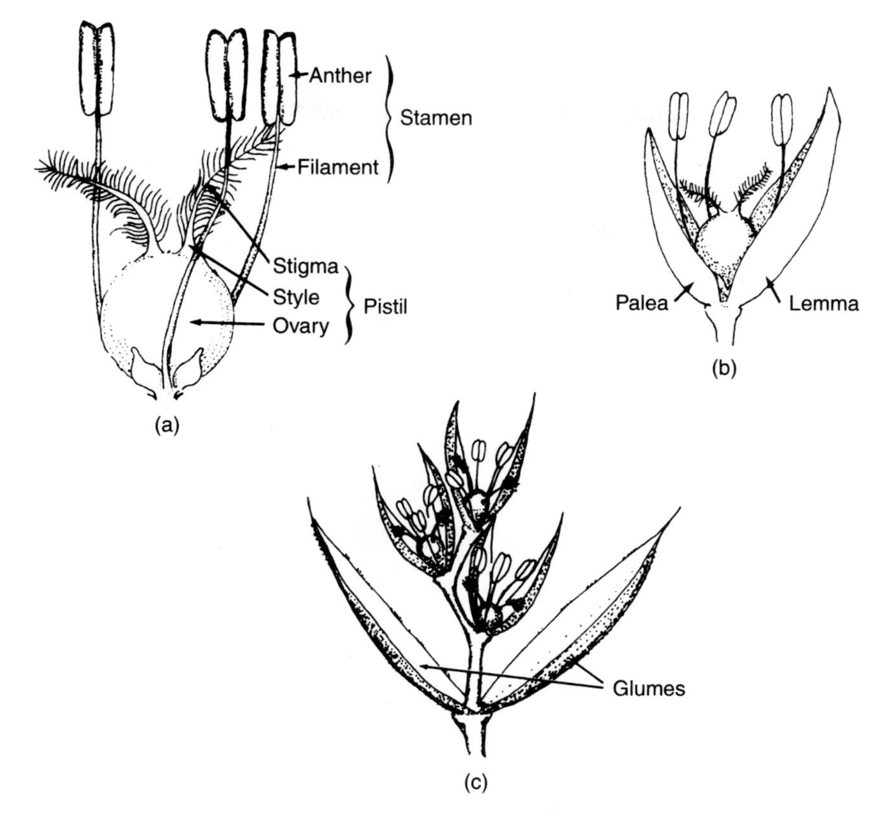

Figure 2.15. The grass flower (a), floret (b), and multiple-floret spikelet (c).

closed floret, or cleistogamy, assures self-pollination and is common in some species, especially annuals.

Once a pollen grain comes into contact with the stigma a pollen tube grows through the style and into the ovule within the ovary (Figure 2.16). Upon entering the embryo sac two sperm nuclei are released. One unites with the egg to form a zygote, and the other unites with two polar nuclei to form the endosperm cell. These sexual unions involving sperm nuclei are termed *double fertilization.* The zygote or fertilized egg forms the embryo, and the fertilized endosperm cell eventually develops into the endosperm, the food source for the embryo during germination. In some grasses sexual reproduction is supplemented by an asexual process called *apomixis.* This involves seed production without actual fusion of male and female gametes in developing the embryo. Consequently, plants resulting from germination of apomictic seed are identical to the female parent. Many Kentucky bluegrasses are highly apomictic and, therefore, can be propagated true-to-type from seed.

Formation of the grass inflorescence occurs in four distinct phases: *maturation* of the plant, *induction* of the flowering stimulus, *initiation* of the stem apex, and *develop-*

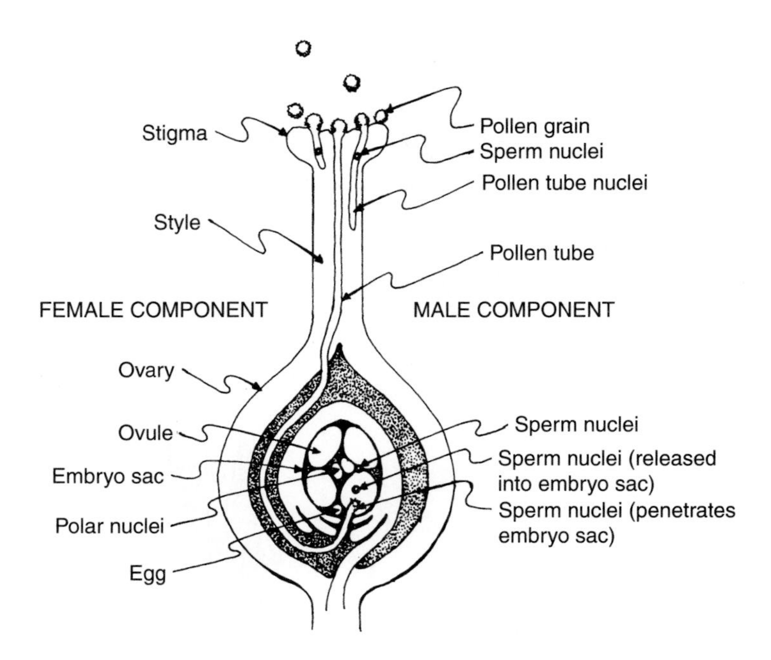

Figure 2.16. Diagram of a grass pistil with germinating pollen grains.

ment of the flowers. Until maturation a juvenile plant is insensitive to environmental conditions that later promote flowering. During induction physiological changes take place in the plant in response to specific environmental conditions. Two types of induction processes are known to exist: low-temperature induction, called *vernalization,* which occurs in the growing point, and photoperiodic induction, which takes place in the leaves. Cool-season (festucoid) grasses have a vernalization requirement as the first prerequisite for flowering. The most effective vernalization temperatures are within a range from 32° to 50°F. This is a reversible process, and devernalization can take place under high-temperature conditions. Short days can substitute for vernalization in some grasses, including perennial ryegrass and colonial bentgrass, while in creeping bentgrass, only short days are required for induction. Warm-season (panicoid-eragrostoid) grasses have no vernalization requirement.

Photoperiodic induction involves the production of a flowering stimulus in the leaves and its translocation to the stem apex under certain day lengths. Cool-season grasses are typically long-day plants in that flowering only occurs after exposure to day lengths greater than a critical number of hours. Warm-season grasses are predominantly short-day plants; flower induction occurs under day lengths shorter than a critical number of hours. Some warm-season grasses are day-neutral plants that flower at any naturally occurring day length if other requirements are satisfied.

The third phase of inflorescence formation is initiation. This involves the transformation of the stem apex from a vegetative to a flowering axis. The first detectable

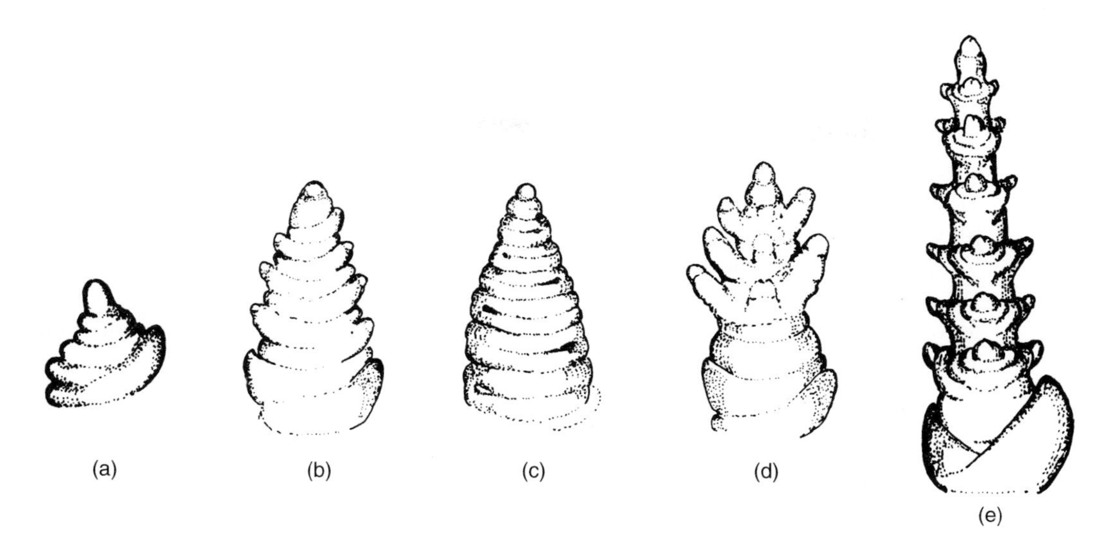

Figure 2.17. Various stages of inflorescence initiation and development, including (a) vegetative, (b) elongation, (c) double ridge, (d) early spikelet, and (e) late spikelet formation stages.

response during initiation is the rapid elongation of the stem apex (Figure 2.17). At this time, leaf primordia form in rapid succession, and the overall length of the stem apex is dramatically increased. The second response is the development of lateral buds in the axils of leaf primordia, resulting in the formation of double ridges along the stem apex.

Finally, the fourth phase of inflorescence formation, called development, includes all the events leading to the formation of branches, spikelets, and florets, and the elevation of the inflorescence above the leaves.

Once the stem apex transforms from the vegetative to the flowering state, no further leaf primordial production occurs. Therefore, removal of the inflorescence by mowing does not remove future leaf-producing structures but, rather, it may stimulate the development of subsidiary tillers and thus enhance leaf production. The future destiny of the flowering shoot, however, is terminal, and perenniality in the turfgrass community depends on the growth of nonflowering shoots and new plants from intravaginal and extravaginal branching. Unless the specific intent is seed production, extensive inflorescence development in a turf is generally undesirable because of the temporary reduction in shoot density that may occur following the death of flowering shoots and the disruptive effect of seedheads on the esthetic appearance of the turf. Furthermore, it is unlikely that inflorescences produced in most turfs will be of any value in yielding new seedling plants since at least several weeks are required for the production of viable seed. Under regular mowing seed maturation is interrupted before viable seeds are produced within newly elevated inflorescences. A notable exception to this is annual bluegrass, which produces viable seed very quickly, resulting in a large reservoir of seed that can germinate and add new plants when conditions are favorable.

SEASONAL GROWTH VARIATION

Cool-season turfgrasses typically show a bimodal growth pattern. Shoot growth begins with a strong spring flush that eventually slows and may even stop entirely during summer, then resumes with moderate vertical growth and vigorous tillering during the fall (Figure 2.18). Because maximum root growth occurs at slightly lower temperatures than shoot growth, it usually begins earlier in spring and later in the fall. Summer rooting is slow and shallow in most cool-season turfgrasses, especially in situations where low levels of carbohydrate reserves may be present in storage organs. This is highly likely where low light limits photosynthetic production or where high rates of nitrogen fertilization encourage excessive shoot growth.

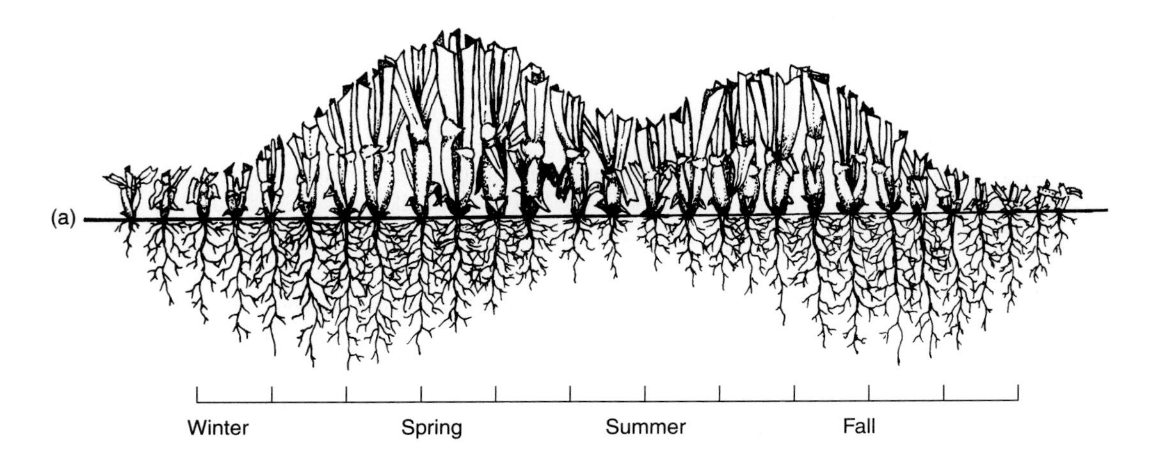

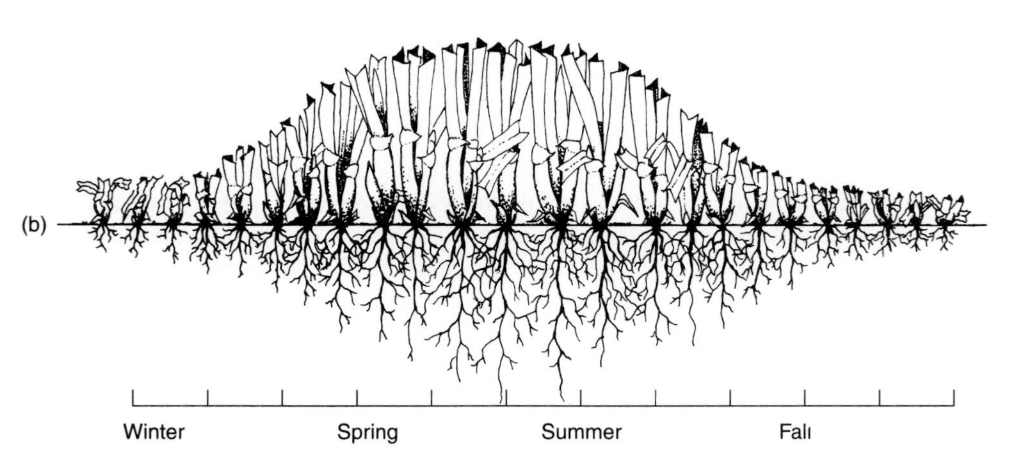

Figure 2.18. Illustrations of seasonal shoot and root growth of cool-season (a) and warm-season (b) turfgrasses.

BIOENERGETICS AND METABOLISM

Plants harvest energy from the sun and convert it to metabolically useful forms. Energy is needed to fuel the metabolism and growth of plants; this involves synthesizing large molecules from smaller ones and moving molecules within the plant. Metabolism is the sum of the chemical and energy transformations that occur in living organisms. These transformations occur in stepwise sequences, called metabolic pathways, in which the product of one transformation becomes the substrate for another. Each transformation in a metabolic pathway rearranges atoms into new molecules, and each one either absorbs or releases energy. The potential energy of a molecule is contained in its chemical bonds. When these bonds break, the energy released can be used to do work, including the formation of other bonds. For example, when a molecule of glucose ($C_6H_{12}O_6$) is oxidized by reaction with six molecules of oxygen gas (O_2) into six molecules each of carbon dioxide (CO_2) and water (H_2O), the rearrangement of chemical bonds results in the release of 686 kcal of energy. This "respiratory" energy is essential for the work the plant must do to grow, compete, and survive.

Many energy transformations in living organisms involve oxidation and reduction of carbon. Reduced carbon contains more energy than oxidized carbon. Organisms extract energy from energy-rich molecules, including starch and fats, via a series of *catabolic* reactions collectively called respiration. They use this energy to drive *anabolic* reactions such as building cell walls or replicating DNA. An important hub through which energy passes during metabolism is ATP (adenosine triphosphate). ATP is a complex organic molecule containing three phosphate (HPO_4^{-2}) groups in series. ATP molecules contain much energy that is released when the terminal phosphate group is cleaved from the molecule. When cells need energy to do work they consume some of the accumulated ATP by converting it to ADP (adenosine diphosphate), inorganic phosphate, and energy.

Starch and fats also store energy. Depending on the size of these molecules, the amount of energy stored can be substantial. Since the bonds connecting carbon with hydrogen and oxygen, as well as with other carbons, in these molecules are very stable, the energy contained in them is not easily available. In contrast, the energy contained in ATP is readily available because the terminal phosphate bond is unstable and easily broken. Generally, ATP is created through an array of catabolic reactions in which energy from stable bonds in organic molecules is transferred to the relatively unstable bonds connecting an additional (terminal) phosphate to ADP. Conversely, ATP is consumed in anabolic reactions in which energy is required for the synthesis of relatively complex molecules from smaller, simpler ones.

Biochemistry

Plant biochemistry is the study of the activities of molecules occurring in plant cells. Biochemicals range from small, simple molecules, called *monomers,* to very large, complex molecules formed by linking many small molecules together into *polymers.* For example, cellulose and starch are polymers of glucose, a simple sugar. Large enzymes and other proteins are made by polymerizing amino acids, and DNA is made by linking many nucleotide molecules together. The four principal types of biochemicals occurring in plants are carbohydrates, proteins, nucleic acids, and lipids.

Carbohydrates

Carbohydrates are polymers of glucose and other sugars in which the individual units are linked together by carbon-oxygen (C-O) bonds. Carbohydrate polymers are called *polysaccharides.* If one molecule of glucose is linked with one molecule of fructose, it forms a dimer (two monomers linked together), or disaccharide, called sucrose; this is the major form of carbohydrate that moves in plants. Carbohydrates are typically divided into two types: structural and nonstructural. The structural carbohydrates are large polymers that hold cells and organisms together. Cellulose is the most abundant structural carbohydrate, making up about half of the substance of cell walls. Several thousand glucose units are linked together to form long chains. A thousand or more of these chains are twisted together to form *microfibrils,* which are like tiny cables of very high tensile strength. Several dozen microfibrils are intertwined to make *fibrils.* Layers of fibrils are cemented together into strong, three-dimensional grids by other structural carbohydrates called pectins and hemicelluloses.

Some carbohydrates are stored in various plant organs (primarily stems) for future consumption. These nonstructural carbohydrates are typically referred to as *carbohydrate reserves.* Cool-season grasses store carbohydrates principally as fructosans. These are relatively short-chain polymers, called oligosaccharides, containing from 26 to 260 6-carbon units of fructose and glucose. Carbohydrate reserves in warm-season grasses occur primarily as long-chain glucose polymers called starch; the two forms are amylose, containing from 50 to 1500 glucose units, and amylopectin, containing from 2000 to more than 200,000 glucose units.

Other forms of nonstructural carbohydrates include the monosaccharide sugars glucose and fructose, and the disaccharides sucrose and maltose. However, these are regarded primarily as intermediary metabolites, and their concentrations change little compared to the fructosans and starches. Carbohydrate reserves are essential for survival and for the production of new plant tissue during periods when carbohydrate use exceeds the supply from photosynthesis. The amounts of nonstructural carbohydrates in the storage organs of turfgrasses vary depending on season and cultural practices. Accumulation is greatest during periods of minimal shoot growth and high light intensities, while depletion is usually associated with periods of rapid shoot growth. Thus, environmental and cultural factors conducive to high growth rates also reduce the level of carbohydrate reserves. If these reserves are allowed to deplete, plants may be unable to recover from injury or disease, regardless of fertilization or irrigation practices. One indication of inadequate carbohydrate reserves is the premature loss of roots during periods of high respiration and vigorous shoot growth. This frequently occurs during summer in intensively cultured cool-season turfs. Since roots depend on the photosynthesizing leaves for carbohydrates, conditions favoring rapid use of carbohydrates tend to favor shoot growth over root growth. This is termed *sink priority.* The source of carbohydrates is the photosynthesizing leaves, and the sinks (locations to which carbohydrates are translocated) include immature leaves, stems, and roots. Since the leaves are closest to the carbohydrate source, they have priority over other plant organs for carbohydrates that are in limited supply. Thus, root growth may cease and much of the root system may be lost under these conditions. Ultimately, the entire plant may die if the roots are unable to supply sufficient moisture and nutrients, especially during stress periods.

Proteins

After carbohydrates, proteins make up most of the biomass of plant cells. Proteins are polymers of amino acids in which the individual units are linked together by carbon-nitrogen (C-N) bonds. Protein polymers are called *polypeptides.* Like carbohydrates, proteins are important in cell structure and as storage reserves. Many proteins are also enzymes that catalyze biochemical reactions. As twenty different amino acids occur in living organisms, the number of possible sequences in polypeptide chains allows for an enormous diversity of proteins and determines their primary structure. Proteins typically form complex three-dimensional shapes due to the formation of various types of bonds between constituents along the chains. The formation of these bonds accounts for the secondary, tertiary, and quaternary structures of protein molecules.

Structural proteins include those occurring in cell walls (2–10% protein) and membranes (50–75% protein). Storage proteins occur mostly in seeds and are used as a source of nutrition for the early development of seedlings. Most of the proteins in a living cell are enzymes. Enzymes serve as catalysts for biochemical reactions. A specific site on each enzyme, called an *active site,* binds to one or more reactants to enable a reaction to proceed many times faster than it would in the absence of the enzyme.

Nucleic Acids

Nucleic acids—composed of long chains of nucleotides—are the most complex biological polymers and include DNA (deoxyribonucleic acid) and RNA (ribonucleic acid). The structure of DNA is a two-stranded spiral called a double helix, while RNA occurs as a single strand. Nucleic acids are unique because they can replicate themselves. Also, DNA can make RNA, which guides the assembly of proteins; thus, nucleic acids form the molecular foundation for all living organisms in that they carry the genetic codes that determine the unique features of each genotype.

Lipids

Lipids, or fats, are water-repellent hydrocarbons that include oils, phospholipids and waxes. Oils are fats that are liquid at room temperature. They are liquids because they are unsaturated; that is, they have double bonds between one or more pairs of carbon atoms that provide molecular rigidity and, thus, prevent molecules from packing tightly into a solid. Oils are most abundant in seeds. Phospholipids are lipids in which a phosphate group is part of the composition of the molecule; they occur mostly as structural components of membranes. Waxes comprise the outermost layer (cuticle) of leaves and are called epicuticular wax. Waxes are harder and more water-repellent than other fats. Another example is suberin, which occurs in some root cells. Waxes function as barriers to water movement into or out of cells.

Photosynthesis and Respiration

Turfgrass growth and development are dependent upon a net increase in dry matter, which is the balance between CO_2 uptake (photosynthesis) and CO_2 evolution (respiration). During growth, daily respiration of plants is usually between 25 and 30% of total photosynthesis, so the plant increases in dry weight. Light reactions transform light energy to the short-term chemical energy of $NADPH_2^+$ (nicotinamide adenine dinucleotide phosphate) and ATP. These compounds are then used to reduce CO_2 to stable organic compounds from which dry weight results. Photosynthesis and respiration are the most important energy transformations in plants. The energy-requiring process is photosynthesis. During this process, light energy absorbed by chloroplasts is used to release O_2 and reduce CO_2, a low-energy molecule, to carbohydrates, which are high-energy molecules (Figure 2.19). The energy-releasing process is cellular respiration. During this process, carbohydrates are oxidized to CO_2 and H_2O.

Photosynthesis

The combination of carbon, hydrogen, and oxygen from carbon dioxide and water yields simple sugars through the process of photosynthesis in green plant tissue. This process is often represented by the formula:

$$6CO_2 + 12H_2O \rightarrow C_6H_{12}O_6 + 6H_2O + 6O_2$$

The water, absorbed principally from the soil but also from atmospheric water vapor, is split into hydrogen (H) and oxygen (O) atoms through a light-activated process called *photolysis.* This is also referred to as the "light reaction" in photosynthesis. Hydrogen ions (H^+) and the electrons (e^-) generated from oxygen ions (O^{-2}) are then used to reduce CO_2 to simple sugars. This is sometimes called the "dark reaction" because it is light-independent.

Visible light, the source of energy used by the plant for photosynthesis, is part of the radiant energy spectrum. At wavelengths above the visible portion of the spectrum (greater than 700 nm), the photons do not have sufficient energy to excite electrons in the light-absorbing pigments. Below the visible portion of the spectrum (less than 400 nm), the photons have too much energy and, if absorbed by chlorophyll, cause pigment degradation. When chlorophyll absorbs a photon, the energy lifts an electron from a lower energy state to a higher (excited) state. When in this excited state chlorophyll can donate electrons to other molecules as well as accept them from other molecules. The energy from absorbed photons catalyzes the removal of electrons from H_2O. Thus, energy is harnessed for a

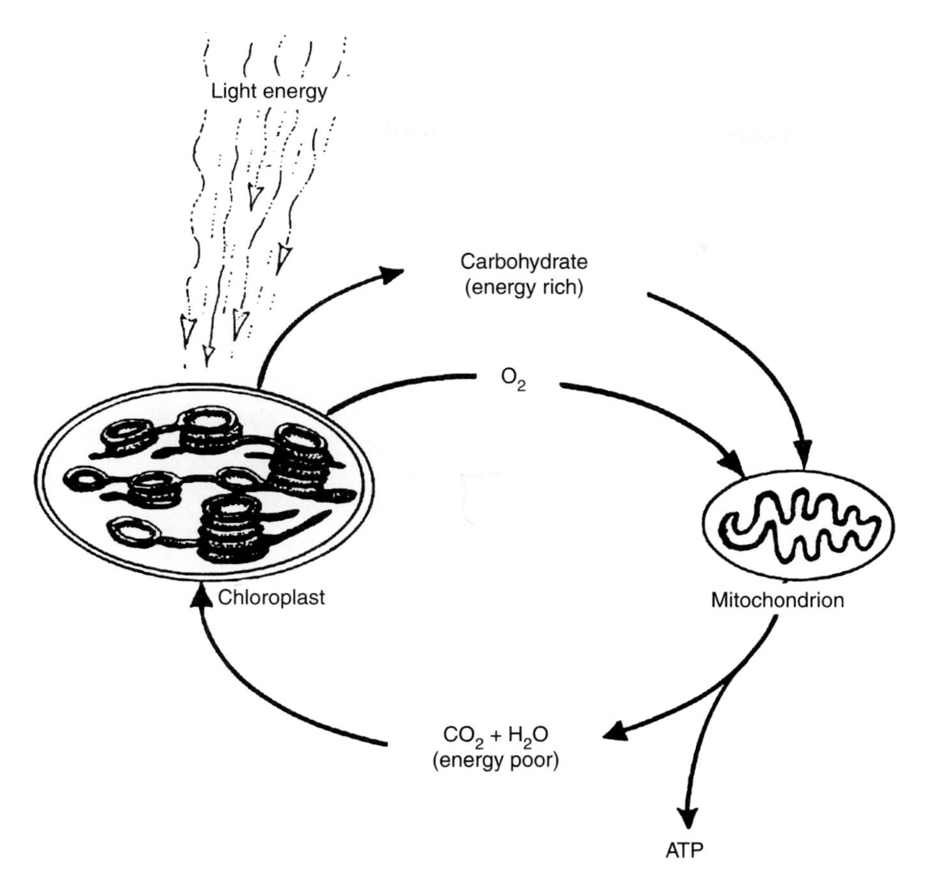

Figure 2.19. Solar energy is captured and stored in carbohydrates through photosynthetic fixation of atmospheric carbon dioxide within chloroplasts. Subsequently, this energy may be released through respiration in the mitochondria and carried in phosphoanhydride bonds of ATP, a high-energy compound that can release small amounts of energy when it loses one of its phosphate groups and converts to ADP.

process called *photophosphorylation*—the formation of ATP from ADP and inorganic phosphorus. This energy also provides hydrogen ions to reduce $NADP^+$ to $NADPH_2^+$, another energy-rich intermediate compound important for reducing CO_2 to simple sugars.

Within the visible portion of the radiant energy spectrum, principally blue and red light are involved in photosynthesis; green light near the center of the visible spectrum is reflected, resulting in the green color of chlorophyll and of plants containing chlorophyll.

The chloroplast, a lens-shaped, subcellular organelle measuring 1 to 10 μm across, is the photosynthetic apparatus of the plant (Figure 2.20). Within the chloroplast are two key areas: the thylakoids and the stroma. The thylakoids are membranes in which chlorophyll and other photosynthetic pigments are concentrated, and this is where the

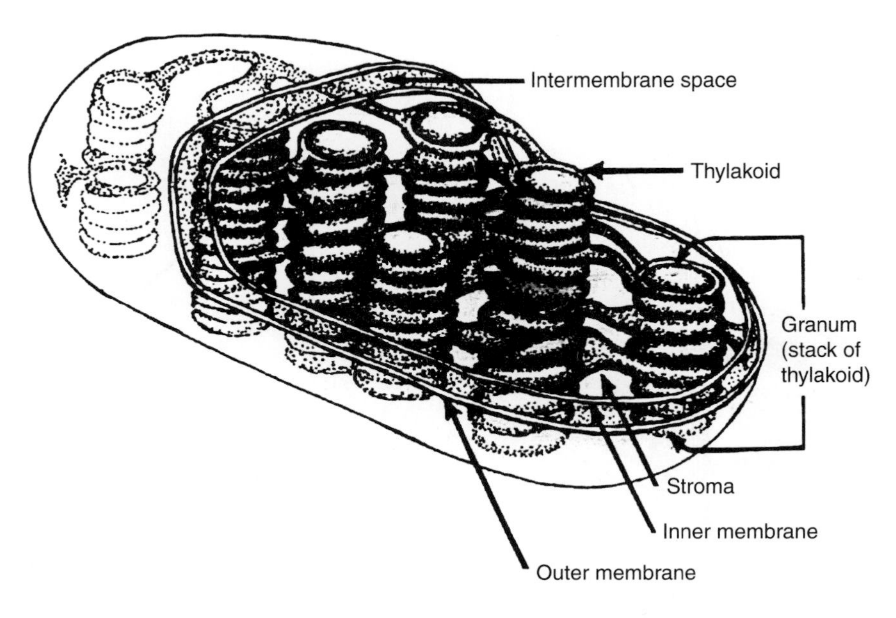

Figure 2.20. Structure of chloroplast showing the two membranes (inner and outer) enclosing a complex series of grana composed of stacks of thylakoids and the fluid stroma between the grana.

light reactions of photosynthesis, including photophosphorylation and photolysis, occur. The stroma is a fluid area where the dark reactions of photosynthesis occur.

The initial products of photosynthetic fixation of carbon vary depending on the specific pathway utilized by the plant. Cool-season grasses typically fix carbon by the Calvin (C_3) cycle in which a five-carbon molecule, RuBP (ribulose 1,5-bisphosphate), functions as the acceptor molecule and initial reductant for CO_2 to form an unstable intermediate that quickly splits to form two three-carbon molecules of PGA (3-phosphoglyceric acid). These then undergo conversions to form the six-carbon sugar, glucose. Warm-season grasses use the Hatch and Slack (C_4) pathway for initially fixing carbon into a four-carbon intermediary compound (oxaloacetate) prior to fixation via the Calvin cycle.

In general, C_4 (warm-season) plants are more photosynthetically efficient than C_3 (cool-season) plants. The CO_2 compensation point (concentration at which photosynthesis and respiration proceed at the same rate) is 5 ppm or less in C_4 plants and between 30 and 70 ppm in C_3 plants. Presumably this is due, at least in part, to the absence or low rates of photorespiration (increased respiratory activity in the presence of light) in C_4 plants compared to C_3 plants, which have relatively high photorespiration rates. Under normal conditions the CO_2 concentration of air is 300 ppm; therefore, the comparison of CO_2 compensation points of C_3 and C_4 plants is a basis for concluding that C_4 plants are more photosynthetically efficient. Also, in C_4 plants, CO_2 fixation increases with increasing temperature and light intensity, while C_3 plants show no positive photosynthetic response to light intensities above 3000 footcandles. (On a clear day, light intensities may reach 10,000 to 12,000 footcandles.) The photosynthetically derived carbohydrates are used by the plant to create other biochemicals as building blocks for growth, and in respiration to provide energy for metabolic activities.

Respiration

All organisms, including plants, harvest energy from stored chemicals through respiration. This process involves the breakdown of complex molecules into simpler chemicals; ultimately, the products of respiration are CO_2 and H_2O:

$$C_6H_{12}O_6 + 6H_2O + 6O_2 \rightarrow 6CO_2 + 12H_2O + \text{energy}$$

Respiration begins with the conversion of storage compounds to hexose (six-carbon) sugars, including glucose, which can then be shunted into a biochemical pathway called *glycolysis*. In glycolysis the chemical bond energy from each molecule of a hexose sugar is used to bond two phosphates (HPO_4^{-2}) to two molecules of ADP to make two molecules of ATP, which serve as mobile energy carriers. Additionally, two molecules of NAD^+ are reduced to two molecules of $NADH_2^+$, which provide the reducing power for subsequent chemical reactions. The chemical products of glycolytic breakdown are two three-carbon molecules of pyruvic acid, which still have most of the energy contained in the original hexose sugar. The synthesis of additional ATP from the energy of pyruvic acid occurs in the second phase of respiration, the Krebs cycle.

Pyruvic acid is transported from the cytoplasm of the cell into the mitochondrion, a subcellular organelle consisting of a smooth outer membrane and a convoluted inner membrane (Figure 2.21). This arrangement of membranes creates two compartments within the mitochondrion: the intermembrane space between the two membranes, and the matrix enclosed by the inner membrane. Many of the enzymes that catalyze the reactions of respiration are bound to the mitochondrial membranes.

The pyruvic acid transported into the mitochondrion is not used in the Krebs cycle directly; instead, it first loses a molecule of CO_2. The remaining two-carbon molecule of acetic acid is attached to a coenzyme to form acetyl coenzyme A, the compound that actually enters the Krebs cycle. Concurrently, another molecule of NAD^+ is reduced to $NADH_2^+$.

In the Krebs cycle, the two-carbon acetyl coenzyme A bonds with the four-carbon oxaloacetic acid to form the six-carbon citric acid. In the subsequent seven steps in the cycle, three molecules of NAD^+ are reduced to $NADH_2^+$, one molecule of ubiquinone is reduced to ubiquinol, one molecule of ATP is produced from ADP and inorganic phosphorus, two molecules of CO_2 are evolved, and oxaloacetic acid—the initial reactant in

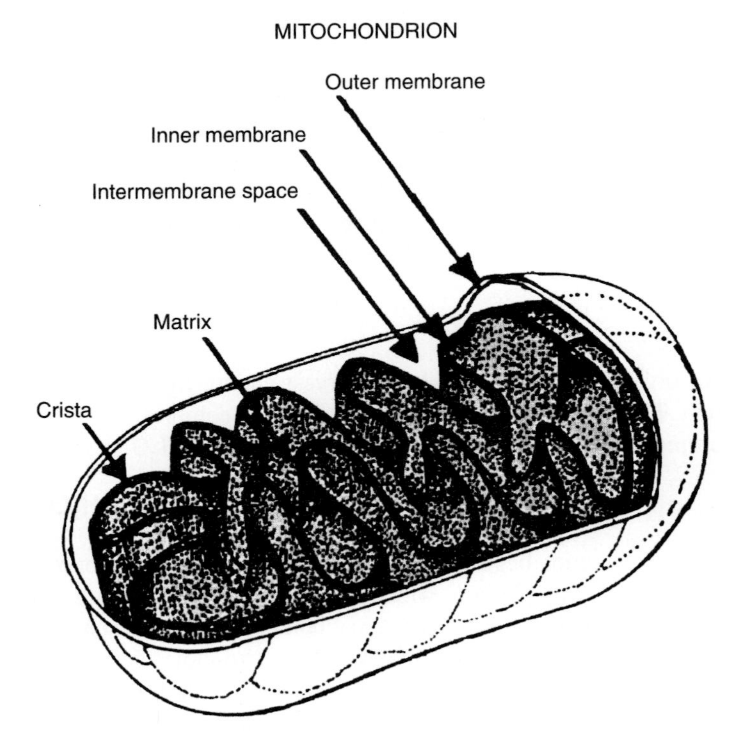

Figure 2.21. Structure of mitochondrion showing the outer membrane and the convoluted inner membrane that forms cristae separating the intermembrane space from the internal matrix.

the cycle—is regenerated. The energy in $NADH_2^+$ and ubiquinol is harvested in the third phase of respiration, called oxidative phosphorylation.

Oxidative phosphorylation involves a complex series of oxidation-reduction reactions of electron transport that enable a cell to use the energy in $NADH_2^+$ and ubiquinol to phosphorylate ADP to ATP. Since three ATPs are typically produced from each $NADH_2^+$ and two ATPs from each ubiquinol in this process, and counting the four ATPs generated directly (two from glycolysis, and two from Krebs cycle), a total of thirty-eight ATPs are produced from a single molecule of glucose. As it is generally accepted that two ATPs are required to transport materials into the mitochondrion, the net production is actually thirty-six. As each ATP contains 7.3 kcal of potential energy, 36 of them would add up to 263 kcal. Since the amount of energy actually contained in a molecule of glucose is 686 kcal, the efficiency with which this energy is harvested through respiration would be 263/686 or 38%. Because of the complexity of these processes, actual respiratory efficiency probably ranges from 22 to 38%. The remaining energy is given off as heat.

QUESTIONS

1. Illustrate and label all components of a mature, rhizomatous turfgrass plant in flower.
2. Differentiate between seminal and adventitious turfgrass **roots.**
3. Illustrate a turfgrass **floret** and label all parts.
4. Illustrate and explain the process of turfgrass seed **germination.**
5. Illustrate and explain **leaf formation** in a turfgrass aerial shoot.
6. Differentiate among the types of **stems** observed in turfgrasses.
7. Illustrate and explain the development of a turfgrass **rhizome.**
8. What is **tillering** and how do tillers contribute to the growth and longevity of a turfgrass plant?
9. Illustrate a turfgrass **root** in cross section and explain water movement into the plant.
10. What are **spikelets** and how are they arranged in the following types of inflorescence: racemes, panicles, and spikes?
11. Illustrate and label a turfgrass **flower.**
12. Explain the **flowering process** in a turfgrass plant.
13. Explain the significance of **nonstructural carbohydrates** in turfgrasses.

CHAPTER 3

Turfgrass Species

All grasses are members of a single family of plants called the *Gramineae,* or *Poaceae.* Within this family, there are six subfamilies containing a total of 25 tribes, 600 genera, and 7500 species. However, only a few dozen of these species form contiguous plant communities tolerant of mowing and traffic and, therefore, are adapted for use as turfgrasses. All turfgrass species are within the three primary subfamilies: Festucoideae, Panicoideae, and Eragrostoideae. Structural features that provide a basis for grouping species within these subfamilies are listed in Table 3.1.

The festucoid turfgrasses are usually referred to as cool-season grasses. Optimum growth of these grasses occurs within a temperature range of 60° to 75°F, and their environmental adaptation is limited primarily by the intensity and duration of seasonal heat and drought stresses. Turfgrasses within the panicoid and eragrostoid subfamilies are called warm-season grasses, and they grow best at temperatures between 80° and 95°F. Warm-season grasses are limited in their poleward adaptation by the intensity and duration of cold temperatures.

TURFGRASS CLIMATIC ADAPTATION

One contemporary scheme of climate classification includes six climatic groups—five humid groups distinguished by temperature, and a dry group in which moisture is the dominant distinguishing feature—and a series of climatic types that further divide these groups based on moisture differences (Trewartha, 1968). The climatic groups include tropical, subtropical, temperate, subarctic, polar, and dry (see Figure 3.1 and inside front cover).

Table 3.1. Morphological and Anatomical Features Separating Festucoid, Eragrostoid, and Panicoid Subfamilies

	Festucoideae	Eragrostoideae	Panicoideae
Roots	Alternately long and short epidermal cells; only short cells (tricoblasts) give rise to root hairs.	All epidermal cells similar; each capable of giving rise to root hairs.	All epidermal cells similar; each capable of giving rise to root hairs.
Stems	Culm internodes typically hollow, surrounded by vascular bundles. No meristematic swelling at base of internodes; swelling apparent at base of leaf sheaths.	Culm internodes typically solid with vascular bundles scattered within pith. Meristematic swelling at base of internodes; very little or no swelling at base of leaf sheaths.	Culm internodes typically solid with vascular bundles scattered within pith. Meristematic swelling at base of internodes; very little or no swelling at base of leaf sheaths.
Leaves	Vascular bundle sheath double, with inner sheath of small, thick-walled cells and outer sheath of large cells. Mesophyll cells loosely arranged with large intercellular spaces. Microhairs absent. Ligules membranous.	Vascular bundle sheath with an outer sheath of large cells and, in some grasses, an inner sheath of small, thick-walled cells. Mesophyll cells radially arranged around vascular bundles. Two-celled microhairs present. Ligules usually a fringe of hairs.	Vascular bundle sheath typically a single sheath of large cells. Mesophyll cells with small intercellular spaces. Two-celled microhairs present. Ligule membranous.
Inflorescence	Spikelet with one to several fertile florets. Lodicules elongate, pointed at tip.	Spikelets with one to several fertile florets. Lodicules small, wedge shaped.	Spikelet with one fertile floret and one reduced floret below. Lodicules short, truncate.
Embryo	Mesocotyl internode absent. Epiblast present. Embryo small, about one-fifth the length of the caryopsis.	Mesocotyl internode present. Epiblast present. Embryo large, about one-half the length of the caryopsis, or larger.	Mesocotyl internode present. Epiblast absent. Embryo large, about one-half the length of the caryopsis, or larger.
Cytology	Basic chromosome number: $x = 7$. Chromosomes typically large.	Basic chromosome number: $x = 9$ or 10. Chromosomes small.	Basic chromosome number: $x = 9$ or 10. Chromosomes small to medium size.

Source: F. W. Gould, *Grass Systematics,* New York: McGraw-Hill Book Company, 1968.

Tropical (A) A humid, frost-free belt extending from the equator to approximately 20° north and south latitudes is the tropical zone. While they are most evident in equatorial South America and Africa, tropical climates also occur in northern Australia, southern Asia, and North America, including the southern tip of Florida in the United States. In marine areas, the average temperature for the coolest month is 65°F or above. The two

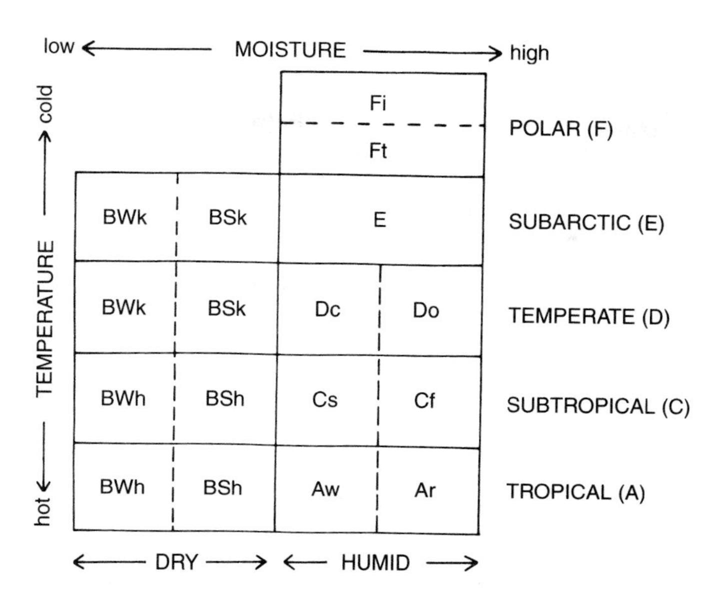

Figure 3.1. Trewartha's climatic classification based on temperature and moisture regimes. Ar = tropical wet (rain forest); Aw = tropical wet and dry (savanna); Cf = subtropical humid; Cs = subtropical dry summer (Mediterranean); Do = temperate oceanic; Dc = temperate continental; E = subarctic (boreal); Ft = polar tundra; Fi = polar icecap; BS = semiarid (steppe); and BW = arid (desert). Suffixes include k (cold) and h (hot) for B group; a (hot summer) and b (cool summer) for C and D groups. ***Source:*** G. T. Trewartha, *An Introduction to Climate,* (New York: McGraw-Hill Book Company, 1968).

main climatic types within this group are *tropical wet* (Ar) or "rainy tropics" with a wet season ten to twelve months in length, and *tropical wet and dry* or "monsoon tropics" with two subtypes: one with a winter dry season of more than two months (Aw) and a rarely occurring one with a summer dry season of more than two months (As). The rainy tropics support tropical rain forests composed of dense stands of broadleaf evergreen trees. The monsoon tropics occur in the transition zones between the perennially wet rainy tropics and the perennially dry tropical deserts (see *dry* [B] climates).

Subtropical (C) Extending poleward from the tropical zone to about 40° latitude are humid regions that have occasional freezing temperatures in continental areas but eight months or more with average temperatures above 50°F. These regions are within the subtropical climatic group. The two principal climatic types within this group are *subtropical humid* (Cf) or humid subtropics, and *subtropical dry summer* (Cs) or summer-dry subtropics (also called a Mediterranean climate because it is most extensive in lands surrounding the Mediterranean Sea). These, as well as the less-commonly occurring *subtropical dry winter* (Cw), are further divided into subtypes with hot (a) or cool (b) summers. The 81°F isotherm for the warmest month was selected by the author to distinguish between subtropical a and b subtypes. The subtropical humid climate typically occurs on the eastern side of a continent. It may have no distinctly dry season and, typically, summer is the wettest season. The subtropical dry-summer climate, in contrast, is characterized by a rainy winter season, and it typically occurs on the western side of a continent. The wettest parts of the humid subtropics may have vegetation similar to that of the tropical rain forest; however, the interior boundaries of these regions are increasingly dry as one proceeds away from the coasts and toward the continental interiors. Trees generally become smaller and more widely spaced. In the interior boundaries, where annual rainfall may average as little as 20 inches, trees tend to give way to grasses. Natural vegetation in the summer-dry subtropics includes woody plants with thick barks and small, stiff leaves that reduce transpiration during the hot, dry summers. A low cover of drought-resistant evergreen shrubs and small trees, called *chaparral* in California and *maquis* in France, is common at lower elevations that are not under cultivation.

Temperate (D) Climates with average temperatures of 50°F or more for four to seven months of the year are termed temperate. The two climatic types within this group are: *temperate oceanic* (Do) and *temperate continental* (Dc). As in the subtropical group, these may be further characterized as having no dry season (f), a winter dry season (w), or a summer dry season (s). Additionally, hot (a) and cool (b) summer subtypes can occur. The 72°F isotherm for the warmest month was selected by the author to distinguish between temperate a and b subtypes. Where Do and Dc climates are contiguous, the boundary separating them is usually the 32°F isotherm for the coolest month; thus, mild winters are characteristic of the oceanic type, and relatively severe winters occur in the continental type. The oceanic type typically occurs on the western side of a continent or on islands, while the continental type is common on the leeward eastern side of a continent or inland. Temperate continental climates occur only in the northern hemisphere because there are no major land masses at these latitudes in the southern hemisphere. Vegetation occurring in temperate continental climates ranges from evergreen and deciduous forests in the more humid coastal areas to grasslands (with trees along water courses) in the interior where annual rainfall averages 20 inches or less. Temperate oceanic climates are found in western Europe, the coastal margins of North and South America, the southern tip of Australia, and the islands of Tasmania and New Zealand. Forest, including the coastal redwoods of northern California, which reach heights of more than 300 feet, is the natural vegetation of these mild, moist areas.

Subarctic (E) The subarctic, or boreal, climate has from one to three months with average temperatures above 50°F. It is characterized by short, cool summers and long, cold winters. Rainfall throughout the year is relatively low compared to that in the warmer D, C, and A climates. Subarctic lands are found only in the northern hemisphere, where they form a broad belt between the northernmost temperate zones and the Arctic with its polar climate. Midwinter days may last only a few hours, and the sun remains low on the horizon; midsummer days are long, with the daylight period extending past midnight in the northernmost areas. The long hours of summer daylight help compensate for the brevity of the growing season. Slow-growing coniferous forest is the predominant natural vegetation of the subarctics.

Polar (F) The summerless polar climate occurs at the highest latitudes, where average temperatures for all months are below 50°F. The two types within this group include *polar tundra* (Ft), in which the warmest month is above 32°F, and *polar icecap* (Fi), which has temperatures below 32°F for all months. Polar climates occur in Greenland and the arctic fringes of North America and Eurasia in the northern hemisphere, and the icebound continent of Antarctica in the southern hemisphere. Toward the poles, the year is divided into months of darkness and months of continuous low-angle sun and continuous daylight. The "bush tundra" is made up of stunted trees only a few feet high while, in the more poleward areas, low-growing sedges, mosses, lichens, and flowering plants comprise the "grass tundra."

Dry (B) A dry climate is one in which evaporation exceeds precipitation. The boundary between humid and dry climates occurs where annual evaporation and annual precipitation are exactly equal. This group includes the *arid* or desert climatic type (BW, from the German *Wuste*), and the *semiarid* or steppe climatic type (BS). The boundary between arid and semiarid climatic types occurs where annual precipitation is half the amount occurring at the humid/semiarid boundary, usually less than 10 inches. The dry climates are further divided into two subtypes: *cold* (k), which occurs in temperate and subarctic zones, and *hot* (h), occurring in subtropical and tropical zones. As in the subtropical and temperate groups, these may be further characterized as having a winter dry season (w) or a summer dry season (s). The hot deserts are concentrated between latitudes 15° and 30° north and south of the equator. While most occur on the western margins of continents, the Sahara Desert cuts a 1200-mile swath across Northern Africa from the Atlantic Ocean to the Red Sea, 3500 miles to the east. This great dry region continues through the Sinai Peninsula into the deserts of Southwest Asia and the Arabian Peninsula. These deserts are the hottest lands on Earth. In Death Valley, California, a temperature of 134°F has been recorded in the shade. Other deserts, as well as steppes, occur across vast expanses of land in the middle latitudes. The lack of precipitation in these regions is due to two factors: the long distance from oceanic sources of moisture (i.e., "continentality") and the rain shadow effect of high mountains. Because of the low moisture, diurnal temperatures fluctuate more widely than in humid climates. Vegetative cover in dry climates ranges from sparse stands of cacti and other drought-resistant perennials or drought-escaping annuals in the hot deserts to denser stands of drought-resistant shrubs and short (prairie) grasses of the middle-latitude steppes.

Cool-season turfgrasses are generally adapted to temperate and subarctic climates, while warm-season turfgrasses are used primarily in tropical and subtropical areas. The poleward adaptation of individual turfgrasses depends primarily on their respective cold tolerances. In the United States, some tall fescues and perennial ryegrasses may not be adapted to the northernmost latitudes, while the bentgrasses and bluegrasses are widely used throughout much of Canada. St. Augustinegrass, which is well adapted to tropical and warm subtropical-humid climates, will usually not persist in the northernmost portions of Arkansas, Alabama, and eastern Georgia, where colder winters occur and the climate shifts to the cool subtropical-humid subtype. The boundary between the temperate and subtropical climates in the eastern United States marks the approximate location of what has been called the *transition zone.* Here some cool-season (primarily Kentucky bluegrass) and warm-season (certain bermudagrasses) turfgrasses encounter the limits of their southern and northern adaptation, respectively. Tall fescue and some zoysiagrasses are uniquely adapted to the transition zone.

In northern Europe, the temperate oceanic climate of the west is the same as that of the U.S. Pacific Northwest east of the Cascade Mountains. In both locations, colonial bentgrass and fine fescues are well adapted. Kentucky bluegrass can be used, but it is somewhat more prone to disease than in the temperate continental regions. Perennial ryegrass is well adapted to a temperate oceanic climate, and improved turf-type cultivars are being used increasingly for athletic field turfs in the United Kingdom, the Netherlands, France, and Germany. This is a dramatic departure from the traditional use in Europe of mixtures containing timothy and crested dogtail for soccer fields and other intensively trafficked sports turfs.

In east-central Africa, the climate is tropical except at high elevations. In the lowlands, only warm-season turfgrasses are used; at altitudes above 6000 feet, cool-season turfgrasses provide excellent turfs even though this area is astride the equator. An analogous situation exists in the southeastern United States due to the Appalachian Mountains. Although they are not very high, the effect of these mountains on local climate allows the use of cool-season turfgrasses in upland portions of Tennessee, North Carolina, and Georgia.

The most dramatic effect of high-altitude areas on climatic conditions is evident east of the U.S. Rocky Mountain range. For a large stretch of land extending across much of Oklahoma, Kansas, Nebraska, and the Dakotas, the climate is semiarid, and diurnal temperatures can fluctuate quite widely. Turfgrasses well adapted to this region include the wheatgrasses in the north, and buffalograss and the gramagrasses throughout most of the Great Plains. Under regular irrigation, many other cool- and warm-season grasses can be sustained within their respective zones of adaptation above or below the transition zone.

CHARACTERIZATION OF TURFGRASSES

Each turfgrass is identified in the text by its Latin binomial. An example is *Poa pratensis*—the Latin name for a turfgrass species known in the United States as Kentucky bluegrass. *Poa* is the Latin designation for the bluegrass genus and *pratensis* is the particular species within this genus. In this example, "Kentucky bluegrass" is the common name. While common names may vary from place to place (for example, Kentucky bluegrass is known as fine-leaved meadowgrass in the United Kingdom), the Latin name is supposed to be universally accepted. There are exceptions, however, depending on the taxonomic authority one uses as the basis for establishing the identity of a particular species. The authority may be added to the Latin binomial immediately following the species designation. For example, Carl Linnaeus, a botanist in the eighteenth century and the father of plant taxonomy, is the authority for many species and is identified simply as **L.**; thus, Kentucky bluegrass is universally known by its Latin name: *Poa pratensis* **L.** Subsequent references to a species are usually made in an abbreviated version: *P. pratensis.*

Sometimes, the Latin binomial is followed by an infraspecies designation when important differences exist within a species but are not sufficient to warrant separation into different species. Categories of infraspecific names, by rank, in common use are subspecies (ssp.), variety (var.), and forma (f.). For example, Chewings fescue is *Festuca rubra* **L. ssp.** *comutata* **Gaud.,** while creeping red fescue is *Festuca rubra* **L. ssp.** *trichophylla* **Gaud.** Charles Gaudichaud-Beaupre, a botanist in the nineteenth century, made this distinction based on morphological characteristics: Chewings fescue has a noncreeping, bunch-type growth habit, while creeping red fescue is rhizomatous. If an infraspecific group is maintained only by cultivation, it is referred to as a cultivated variety or cultivar (cv.). Baron Kentucky bluegrass would be designated as *Poa pratensis* **L. cv.** Baron. An alternative would be to use single quotes: *Poa pratensis* **L.** 'Baron.'

Multiple Latin names occur where different authorities are identified as being responsible for naming and describing the species. Redtop is identified by the Latin names *Agrostis alba* **L.** and *Agrostis gigantia* **With.** (William Withering, also a botanist in the eighteenth century), reflecting the two different authorities associated with this species. Sometimes, two different authorities are credited with identifying a species, as in manilagrass or *Zoysia matrella* **[L.] Merr.** This indicates that the species was first described by Linnaeus in a different genus, as a separate species, or at a different rank. It was later modified by Elmer Drew Merrill, an early twentieth century botanist. Another rule that applies to authorities is evident in the Latin name of mascarenegrass or *Zoysia tenuifolia* **Willd. ex Trin.** The term **ex** means that Carl Ludwig Willdenow first proposed the name, but that Carl Bernhard von Trinius later provided the valid description.

The characteristics and organization of turfgrass species within their respective genera, tribes, and subfamilies follow.

Festucoideae Subfamily

Festucoids are cool-season grasses occurring mostly in temperate and subarctic, and sometimes in subtropical, climates. They are usually long-day plants in which floral initiation must be accompanied by cool nights and preceded by vernalization. Within their one- to several-flowered spikelets, disarticulation occurs above the glumes, which remain attached to the plants after the florets are shed, and between individual florets. Inflorescences are usually panicles, but occasionally are racemes or spikes, which are laterally compressed due to lengthwise folding of the floral bracts. Carbon fixation in photosynthesis occurs principally through the Calvin (C_3) cycle; therefore, the Festucoideae are sometimes referred to as C_3 grasses. This subfamily contains nine tribes, three of which (Festuceae, Aveneae, and Triticeae) include turfgrass genera.

Festuceae Tribe

Genera within the Festuceae have two- to several-flowered spikelets; one or both of the glumes are shorter than the lowermost lemma. There are three major (*Festuca, Poa, Lolium*) and three minor (*Bromus, Cynosurus, Puccinellia*) turfgrass genera within this tribe.

Fescues (*Festuca* **L.**)

Of the approximately one hundred fescue species, six are commonly used as turfgrasses. These are divided into two subgeneric types based on leaf texture: the coarse fescues (*Festuca arundinaceae*—tall fescue, *Festuca elatior* or *pratensis*—meadow fescue); and the fine fescues (*Festuca rubra*—creeping red fescue, *Festuca rubra* **ssp.** *commutata*—Chewings fescue, *Festuca ovina*—sheep fescue, *Festuca longifolia*—hard fescue).

Creeping red fescue (*Festuca rubra* **L. ssp.** *rubra* and *F. rubra* **L. ssp.** *trichophylla* **Gaud.** or **ssp.** *litoralis* **[Meyer] Auquir**). There are two distinct types of creeping red fescue, one that is a strong creeping type with fifty-six chromosomes (**ssp.** *rubra*) and a slender creeping type with short rhizomes and forty-two chromosomes (**ssp.** *trichophylla* **Gaud.** or **ssp.** *litoralis* **[Meyer] Auquir**).

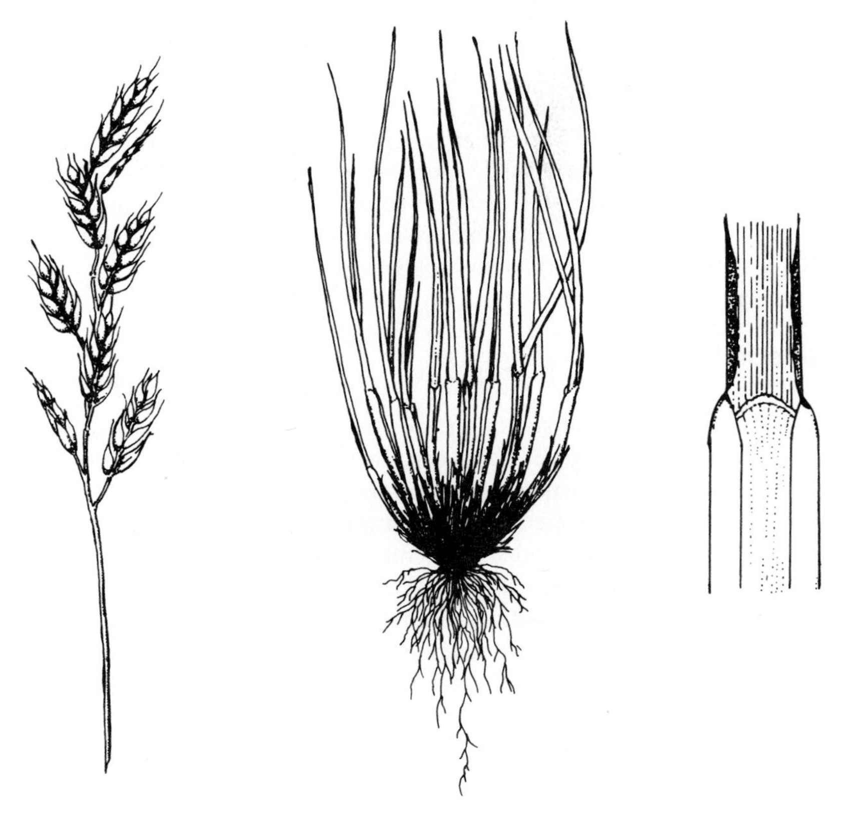

Description: Vernation—folded; ligule—membranous, 0.5 mm long, truncate; auricles—absent; collar—narrow, continuous, without hairs; blade—1.5 to 3 mm wide, deeply ridged on adaxial surface, abaxial surface and margins smooth; inflorescence—contracted panicle.

Adaptation and Use: Well-drained, moderately shaded sites and droughty infertile acid soils. Intolerant of wet conditions and high fertility. Used in seed mixtures with Kentucky bluegrass or colonial bentgrass in cooler temperate and subarctic climates and with perennial ryegrass for overseeding warm-season turfs in subtropical climates for winter play and color.

Cultural Intensity: Low to moderate. Mowing heights of 1.5 to 2 inches preferred. Fertilization minimal: 2 lb N/1000 ft^2/yr or less.

Cultivars: New cultivars have been sought to improve the close-mowing tolerance, disease resistance, and heat hardiness of the creeping red fescues. Improved cultivars include the strong creeping (fifty-six chromosomes) types Boreal, Ensylva, and Shademaster; and the slender creeping (forty-two chromosomes) types Dawson and Seabreeze. The inclusion of *Neotyphodium typhinum* and *Epichloe typhina* (formerly *Acremonium*) endophytes in creeping red fescue cultivars has been shown to improve drought resistance, insect tolerance, and resistance to some diseases.

Chewings fescue (*Festuca rubra* **L. ssp.** *commutata* **Gaud.**) This is a noncreeping, bunch-type grass that is otherwise similar to creeping red fescue.

Adaptation and Use: Compared to creeping red fescue, Chewings forms a somewhat denser turf, especially under the closer mowing heights (about 1 inch) practiced in northwestern Europe. In continental regions, the preferred mowing height is 1.5 to 2 inches. Chewings is somewhat less tolerant of temperature extremes than creeping red fescue; otherwise, the two are similar in their environmental adaptation.

Cultural Intensity: Same as creeping red fescue.

Cultivars: Improved cultivars include Banner, Bargena, Jamestown, Koket, Longfellow, Shadow, and Tiffany. The inclusion of *Neotyphodium typhinum* and *Epichloe typhina* (formerly *Acremonium*) endophytes in Chewings fescue cultivars has been shown to improve drought resistance, insect tolerance, and resistance to some diseases.

Sheep fescue (*Festuca ovina* **L.**) A noncreeping, bunch-type grass with stiff leaves and twenty-eight chromosomes.

Description: Vernation—folded; ligule—membranous, 0.3 mm long, rounded; auricles—absent; collar—broad, divided, without hairs; blade—1 to 2 mm wide, ridged on adaxial surface, abaxial surface and margins smooth; inflorescence—contracted panicle.

Adaptation and Use: Well-drained, droughty, sandy, or gravelly acid soils of low fertility. Used for soil stabilization.

Cultural Intensity: Very low. Not adapted to intensive culture.

Cultivars: Quatro is one of the few improved cultivars reported for this species.

Hard fescue (*Festuca longifolia* **Thuill.**) A noncreeping, bunch-type grass similar to sheep fescue, but with tougher, wider leaves and forty-two chromosomes.

Adaptation and Use: Less drought tolerant, but more tolerant of moist, fertile soils than sheep fescue. Used primarily for soil stabilization.

Cultural Intensity: Low. Similar to sheep fescue.

Cultivars: Biljart, Reliant, Scaldis, and Silvana are cultivars with improved drought, heat, and disease tolerance. The inclusion of *Neotyphodium typhinum* and *Epichloe typhina* (formerly *Acremonium*) endophytes in hard fescue cultivars has been shown to improve drought resistance, insect tolerance, and resistance to some diseases.

Tall fescue (*Festuca arundinacea* **Schreb.**) A coarse-textured, bunch-type grass that is considered a good utility turfgrass and also a weed when it occurs in finer-textured turfs.

Description: Vernation—rolled; ligule—membranous, 0.4 to 1.2 mm long, truncate; auricles—short, blunt, pubescent; collar—broad, divided, short hairs on margins; blade—5 to 10 mm wide, flat, stiff, ridged on adaxial surface, abaxial surface smooth, keeled, and somewhat glossy, margins scaly; inflorescence—contracted panicle.

Adaptation and Use: Adapted to a wide range of soil conditions. Heat and drought stress tolerance especially good for a cool-season grass; cold tolerance is relatively poor. Widely used as a utility turfgrass in warm temperate and cool subtropical climates; an increasingly important lawn species in the cool subtropical climatic belt.

Cultural Intensity: Low to moderate. Mowing range—above 1.5 inches; minimal fertilization in late spring and summer, optimum response from early spring and fall fertilization; irrigation required under semiarid conditions for survival and under all conditions for optimum growth and quality.

Cultivars: Alta and Kentucky-31 are the most widely used cultivars. Many fine-leaved cultivars provide high-quality lawn turf under proper management; these include Bonsai, Jaguar, and Rebel. The inclusion of *Neotyphodium coenophialum* (formerly

Acremonium) endophyte in tall fescue cultivars has been shown to improve drought resistance, insect tolerance, and resistance to some diseases.

Meadow fescue (*Festuca elatior* **L.** or *F. pratensis* **Huds.**) Similar to tall fescue in general appearance, growth habit, and environmental adaptation; meadow fescue is more widely used in Europe than in the United States.

Description: Vernation—rolled; ligule—membranous, 0.2 to 0.5 mm long, truncate; auricles—short, blunt, without hairs; collar—broad, continuous; blade—3 to 8 mm wide, flat, ridged on adaxial surface, abaxial surface smooth and somewhat glossy, margins scaly; inflorescence-contracted panicle.

Adaptation and Use: Similar to tall fescue, but less persistent under drought and heat stresses.

Cultural Intensity: Same as for tall fescue.

Cultivars: No improved turf-type cultivars are currently available.

Bluegrasses (*Poa* **L.**)

There are more than 200 species of bluegrasses, including the most widely adapted cool-season grasses. Four species are commonly found in turf, including Kentucky bluegrass

(*Poa pratensis*), the most widely used perennial turfgrass in temperate and subarctic climates; Canada bluegrass (*Poa compressa*), which is native to North America; rough bluegrass (*Poa trivialis*), a specialty grass for moist shaded environments; and annual bluegrass (*Poa annua*), which, although rarely planted intentionally, occurs in intensively cultured turfs in subarctic, temperate, and subtropical climates. The most distinguishing vegetative features of the bluegrasses are the boat-shaped leaf tip and the parallel light lines occurring on either side of the central vein of the leaf blade.

Kentucky bluegrass (*Poa pratensis* **L.**) This is a highly variable, rhizomatous species with cultivars that differ in color, texture, density, close-mowing tolerance, disease resistance, and other parameters.

Description: Vernation—folded; ligule—membranous, very short, 0.2 to 0.6 mm long, truncate; auricles—absent; collar—broad, divided; blade—V-shaped or flat, parallel sided, smooth on both surfaces, two light lines astride central vein, boat-shaped tip; inflorescence—open, pyramidal panicle.

Adaptation and Use: Well-drained, moist, neutral to slightly acid, fertile soils on sunny or slightly shaded sites. Used throughout subarctic and temperate climates and at high altitudes in tropical and subtropical climates. Widely used for lawns, golf turf (except greens), athletic fields, and other general-purpose turfs.

Cultural Intensity: Low to high, depending upon cultivar. Mowing tolerance range—0.75 to 2.5 inches; fertilization requirement—2 to 6 lb N/1000 ft^2/yr or higher where clippings are removed; irrigation as necessary to prevent wilt and sustain density.

Cultivars: Many improved cultivars and intensive breeding efforts underway throughout Europe and the United States. Cultivars adapted to intensive culture include Touchdown, A-34, and Brunswick. For moderate cultural intensities, the list is long and includes A-20, Adelphi, Baron, Bonnieblue, Cheri, Eclipse, Enmundi, Georgetown, Glade, Majestic, Merit, Midnight, Parade, Plush, Rugby, and Victa. The common types, including Kenblue, Park, Delta, and South Dakota Certified, are adapted to low intensities of culture. Important criteria for selecting cultivars for moderate to high cultural intensities include resistance to leaf spot (*Helminthosporium* melting out), striped smut, and patch diseases.

Canada bluegrass (*Poa compressa* **L.**) A bluish green, weakly rhizomatous grass that forms an open, stemmy turf of low quality; Canada bluegrass is adapted to cool temperate and subarctic climates.

Description: Vernation—folded; ligule—membranous, 0.2 to 1.2 mm long, truncate; auricles—absent; collar—narrow, divided; blade—flat or V-shaped, smooth on both surfaces, tapering toward boat-shaped tip, dull bluish green and occasionally with reddish margins, 1 to 3 mm wide, two light lines astride central vein; inflorescence—narrow panicle.

Adaptation and Use: Acid, droughty, infertile soils. Used for soil stabilization.

Cultural Intensity: Very low.

Cultivars: Canon is reported as having better disease resistance, leafiness, cold tolerance, and spring green-up than the common type.

Rough bluegrass (*Poa trivialis* **L.**) A fine-textured, usually yellow green, stoloniferous grass, rough bluegrass is extremely limited in its adaptation; however, it may be an important turfgrass for sites where alternative species do not survive.

Description: Vernation—folded; ligule—membranous, 2 to 6 mm long, pointed or notched; auricles—absent; collar—broad, divided; blade—flat, 1 to 4 mm wide, slightly tapering toward boat-shaped tip, smooth on both sides, glossy abaxial surface, light lines not prominent (sheath base has an "onion-skin" appearance); inflorescence—open panicle.

Adaptation and Use: Damp or moist, fertile soils on moderately to heavily shaded sites. Occurs as a weed on sunny sites where it thins and turns brown during midsummer stress periods. Not compatible with other turfgrasses in lawn seed mixtures. Used for overseeding bermudagrass greens for winter play. Has good cold tolerance, but poor heat and drought tolerance.

Cultural Intensity: Moderate. Mowing heights of 0.5 to 1 inch optimum; fertilization—2 to 4 lb N/1000 ft^2/yr; irrigation necessary to sustain quality during drought periods. Susceptible to injury from 2,4-D and related herbicides.

Cultivars: Commercially available cultivars include Colt, Lazer, and Sabre.

Annual bluegrass (*Poa annua* **L.**) Widely distributed as a winter annual in both cool- and warm-season turfs, it is also sustained as a perennial turfgrass on intensively cultured sites in cool temperate and subarctic climates. Annual bluegrass can form a very dense, bunch-type or weakly stoloniferous turf under close mowing.

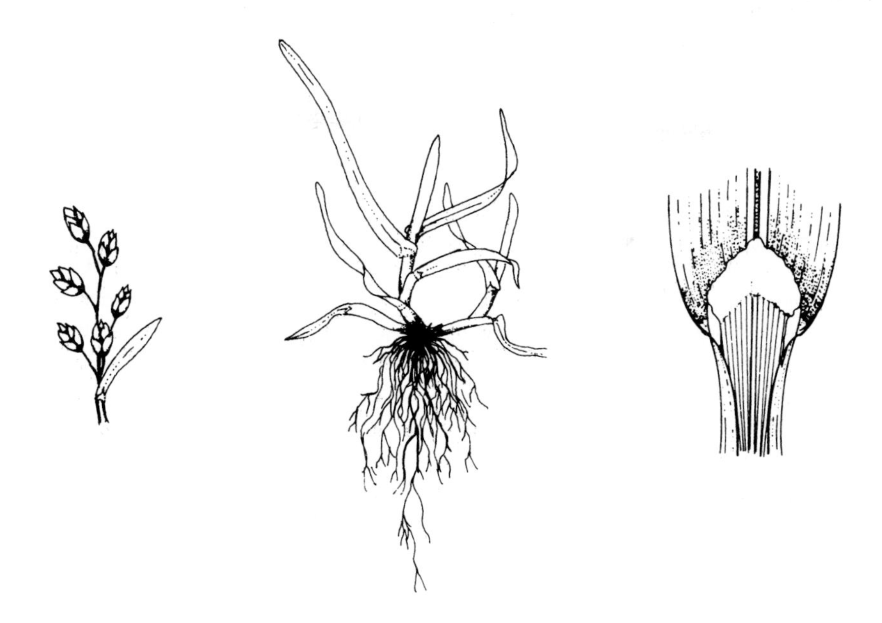

Description: Vernation—folded; ligule—membranous, 0.8 to 3 mm long, pointed; auricles—absent; collar—broad, divided; blade—flat or V-shaped, 2 to 3 mm wide, parallel sided or slightly tapering toward boat-shaped tip, smooth on both sides, light green at both ends of the growing season or in winter, multiple light lines appear parallel to veins; inflorescence—small, open panicle, seed heads apparent throughout most of the growing season but particularly abundant during early spring to midspring.

Adaptation and Use: Although relatively intolerant of heat, cold, and drought stresses, annual bluegrass grows vigorously during spring and fall periods in temperate climates and during cool weather in subtropical climates. It is rarely planted intentionally; yet it often becomes the major component of some intensively cultured turfgrass communities, especially at high latitudes. Best adapted to moist, fertile, neutral to slightly acid, well-drained soils and cool, shaded environments.

Cultural Intensity: High. Mowing heights of 1 inch or lower optimum; fertilization—2 to 6 lb N/1000 ft^2/yr; irrigation often, perhaps daily, during droughty periods and supplemented with midday syringing in hot weather. Fungicide treatments are necessary to control dollar spot, brown patch, *Pythium,* anthracnose, snow molds, and other diseases.

Cultivars: None; however, many botanical varieties varying in density, color, texture, growth habit, and persistence are recognized. These occur primarily within two subspecies groups: *P. annua.* **var.** *annua* **L.,** a winter annual, upright-growing biotype; and *P. annua* **var.** *reptans* **Hausskn,** a perennial, prostrate-growing biotype that persists under close mowing and frequent irrigation.

Ryegrasses (*Lolium* **L.**)

The ryegrasses include about ten species that are distributed primarily within the temperate climatic zone. The two species used as turfgrasses are perennial (English) ryegrass (*Lolium perenne*) and annual (Italian) ryegrass (*Lolium multiflorum*). Under evaluation are several "intermediate" ryegrasses—interspecific hybrids of annual and perennial ryegrasses. Traditionally, the ryegrasses have been used as nursegrasses in seed mixtures because of their rapid germination rate and vigorous seedling growth. With the development of improved turfgrass cultivars, the contemporary role of ryegrass is changing.

Perennial ryegrass (*Lolium perenne* **L.**) A cool-season, bunch-type grass, perennial ryegrass may behave as an annual, short-lived perennial, or perennial, depending on environmental conditions.

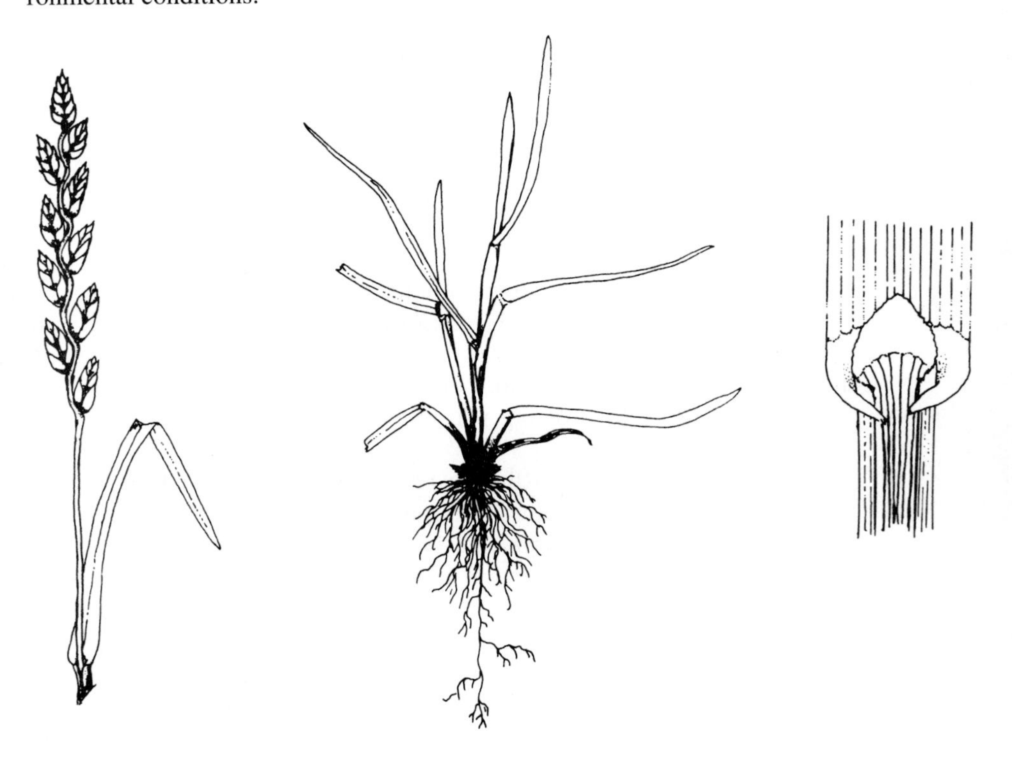

Description: Vernation—folded; ligule—membranous, 0.5 to 2 mm long, truncate to rounded; auricles—small, soft, clawlike; collar—broad, divided; blade—flat, 2 to 5 mm wide, ridged on adaxial surface, glossy and keeled on abaxial surface; inflorescence—flat spike with awnless spikelets.

Adaptation and Use: Moist cool environments without severe winter or summer temperatures. Although adapted to a wide range of soil conditions, perennial ryegrass grows best in neutral to slightly acid, moist soils of moderate to high fertility. It is used as a nursegrass in cool-season turfgrass seed mixtures and in combination with Kentucky bluegrass for intensively trafficked athletic turfs. Football fields may be overseeded with perennial ryegrass to promote recovery in spring and to provide a wear-resistant turf during the playing season. Overseeding is also performed in dormant or semidormant, warm-season turfs for winter play and color.

Cultural Intensity: Moderate. Mowing height—1.5 to 2 inches preferred; mowing quality is sometimes poor because of the tough vascular bundles within the leaf blades; fertilization—2 to 6 lb N/1000 ft^2/yr; irrigation is necessary to ensure survival during extended drought periods.

Cultivars: Improved, turf-type perennial ryegrasses are finer textured, denser, and more persistent, and have better mowing quality than the common type. These include Citation, Derby, Diplomat, Manhattan, Manhattan II, Palmer, Pennant, Pennfine, and Repell. The inclusion of *Neotyphodium lolii* (formerly *Acremonium*) endophyte in perennial ryegrass cultivars has been shown to improve drought resistance, insect tolerance, and resistance to some diseases.

Annual ryegrass (*Lolium multiflorum* **Lam.**) A cool-season annual or short-lived perennial, bunch-type grass, annual ryegrass forms a fine- to coarse-textured turf, depending on seedling rate and density. Its use in seed mixtures is discouraged because of its aggressive seedling growth, which may prevent establishment of the more desirable cool-season turfgrasses.

Description: Vernation—rolled; ligule—membranous, 0.5 to 2 mm long, rounded; auricles—pointed, clawlike; collar—broad, continuous; blade—flat, 3 to 7 mm wide, ridged on adaxial surface, smooth and glossy on abaxial surface; inflorescence—flat spike with awned spikelets.

Adaptation and Use: Similar to perennial ryegrass but less tolerant of temperature extremes. Annual ryegrass is widely used in subtropical climates for overseeding dormant or semidormant, warm-season turfs for winter play and color. It has been partially replaced by the improved, turf-type perennial ryegrasses or mixtures containing several cool-season turfgrass species. In temperate climates, it is occasionally used for establishing temporary lawns in late spring; these can subsequently be reestablished with desirable turfgrasses when favorable environmental conditions exist.

Cultural Intensity: Similar to perennial ryegrass.

Cultivars: No improved cultivars for turfgrass use.

Bromegrass (*Bromus* **L.**)

There are approximately one hundred species of bromegrass worldwide, but only one species, smooth brome (*Bromus inermis*), is used as a turfgrass.

Smooth bromegrass (*Bromus inermis* **Leyss.**) This is a coarse-textured, rhizomatous, cool-season grass that is well adapted for use in semiarid areas within the temperate zone.

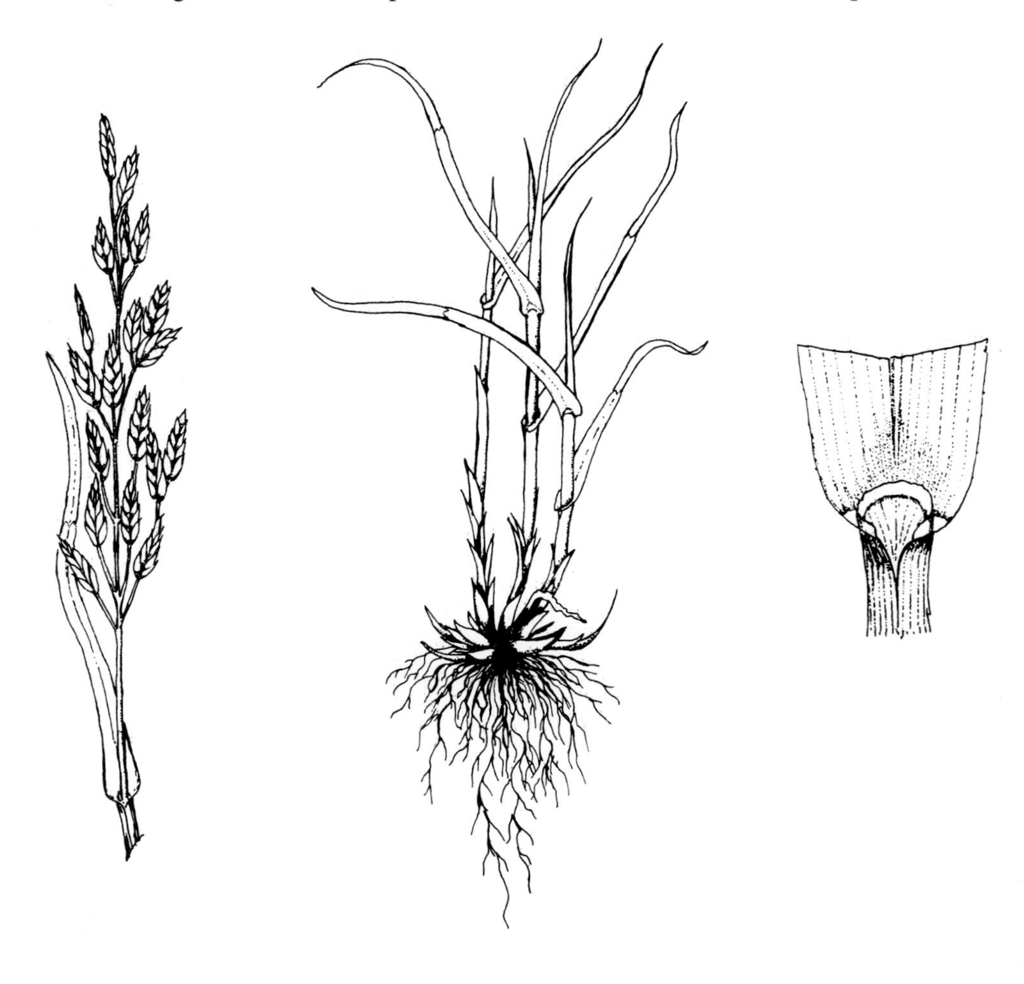

Description: Vernation—rolled; ligule—membranous, 1 mm long, truncate to rounded; collar—broad, divided; blade—flat, 8 to 12 mm wide, smooth on both surfaces; inflorescence—contracted panicle, branches whorled.

Adaptation and Use: Due to its good drought, heat, and cold tolerance, smooth bromegrass is well adapted to the extreme conditions encountered in temperate, semi-arid regions. Since it forms an open, coarse-textured turf that is intolerant of intense traffic or close mowing, its use is limited to soil stabilization along roadsides and other low-use sites.

Cultural Intensity: Very low.

Cultivars: Many cultivars developed for pasture and forage use. Most fall within two subspecies groups: northern types, which are very cold hardy, and the more aggressive southern types, which are more tolerant of heat and drought stresses.

(*Cynosurus* **L.**)

This genus contains four species that occur primarily in Europe. Only one species, crested dogtail (*Cynosurus cristatus* **L.**), is used as a turfgrass. It is a bunch-type grass that appears similar to perennial ryegrass. In Europe, it is sometimes included in seed mixtures to improve the wear resistance of athletic turfs. However, its use is limited because of its poor summer color and intolerance of cold weather.

Alkaligrass (*Puccinellia* **Parl.**)

Of the thirty species within this genus, three have been reported as potentially valuable turfgrasses for use on saline or alkaline soils in temperate climates. These include weeping alkaligrass (*Puccinellia distans* **[L.] Parl.**), Nutall alkaligrass (*Puccinellia airoides* **[Nutt.] Wats** and **Coult.**), and Lemmon alkaligrass (*Puccinellia lemmoni* **[Vasey] Scribn.**). Naturally occurring stands of weeping alkaligrass were observed along roadsides in northern Illinois where salt splash from applications to control highway icing destroyed other vegetation. Fults, a cultivar of this species, is now being produced for commercial distribution.

Aveneae Tribe

Genera within the Aveneae have one- to several-flowered spikelets; glumes are usually longer than the lowermost lemma. There are one major (*Agrostis*) and one minor (*Phleum*) turfgrass genera within this tribe.

Bentgrass (*Agrostis* **L.**)

This genus includes about 125 species that occur in temperate and subarctic climates and at high altitudes in tropical and subtropical zones. Five species are used as turfgrasses, including creeping bentgrass (*Agrostis palustris, A. stolonifera*), colonial bentgrass (*Agrostis capillaris*), dryland bentgrass (*Agrostis castellana*), velvet bentgrass (*Agrostis canina*), and redtop (*Agrostis alba, A. gigantea*). With the exception of redtop, the bentgrasses have been used extensively for greens and other intensively cultured turfs. Morphological features common to these species include ridged leaf blades on the adaxial side, rolled vernation, and single-floret spikelets.

Creeping bentgrass (*Agrostis palustris* **Huds.,** *A. stolonifera* **L.**) A fine-textured, stoloniferous species, creeping bentgrass is the most widely used cool-season grass for golf and bowling greens. In the early 1900s, greens were planted with multiple-species mixtures containing small percentages of creeping bentgrass. Combined efforts of the U.S. Department of Agriculture and the U.S. Golf Association Green Section during the 1920s and 1930s resulted in the development of numerous creeping bentgrass cultivars, some of which are still regarded as valuable turfgrasses.

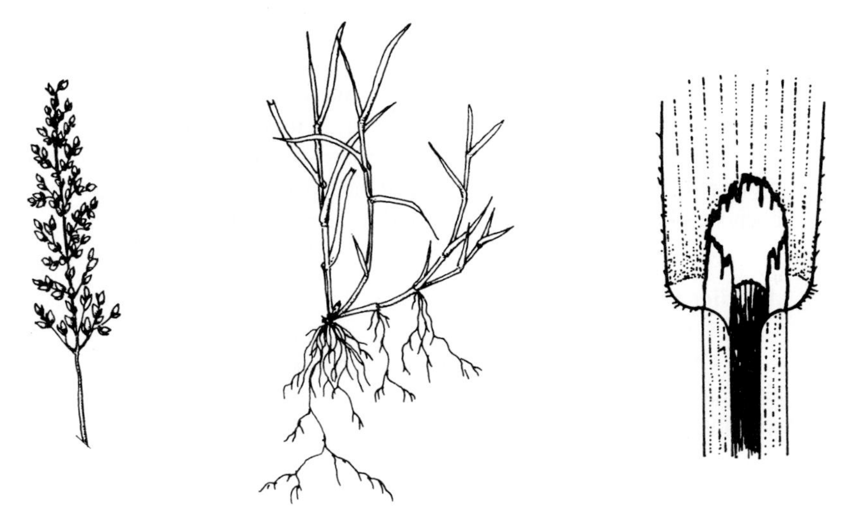

Description: Vernation—rolled; ligule—membranous, 0.6 to 3 mm long, finely toothed or entire, rounded; auricles—absent; collar—narrow to broad, oblique; blade—flat, 2 to 3 mm wide, ridged on adaxial surface, abaxial surface smooth, margins scaly; inflorescence—contracted panicle, pale or purple colored.

Adaptation and Use: Moist, fertile, acid to slightly acid soils. Used for greens, closely mowed trees and fairways, and exotic lawns.

Cultural Intensity: High. Mowing heights of 0.2 to 0.5 inch preferred; fertilization—4 to 8 lb N/1000 ft^2/yr; irrigation frequent; topdressing helpful in controlling thatch; shallow vertical mowing and brushing to control grain; core cultivation and spiking to improve water infiltration and relieve soil compaction; may require fungicide applications for disease control.

Cultivars: Earlier, most cultivars were propagated vegetatively; these include Washington (C-50), Toronto (C-15), Cohansey (C-7), Pennpar, and numerous others found in old established greens. Today, most creeping bentgrass turfs are propagated by seed. Seeded cultivars include Emerald, Penncross, Penneagle, Pennlinks, Prominent, Providence, Putter, Seaside, and SR1020. The vegetatively propagated cultivars tend to be more uniform in appearance.

Colonial Bentgrass (*Agrostis capillaris* **L.**) A fine-textured, bunch-type to weakly creeping (short stolons and rhizomes) grass that is best adapted to temperate oceanic climates. This is very similar in appearance and adaptation to dryland bentgrass (*Agrostis castellana* **Boiss.** and **Reut.**); however, colonial bentgrass is a tetraploid with twenty-eight chromosomes, while dryland bentgrass is a hexaploid with forty-two chromosomes.

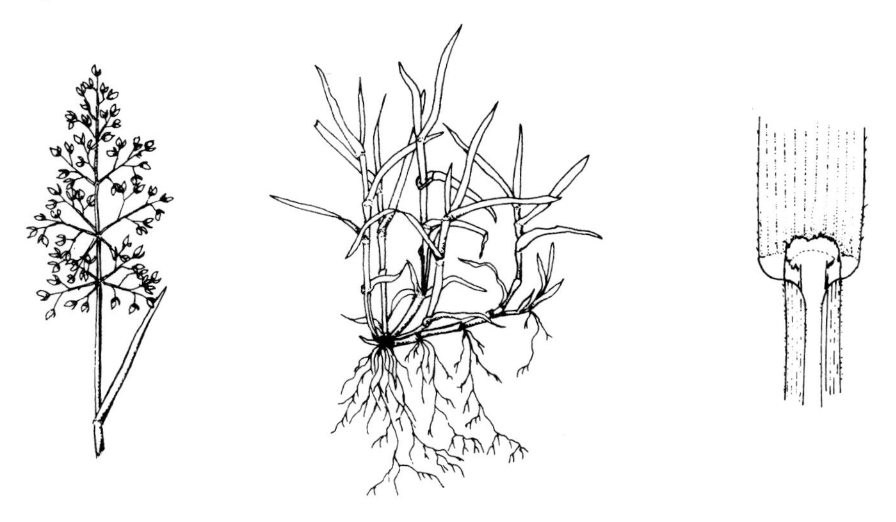

Description: Vernation—rolled; ligule—membranous, 0.3 to 1.2 mm long, truncate; auricles—absent; collar—narrow, oblique; blade—flat, 1 to 3 mm wide, ridged on adaxial surface, abaxial surface smooth; inflorescence—open panicle.

Adaptation and Use: Well-drained, sandy, acid to slightly acid soils of moderate fertility. Its poor heat and drought tolerance limits its use to oceanic climates of the U.S. Pacific Northwest and New England states, northwestern Europe, New Zealand, and locations with similar conditions.

Cultural Intensity: Moderate to high. Mowing heights of 0.3 to 0.8 inch; fertilization—1 to 6 lb N/1000 ft^2/yr; irrigation—frequent during drought periods.

Cultivars: U.S. cultivars include Allure, Astoria, Bardat, Egmont, and Exeter. Breeding efforts in northern Europe resulted in the development of Holfior and Boral. In the United States, a common type of dryland bentgrass is Highland.

Velvet bentgrass (*Agrostis canina* **L.**) A very fine-textured, stoloniferous grass that forms a velvety turf of very high density. Its use is restricted to temperate oceanic climates where, with meticulous culture, it forms turfs of exceptionally high quality.

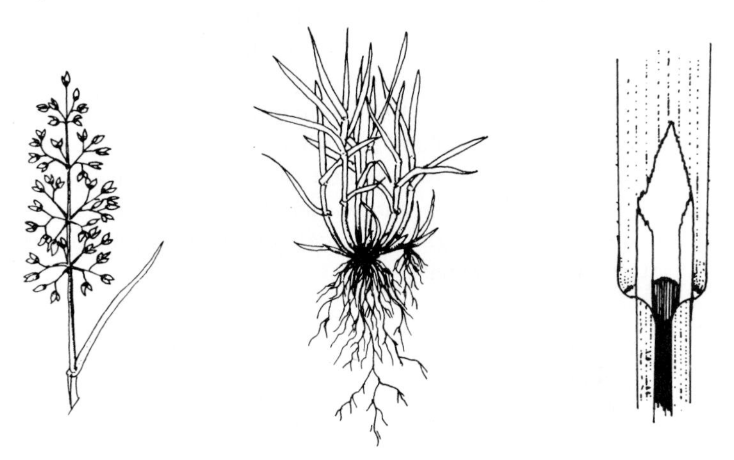

Description: Vernation—rolled; ligule—membranous, 0.4 to 0.8 mm long, pointed; auricles—absent; collar—broad; blade—flat, 1 mm wide, slightly ridged on adaxial surface, smooth on abaxial surface, margins scaly; inflorescence—reddish, spreading panicle.

Adaptation and Use: Well-drained, acid to slightly acid, sandy soils of moderate fertility. It is quite shade tolerant compared to other bentgrasses. Like colonial bentgrass, its use is limited to very mild, temperate oceanic climates. It is occasionally found in old greens as very dense, isolated patches, especially in the shade.

Cultural Intensity: High. Mowing heights of 0.2 to 0.4 inch preferred; fertilization—2 to 4 lb N/1000 ft^2/yr; irrigation—frequent; frequent top dressing necessary to prevent dense thatch accumulation; frequent preventive fungicide applications for disease control. Due to its slow growth rate, severe cultivations should be avoided.

Cultivars: Breeding efforts have been directed toward improved disease resistance, better recovery from injury, and darker color. Currently, the best cultivar is Kingstown.

Redtop (*Agrostis alba* **L.,** *A. gigantea* **With.**) A coarse-textured, grayish green, rhizomatous species, redtop was used extensively as a component of cool-season turfgrass seed mixtures to promote rapid cover. Because of its highly competitive seedling growth and its unsightly appearance and persistence in new turfs, the use of redtop as a nursegrass is discouraged in favor of other species.

Description: Vernation—rolled; ligule—membranous, 1.5 to 5 mm long, rounded; auricles—absent; collar—broad, divided; blade—flat, tapering toward tip, 3 to 10 mm wide, ridged on adaxial surface, smooth on abaxial surface, margins scaly; inflorescence—reddish, spreading panicle.

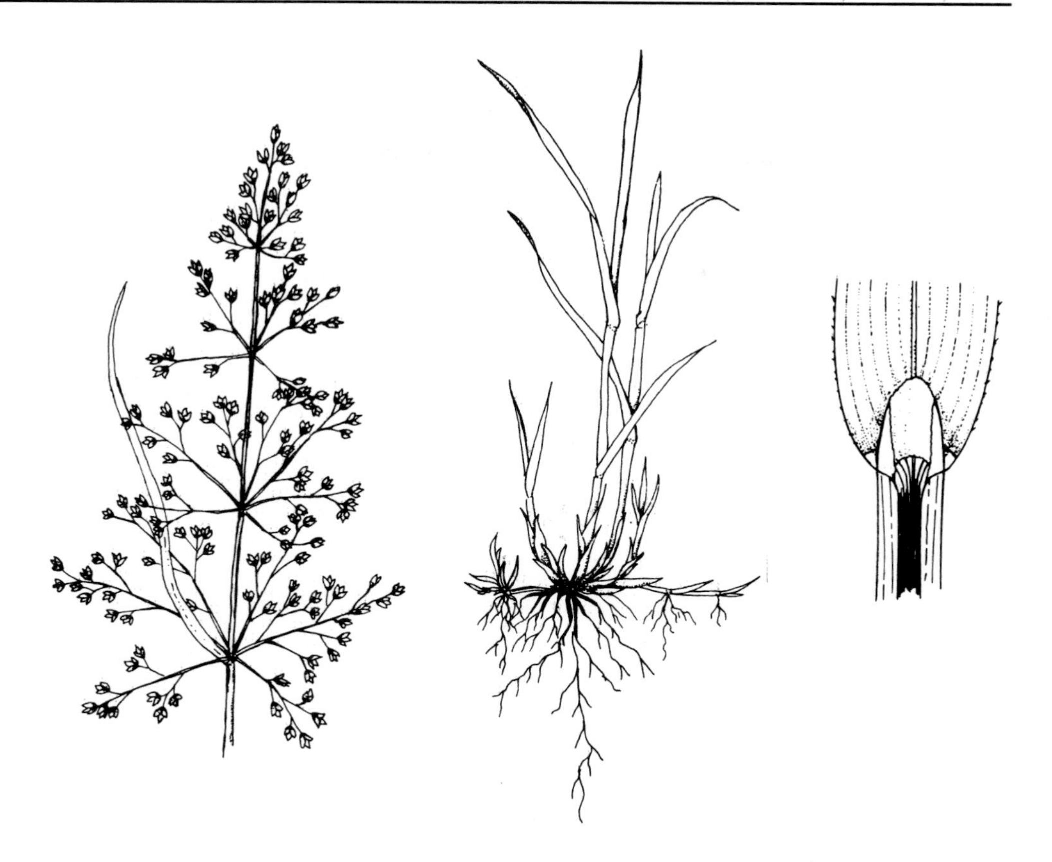

Adaptation and Use: Adapted to a wide range of soil conditions. It is sometimes used in seed mixtures for planting along roadsides in drainage ditches and other wet or poorly drained, infertile sites. When used as a nursegrass, some plants persist in isolated clumps or patches that reduce the uniform appearance of the turf. It is occasionally used as a "repair grass" for greens and other turfs that have been damaged or diseased; however, fine fescues or ryegrasses are preferred choices.

Cultural Intensity: Low to moderate. Due to its intolerance of close mowing, a mowing height of at least 1.5 to 2 inches is necessary for redtop to persist. Although quite tolerant of infertile, droughty soils, it does respond to fertilization and irrigation practices.

Cultivars: None available.

Timothy (*Phleum* **L.**)

Of the ten species within this genus, two are used in subarctic and cool temperate climates as turfgrasses. They are common timothy (*Phleum pratense* **L.**) and turf timothy (*Phleum nodosum* **L.** or *Phleum bertolonii* **D.C.**).

Common timothy (*Phleum pratense* **L.**) A coarse-textured, bunch-type grass, timothy is used primarily in northern Europe in seed mixtures for athletic field turf.

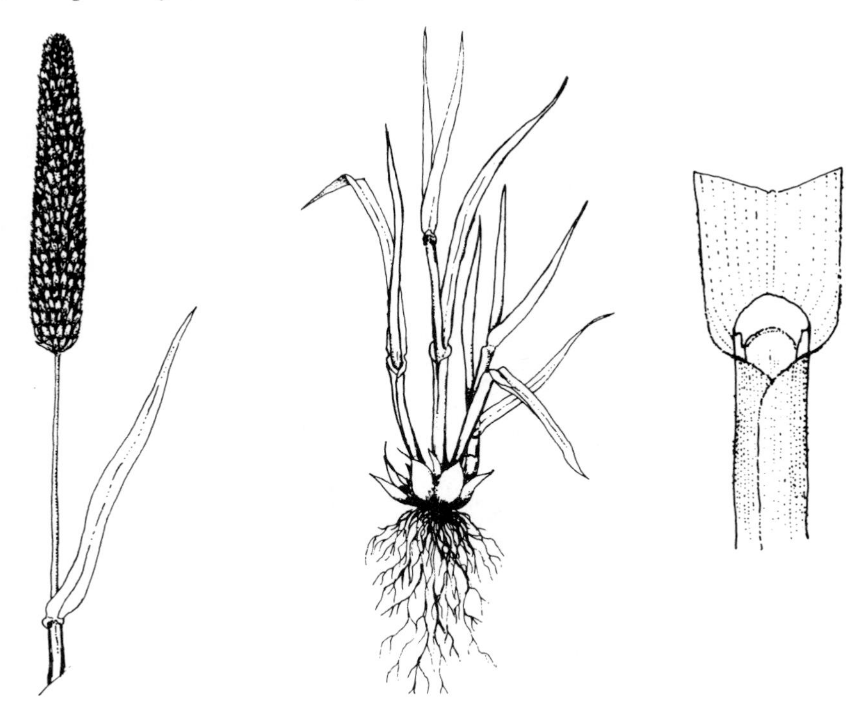

Description: Vernation—rolled; ligule—membranous, 3 to 6 mm long, pointed at center with distinct notches on either side; auricles—absent; collar—broad, continuous, margins sparsely hairy; blade—flat, 4 to 8 mm wide, ridged on adaxial surface, smooth on abaxial surface, margins scaly; inflorescence—cylindrical, tightly contracted panicle.

Adaptation and Use: Moist, fine-textured, well-drained, slightly acid soils with good fertility. Timothy is valued in northern Europe because of its wear resistance, cold tolerance, and winter color; however, its intolerance of heat and drought stresses severely limits its use at other locations. In the United States, it is generally considered a weed in turf, but it is an important forage and pasture grass.

Cultural Intensity: Moderate to high. Mowing heights below 1 inch are tolerated; fertilization—3 to 6 lb N/1000 ft^2/yr; irrigation essential during drought periods.

Cultivars: Breeding efforts in northern Europe are oriented toward developing finer-textured and more persistent cultivars. The diploid species, turf timothy (*P. bertolonii* **D.C.**), is somewhat finer textured than the common type.

Triticeae Tribe

Genera within the Triticeae have two- to several-flowered spikelets; inflorescences are two-sided spikes or spicate racemes. Only one minor turfgrass genus, *Agropyron,* occurs within this tribe.

Wheatgrasses (*Agropyron* **Gaertn.**)

Of the sixty species within this genus, several are important forage grasses on semi-arid rangelands in the temperate and subarctic climatic zones. Those species that have some value as turfgrasses for unirrigated sites include crested (fairway) wheatgrass (*Agropyron cristatum* **[L.] Gaertn.**), western wheatgrass (*Agropyron smithii* **Rydb.**), and desert wheatgrass (*Agropyron desertorum* **[Fisch, ex Link] Schult.**).

Crested Wheatgrass (*Agropyron cristatum* **[L.] Gaertn.**) Also called fairway wheatgrass, this is a coarse-textured, bunch-type, bluish green, cool-season grass that is well adapted for dryland turfs.

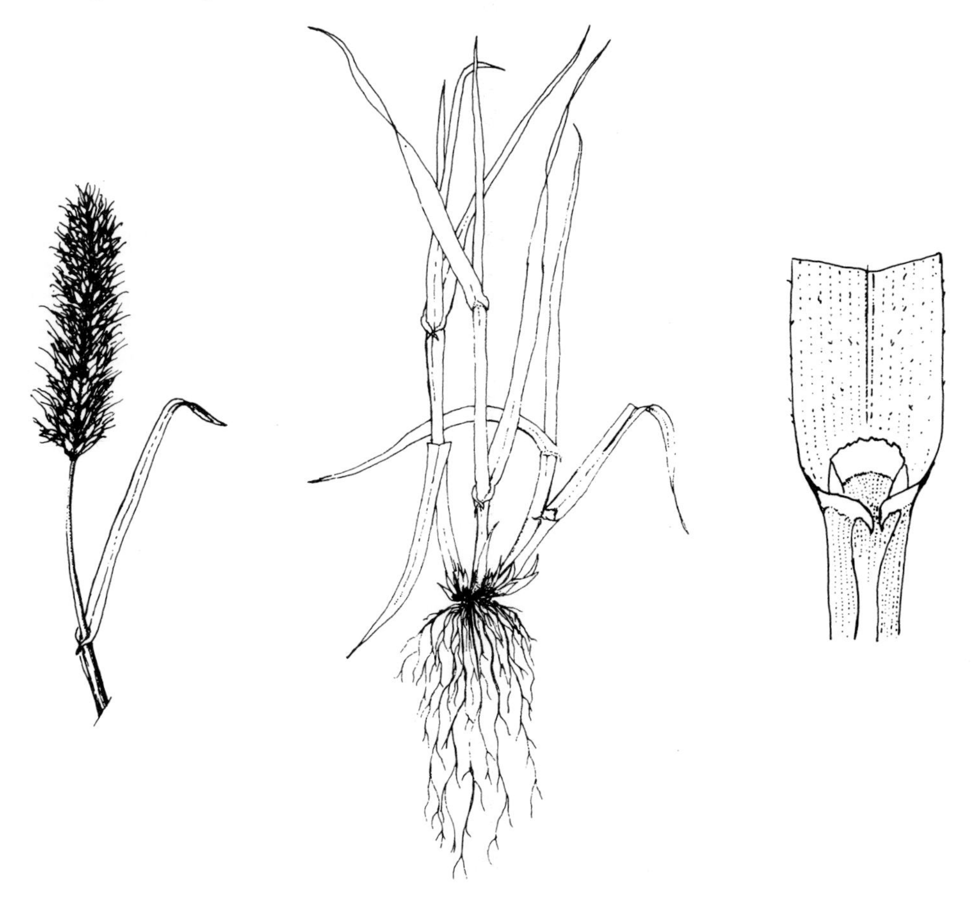

Description: Vernation—rolled; ligule—membranous, 0.1 to 0.5 mm long, truncate margin with short hairs; auricles—narrow, clawlike; collar—broad, divided; blade—flat, stiff, 2 to 5 mm wide, ridged and pubescent on adaxial surface, smooth on abaxial surface; inflorescence—flattened spike, glumes awned.

Adaptation and Use: Adapted to a wide range of soil conditions. Because of its excellent cold and drought tolerance, crested wheatgrass can persist under the extreme conditions of the northern U.S. Great Plains, Canadian prairies, and central Asia. Where adapted, it is used for unirrigated fairways, lawns, and roadside turf.

Cultural Intensity: Low to moderate. Mowing heights of 2.5 to 3.5 inches or higher are tolerated; fertilization—1 to 3 lb N/1000 ft^2/yr; if practiced, irrigation should be minimal to prevent deterioration of the turf.

Cultivars: An improved cultivar is Ephraim. This is reported to have superior cold and drought tolerance. Related species may be used on sites where crested wheatgrass is not well adapted. Western wheatgrass, a rhizomatous species, is better adapted for use in warmer and wetter climates. Desert wheatgrass, a coarser-textured, bunch-type grass, is more vigorous than crested wheatgrass but less suitable for dryland lawns.

Eragrostoideae Subfamily

Eragrostoids are warm-season grasses occurring mostly in tropical, subtropical, and warm temperate climates. Some species are well adapted to semiarid regions within these climatic zones. They are usually short- or intermediate-day plants requiring warm nights and no vernalization at the time of floral initiation. Eragrostoid grasses are festucoid in most spikelet characters but panicoid with respect to chromosome number and size and in most features of the embryo, root, stem, and leaf. Carbon fixation in photosynthesis occurs principally through the C_4 (Hatch and Slack) pathway. This subfamily contains eight tribes, two of which (Chlorideae and Zoysieae) include turfgrass genera.

Chlorideae Tribe

Genera within the Chlorideae have one- to several-flowered spikelets; the inflorescence is a unilateral spike or has several unilateral spicate branches; the ligule is usually a fringe of hairs. There are one major (*Cynodon*) and two minor (*Buchloë, Bouteloua*) turfgrass genera within this tribe.

Bermudagrasses (*Cynodon* L.C. Rich.)

Of the approximately ten species within this genus, several, including some interspecific hybrids, are used as turfgrasses in tropical and subtropical climates. The principal turfgrass species are common bermudagrass (*Cynodon dactylon*), Bradley bermudagrass (*Cynodon bradleyi*), Magennis (hybrid) bermudagrass (*CynodonXmagennisii*), and African bermudagrass (*Cynodon transvaalensis*).

Bermudagrass (*Cynodon dactylon* **[L.] Pers.**) This is a highly variable, warm-season species within which substantial differences exist in color, texture, density, vigor, and environmental adaptation. Lateral growth is by both stolons and rhizomes.

Description: Vernation—folded; ligule—a fringe of hairs, 2 to 5 mm long; auricles—absent; collar—narrow, continuous, hairy on margins; blade—flat, 1.5 to 4 mm wide, smooth or hairy on both surfaces, tapers toward the tip; inflorescence—four or five spicate branches.

Adaptation and Use: Adapted to a wide range of soil conditions. Although bermudagrass is generally not very cold tolerant, the poleward limits of adaptation have been extended with the development of several new cultivars. Its intolerance of shade necessitates the use of alternative warm-season species on sites where trees and other structures restrict sunlight penetration. It is used for lawns, sports fields, and along roadsides.

Cultivars: Breeding efforts are oriented toward improved disease and insect resistance, seed head reduction, increased density, low-temperature color retention, and cold tolerance. Superior cultivars for greens include Tifgreen and Tifdwarf; Tifway is an excellent grass for tees, fairways, and lawns; Midiron is a cold-hardy grass for athletic fields located in the transition zone between temperate and subtropical climates. All these are hybrids between *C. dactylon* and *C. transvaalensis.* Cultivars within *C. dactylon* include Ormond and Royal Cape, which have good low-temperature color retention; Texturf 1F and Texturf 10, which develop relatively few objectionable seed heads and are well adapted for use on sports fields and lawns; and Tiflawn, which has excellent wear tolerance for use on football fields. While all of these cultivars are propagated vegetatively, several new seeded cultivars have been developed, including Cheyenne, Jackpot, Mirage, and Sahara.

Bradley Bermudagrass (*Cynodon bradleyi* **Stent.**) A fine-textured, dull green, stoloniferous grass with hairy leaves. It is used in the lowlands of central Africa for golf greens.

Magennis Bermudagrass (*Cynodon magennisii* **Hurcombe**) A very fine-textured, bright green grass with threadlike stolons. It is a natural hybrid between *C. dactylon* and *C. transvaalensis.* An improved cultivar of this species is Sunturf.

African Bermudagrass (*Cynodon transvaalensis* **Burtt-Davy**) Similar in appearance to *C. magennisii,* this is often puffy grass with soft hairy leaves. It is one of the parents of many fine-textured hybrid bermudagrass cultivars.

(*Buchloë* **Engelm.**)

One of five dioecious genera within the Chlorideae tribe, *Buchloë* has male (staminate) and female (pistillate) flowers occurring on different plants. It contains only one species, buffalograss (*Buchloë dactyloides*), a native of the North American "short grass" prairie.

Buffalograss (*Buchloë dactyloides* **[Nutt.] Engelm.**) A fine-textured, grayish green, stoloniferous species, buffalograss is well adapted to semiarid regions within the temperate and subtropical climatic zones.

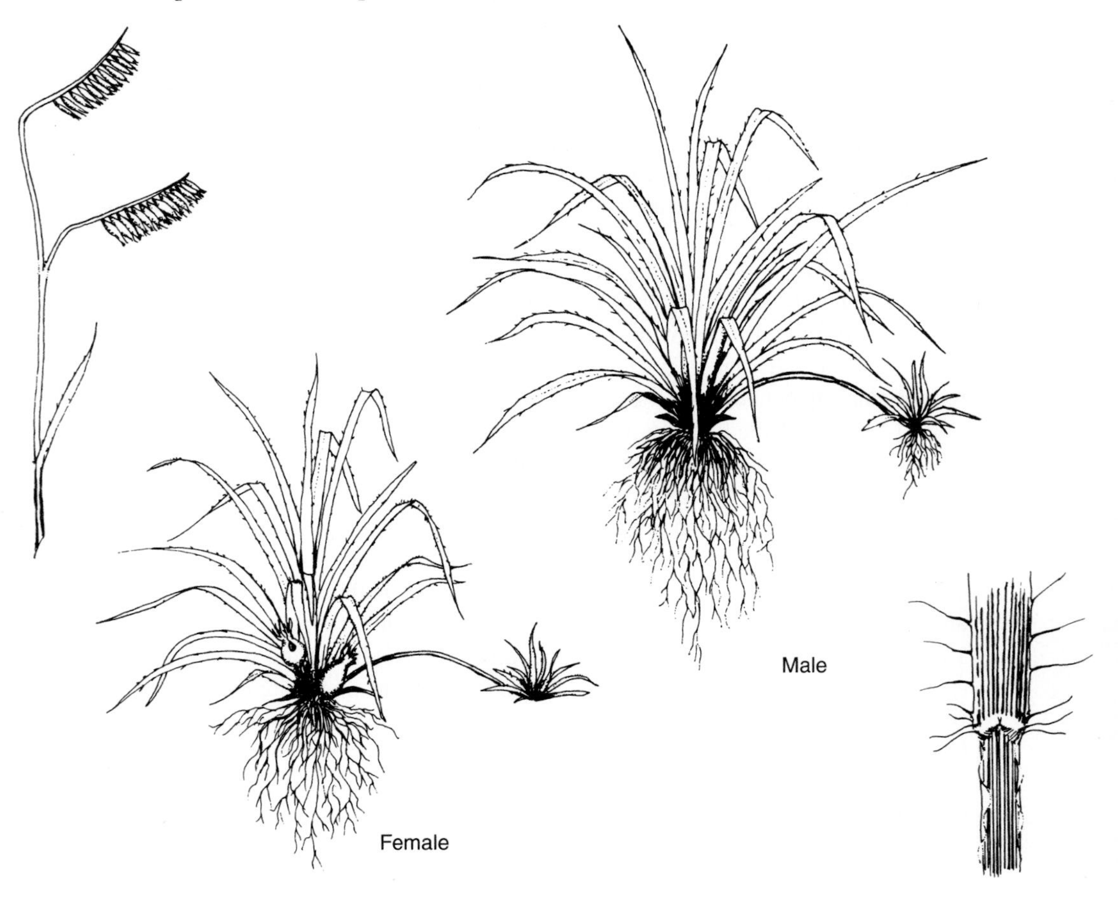

Description: Vernation—rolled; ligule—a fringe of hairs, 0.5 to 1 mm long; auricles—absent; collar—broad, continuous, with long hairs; blade—flat, 1 to 3 mm wide, ridged on adaxial surface, long hairs scattered on both surfaces; inflorescence—dioecious, pistillate heads composed of a burr of female spikelets partially hidden within the leaves, staminate heads with two or three short spikes on slender culms elevated above the leaves.

Adaptation and Use: Adapted to a wide range of soil conditions, but exceptionally well suited to fine-textured, alkaline soils. Because of its excellent hardiness to drought and temperature extremes, buffalograss is used for unirrigated turfs in semiarid regions.

Cultural Intensity: Low. Mowing heights of 0.5 to 1.2 inches for lawns at infrequent intervals due to slow vertical growth; fertilization—0.5 to 2 lb N/1000 ft^2/yr; irrigation minimal or not at all.

Cultivars: Improved cultivars include the seeded types Sharps and Texolea and the vegetatively propagated turf types Prairie from Texas and 609 from Nebraska.

Gramagrasses (*Bouteloua* **Lag.**)

This genus consists of about fifty species that occur in semiarid regions primarily within the subtropical climatic zone. Species that have some value for dryland turf include blue grama (*Bouteloua gracilis*) and sideoats grama (*Bouteloua curtipendula*).

Blue Grama (*Bouteloua gracilis* **[H.B.K.] Lag. ex Steud.**) A fine-textured, grayish green, weakly rhizomatous species, blue grama is an important range grass that can be used along roadsides or for unirrigated utility turfs in semiarid regions.

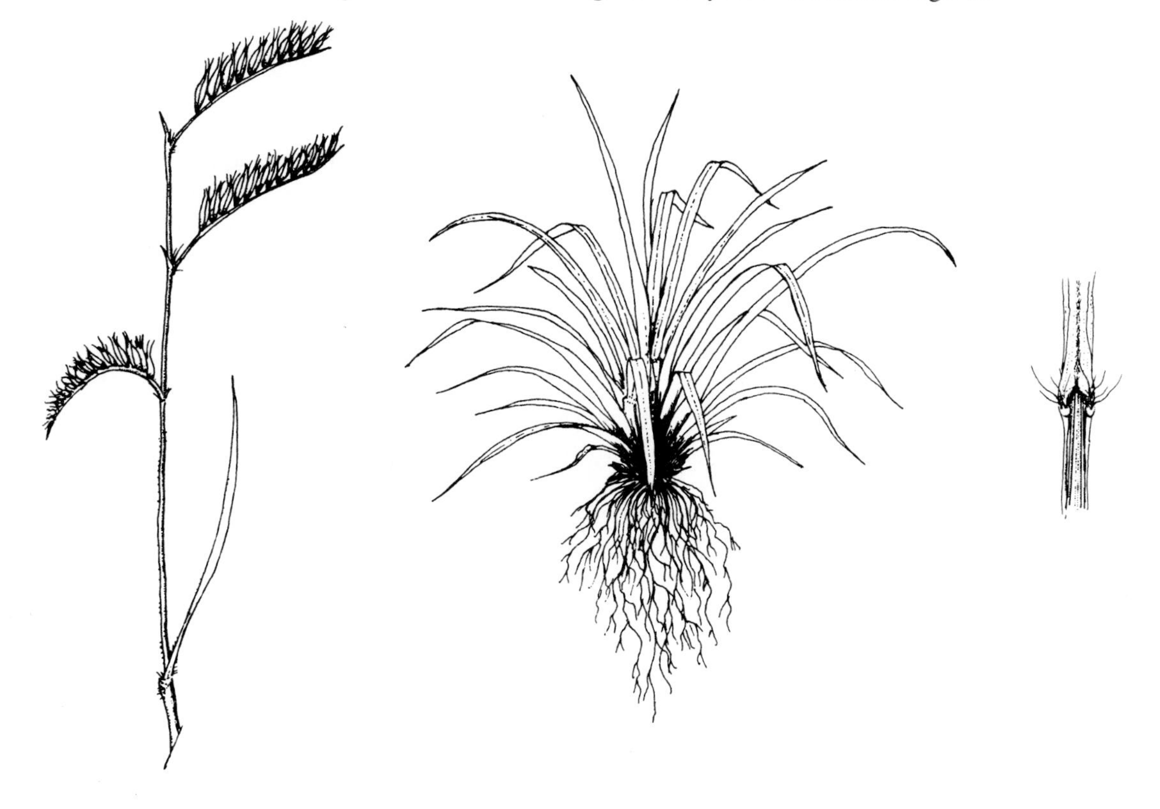

Description: Vernation—folded; ligule—a dense fringe of hairs, 0.1 to 0.5 mm long; auricles—absent; collar—broad, continuous, with long hairs; blade—flat or curled, 1 to 2 mm wide, ridged on adaxial surface, pubescent near the base, smooth on abaxial surface, margins scaly; inflorescence—one to three curled spikes.

Adaptation and Use: Adapted to a wide range of soil conditions, but is well suited for use on fine-textured, upland, alkaline soils. Because of its good heat and drought tolerance, blue grama is well adapted to semiarid regions for use in low-quality, utility turfs.

Cultivars: None specifically for turf, but many forage cultivars.

Sideoats Grama (*Bouteloua curtipendula* **[Michx.] Torr.**) Although more widely distributed than blue grama, sideoats grama is less tolerant of severe droughts and less desirable for use as a utility turfgrass in semiarid regions. The two species are fairly similar in appearance, but sideoats grama has a rolled vernation and long, scattered hairs along the adaxial leaf blade surface.

Zoysieae Tribe

Genera within the Zoysieae have one-flowered spikelets; the inflorescence is a contracted raceme; the ligule is usually a fringe of hairs. There is one major genus (*Zoysia*) within this tribe.

Zoysiagrass (*Zoysia* **Willd.**)

Three of five species within this genus are used as turfgrasses. They are Japanese (Korean) lawngrass (*Zoysia japonica*), manilagrass (*Zoysia matrella*), and mascarenegrass (*Zoysia tenuifolia*). These species are differentiated in terms of aggressiveness, texture, and cold tolerance. All are propagated vegetatively.

Japanese Lawngrass (*Zoysia japonica* **Steud.**) A medium-textured, slow-growing, warm-season grass that spreads by stolons and rhizomes, Japanese lawngrass is adapted to tropical, subtropical, and temperate climates. Because of its outstanding cold tolerance,

it will even persist in a subarctic climate; however, its very short growing season makes it impractical for use in this climatic zone.

Description: Vernation—rolled; ligule—a fringe of hairs, 0.2 mm long; auricles—absent; collar—broad, continuous, with long hairs; blade—flat, 2 to 4 mm wide, stiff, long hairs distributed along smooth adaxial and, sometimes, abaxial surfaces; inflorescence—short, contracted raceme, spikelets laterally compressed.

Adaptation and Use: Adapted to a wide range of soil conditions, but grows best on well-drained, slightly acid, medium-textured soils of moderate fertility. Although quite tolerant of drought, heat, and cold stresses, it is slow to green up in spring, and late-season discoloration begins with the onset of 50° to 55°F temperatures. This results in a straw-colored turf during much of the winter season. Another feature that limits the use of Japanese lawngrass is its slow establishment rate. However, it is being successfully used as a fairway and lawn grass in transitional climates, between temperate and subtropical zones, where some cool-season and other warm-season grasses encounter the limits of their environmental adaptation.

Cultural Intensity: Moderate. Because of the density and toughness of the shoots, mowing should be performed with heavy, reel-type mowers to sustain a uniform surface. Mowing heights of 0.5 to 1 inch are optimum; fertilization—1.5 to 3 lb N/1000 ft^2/yr; irrigation should be performed as needed to sustain color and growth during drought periods.

Cultivars: Meyer is the most widely used cultivar. Interspecific hybrids of *Z. japonica* and *Z. tenuifolia,* called Emerald and El Toro, form very dense, dark green turfs. The hybrid zoysias are not as cold hardy as the *Z. japonica* and, therefore, are not as persistent in temperate and colder climates.

Manilagrass (*Zoysia matrella* **[L.] Merr.**) This is a finer-textured, denser, slower-growing species of *Zoysia* that lacks the cold tolerance of Japanese lawngrass. It is suitable for use as a lawn grass in warm subtropical and tropical climates.

Mascarenegrass (*Zoysia tenuifolia* **Willd. ex Trin.**) This is the finest-textured, densest, and slowest-growing *Zoysia* species. Its cold tolerance is even lower than that of manilagrass. Where used in warm subtropical and tropical climates, it is usually maintained as an unmowed ground cover. Where a very dense thatch is allowed to develop, especially on untrafficked sites, convolutions develop in the surface that disrupt the uniform appearance of the turf.

Panicoideae Subfamily

Panicoids are warm-season grasses occurring mostly in tropical and subtropical climates. They are usually short- or intermediate-day plants requiring warm nights and no vernalization at the time of floral initiation. Within their typically one-flowered spikelets, disarticulation occurs below the glumes, and the entire spikelet (minus glumes) falls when florets are shed. Inflorescences are usually panicles, but occasionally racemes, with spikelets dorsally compressed. Carbon fixation, in photosynthesis, occurs principally through the C_4 (Hatch and Slack) pathway. This subfamily contains two tribes, Paniceae and Andropogoneae, both of which include turfgrass genera.

Paniceae Tribe

Genera within the Paniceae have panicle or raceme inflorescences; the ligule may be membranous, a fringe of hairs, or absent. The four turfgrass genera within this tribe are *Axonopus, Paspalum, Pennisetum,* and *Stenotaphrum.*

Carpetgrass (*Axonopus* **Beauv.**)

Of the seventy species within this genus, only two, common carpetgrass (*Axonopus affinis*) and tropical carpetgrass (*Axonopus compressus*) are used as turfgrasses.

Common Carpetgrass (*Axonopus affinis* **Chase**) A coarse-textured, low-growing, light green, warm-season grass, common carpetgrass spreads by stolons and seed, and is adapted to tropical and warm subtropical climates.

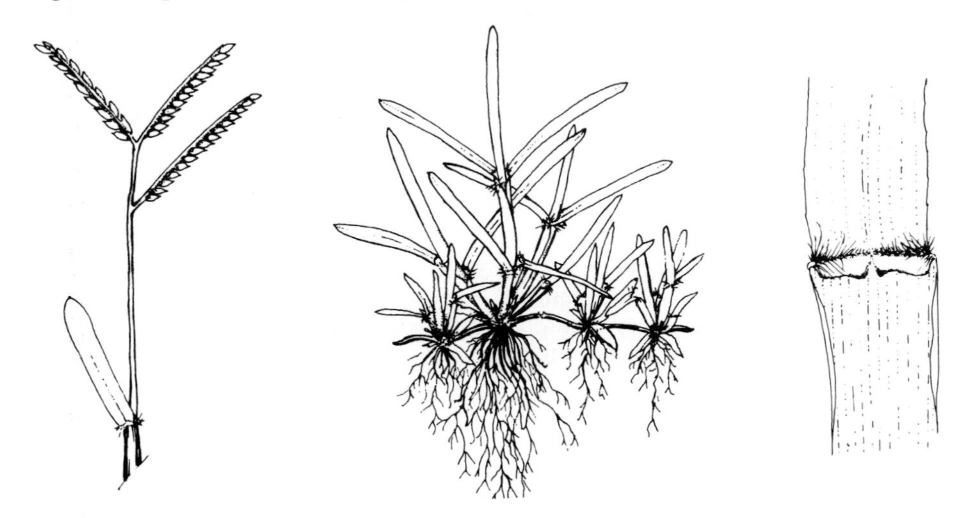

Description: Vernation—folded; ligule—a fringe of hairs fused at the base, 1 mm long; auricles—absent; collar—narrow, continuous, occasionally with hairs; blade—4 to 8 mm wide, margins with short hairs near the tip; inflorescence—three spicate racemes, spikelets widely spaced in two rows on one side of a flattened rachis.

Adaptation and Use: Wet or moist, sandy, acid soils of low fertility. Where adapted, it is used along roadsides and for some lawns. Because of its rapid establishment from seed and minimal cultural requirements, it is especially suited for soil stabilization on steep slopes.

Cultural Intensity: Low. Mowing heights of 1 to 2 inches are optimum; a rotary mower should be used during summer to cut off unsightly seed heads; fertilization—2 lb N/1000 ft^2/yr or less; irrigation is required to sustain carpetgrass on well-drained soils.

Cultivars: None available.

Tropical Carpetgrass (*Axonopus compressus* **[Swarty] Beauv.**) Similar to common carpetgrass in appearance and environmental adaptation; however, it is less cold hardy and its use should be restricted to humid subtropical and tropical climates.

(Paspalum **L.**)

This is a large genus containing 320 species; however, only two, bahiagrass (*Paspalum notatum*) and seashore paspalum (*Paspalum vaginatum*), are used as turfgrasses.

Bahiagrass (*Paspalum notatum* **Flugge**) A tough, coarse-textured, rhizomatous, warm-season turfgrass, bahiagrass is adapted to tropical and warm subtropical climates.

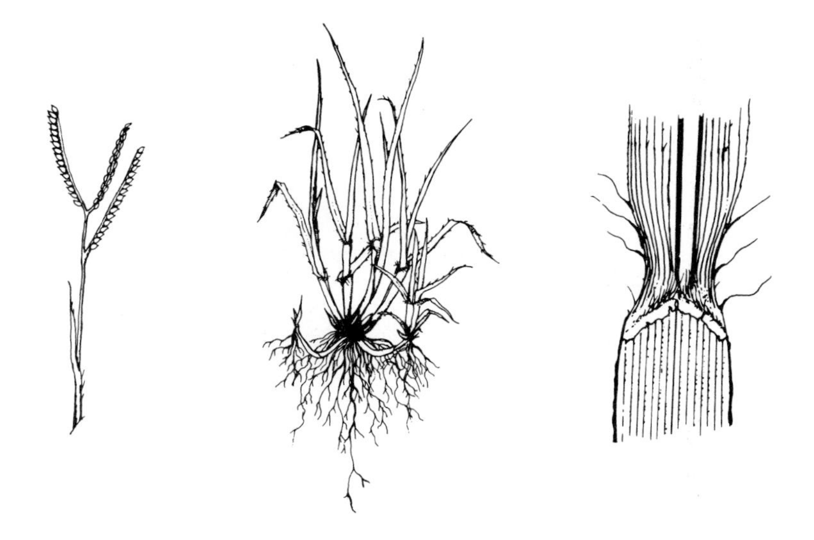

Description: Vernation—rolled; ligule—membranous, 1 mm long, truncate; auricles—absent; collar—broad; blade—flat to folded, 4 to 8 mm wide, margins sparsely hairy toward base; inflorescence—two or three unilateral spikelike branches (racemes).

Adaptation and Use: Adapted to a wide range of soil conditions, but grows best on sandy, slightly acid, infertile soils. Bahiagrass can be propagated by seed; it forms a tough, wear-resistant, open turf and is especially well suited for use along roadsides and for other utility turfs.

Cultural Intensity: Low. Mowing heights of 2.5 to 3.5 inches preferred; a rotary mower should be used during summer to cut off unsightly seed heads; fertilization—1 to 4 lb N/1000 ft^2/yr; irrigation requirement is minimal.

Cultivars: Pensacola is a finer-textured cultivar used extensively along roadsides in Florida. Argentine is preferred for use in lawns because of its superior density, color, disease resistance, and response to fertilization.

Seashore Paspalum (*Paspalum vaginatum* **Swartz.**) Also called sand knotgrass, seashore paspalum is native to Africa and the Americas. It is adapted to tropical and warm subtropical climates. It forms a dense, fine-textured turf of dark green color. Both stolons and rhizomes may be present. Its outstanding feature is its salt tolerance and adaptation to areas where salt accumulation in soil limits the survival of other turfgrasses.

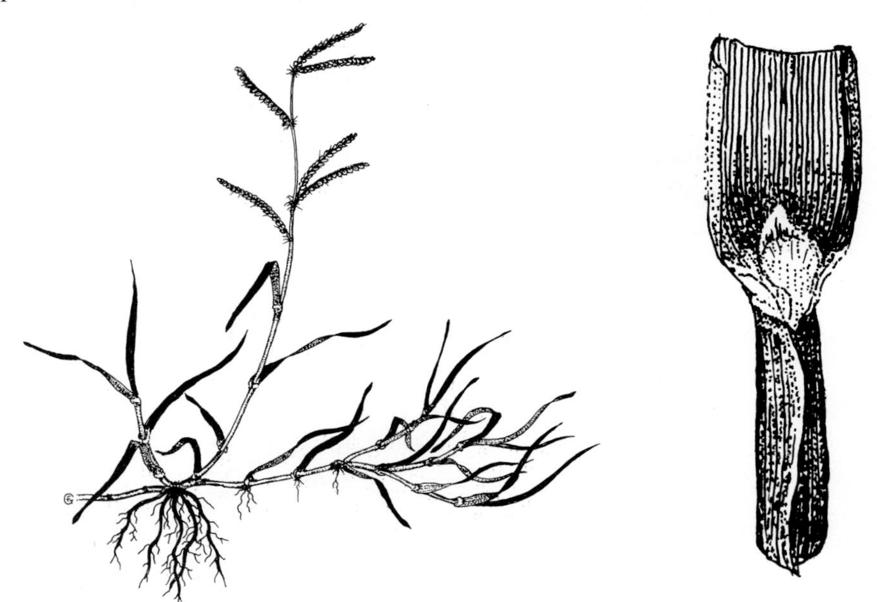

Description: Vernation—rolled; ligule—membranous, 2–3 mm long, pointed; auricles—absent; collar—broad, continuous; blade—flat with margins rolled inward, 2 to 4mm wide, margins sparsely hairy toward base; inflorescence—one or two unilateral spikelike branches (racemes).

Adaptation and Use: Adapted to marshy, brackish conditions and moist, saline soils. It will tolerate waterlogged conditions and periodic flooding. It can be used for utility, lawn, and sports turfs (including greens) on salt-affected sites.

Cultural Intensity: Low to moderate. Mowing heights of 0.15 to 0.5 inch preferred; fertilization—2 to 4 lb N/1000 ft^2/yr; because of its outstanding drought tolerance, its irrigation requirement is less than that of many other warm-season turfgrasses maintained at the same intensity of culture; vertical mowing needed to control puffiness.

Cultivars: Commercially available cultivars include Excaliber (formerly Adalayd) and Futurf from Australia.

(*Pennisetum* **L. Rich.**)

A genus of about eighty species, *Pennisetum* includes one species, Kikuyugrass (*Pennisetum clandestinum*), that invades subtropical and highland tropical turfs and is sometimes cultured as a desirable turfgrass.

Kikuyugrass (*Pennisetum clandestinum* **Hochst ex Chiov.**) A medium-textured, light green, warm-season grass that spreads by vigorous rhizomes and stolons, Kikuyugrass forms a tough, dense, springy turf under close mowing.

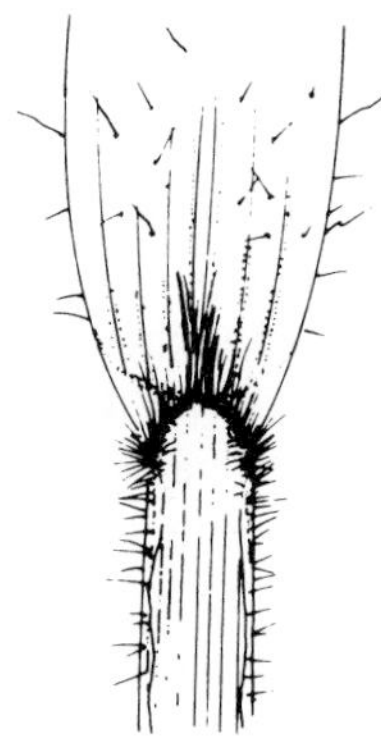

Description: Vernation—folded; ligule—a fringe of hairs, 2 mm long; auricles—absent; collar—broad, continuous; blade—flat, keeled, sparsely hairy on both surfaces; inflorescence—dense, bristly, tightly contracted panicle.

Adaptation and Use: Moist, medium-textured soils of high fertility. It is well adapted to elevations above 6000 feet in the moist tropics of Mexico and central Africa; however, in the warm, damp, tropical lowlands, it becomes severely diseased and does not persist. It is regarded as a serious weed in intensively cultured turfs along coastal regions of southern California. Its cultural requirements have not been adequately characterized.

(Stenotaphrum **Trin.**)

Of the six species within this genus, only one, St. Augustinegrass (*Stenotaphrum secundatum*), is used as a turfgrass in tropical and warm subtropical climates.

St. Augustinegrass (*Stenotaphrum secundatum* **[Walt.] Kuntze.**) A coarse-textured, aggressive, stoloniferous, warm-season grass that is widely used as a lawn grass within its region of adaptation.

Description: Vernation—folded; ligule—a fringe of hairs, 0.3 mm long; auricles—absent; collar—broad, continuous, narrowed to form a short stalk at the base of the blade; blade—flat, 4 to 10 mm wide, smooth on both surfaces, with blunt tip; inflorescence—short, unilateral, spikelike raceme.

Adaptation and Use: Adapted to a wide range of soil conditions, but grows best in moist, well-drained, sandy, slightly acid soils of moderate to high fertility. Because of its poor cold tolerance, it is best adapted to coastal areas with mild winters. Some cultivars are very susceptible to several diseases and to injury from chinchbugs.

Cultural Intensity: Moderate. Mowing heights of 2.5 to 3.5 inches; fertilization—3 to 6 lb N/1000 ft^2/yr; irrigation may be required frequently during drought periods, especially where established in sandy soils with poor water-holding capacity.

Cultivars: Improved density and finer texture is obtainable with Bitter Blue, Floratine, Raleigh, and Seville. Resistance to chinchbugs and to a viral disease, St. Augustine decline (SAD), is reported for Floratam, a coarse-textured cultivar widely used in Florida and other U.S. Gulf Coast states. A similar cultivar, Floralawn, offers increased chinchbug tolerance and drought tolerance.

Andropogoneae Tribe

Genera within the Andropogoneae have spikelets organized in pairs of one sessile and one pedicel, the pediceled spikelet reduced in most genera; the ligule may be membranous, a fringe of hairs, or absent. The one turfgrass genus within this tribe is *Eremochloa.*

(Eremochloa **Buese**)

Of the ten species within this genus, only one, centipedegrass (*Eremochloa ophiuroides*), is used as a turfgrass in tropical and warm subtropical climates.

Centipedegrass (*Eremochloa ophiuroides* **[Munro] Hack.**) A medium-textured, slow-growing, stoloniferous, warm-season grass used as a lawn or utility turf.

Description: Vernation—folded; ligule—membranous with short hairs at the top, 0.5 mm long; auricles—absent; collar—broad, continuous, constricted by the fused keel, pubescent below; blade—flat, 3 to 5 mm wide, smooth, margins hairy toward the base; inflorescence—single, spikelike raceme.

Adaptation and Use: Adapted to a wide range of soil conditions, but especially suited for moist, acid, sandy soils of low fertility. It is used for soil stabilization and for minimum-maintenance turfs.

Cultural Intensity: Low. Mowing heights of 1 to 2 inches; fertilization—1 to 2 lb N/1000 ft^2/yr, iron applications may be necessary to correct chlorosis; irrigation may be required during drought periods to sustain color and growth.

Cultivars: Common type used most extensively. Oklawn is more tolerant of drought and temperature extremes.

TURFGRASS LEAF ANATOMY AND MORPHOLOGY

The anatomy of turfgrass leaf blades (laminae) varies among taxonomic units; however, certain features are common throughout the grass family. The mature lamina of all grasses consists of an adaxial (upper) and abaxial (lower) epidermis enclosing several layers of mesophyll cells interspersed within a network of vascular bundles (Figure 3.2). The epidermis includes specialized elements: stomates, trichomes, epicuticular wax, and an array of differentiated cells. Epidermal cells vary in size and shape. For example, large, thin-walled bulliform cells may be present that function in unrolling or unfolding the leaf in response to increasing turgor pressure and water availability. The stomates are the pores in the epidermis through which gaseous exchange takes place. While occurring on both sides of the leaf, they tend to be found in regions overlying mesophyll cells rather than over the veins. Each stomate consists of two kidney-shaped guard cells surrounding a central pore. Trichomes are epidermal outgrowths, or hairs, that vary widely among species in their size and density. The cuticle, which is composed mostly of cutin, lipids, and waxes, is a layer covering the entire leaf surface. In most turfgrasses it is thin and nearly transparent.

The internal mesophyll cells constitute the primary photosynthetic tissue of the leaf. These vary in size and shape, with many air spaces interspersed among them. The vascular

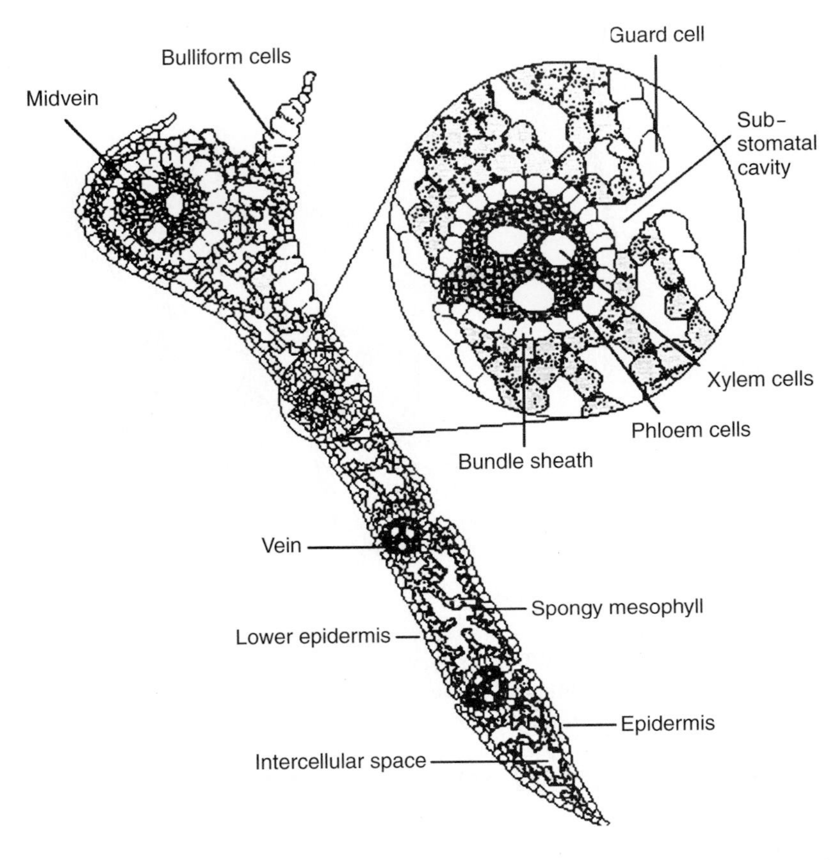

Figure 3.2. Illustration of turfgrass leaf cross-section with key anatomical features labeled.

bundles are surrounded by a bundle sheath. The bundle sheath is formed by a single layer of cells in C_4 (eragrostoid and panicoid) turfgrasses. Beyond this layer of cells is a series of elongated mesophyll cells radiating from the vascular bundles; this is called "Kranz anatomy." In C_3 (festucoid) turfgrasses, a double layer of cells forms the bundle sheath, while mesophyll cells external to the bundle sheath are randomly arranged.

Festucoid Species

The bluegrasses (*Poa* species), as seen in Figure 3.3a, b, and c, have relatively smooth leaf surfaces and very prominent bulliform cells located on the adaxial side on either side of the central vein. Under conditions of moisture stress, these cells lose turgidity, causing the leaf to fold. As a consequence, bluegrasses are said to have a folded vernation. Because of the large central vein (called the midrib), the abaxial side is slightly keeled.

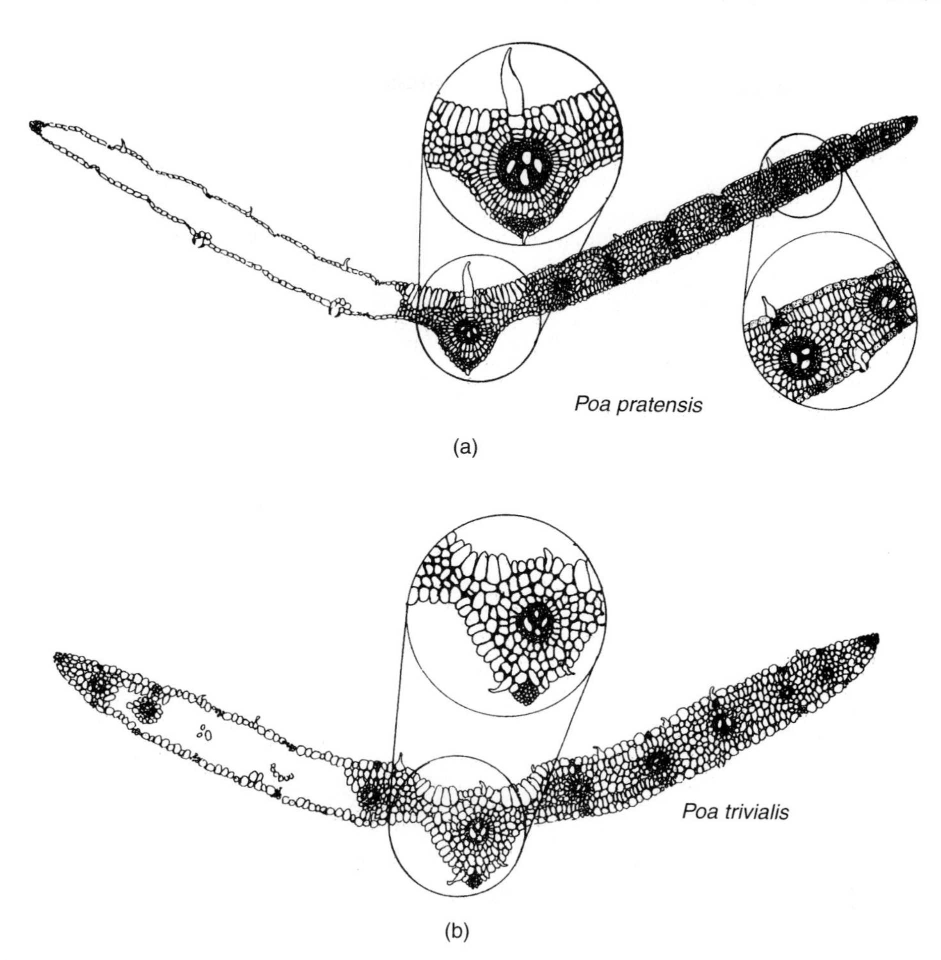

Figure 3.3. Leaf blade cross-sections of (a) *Poa pratensis,* (b) *Poa trivialis,* (c) *Poa annua,* (d) *Agrostis stolonifera,* (e) *Agrostis capillaris,* (f) *Agrostis canina,* (g) *Agrostis gigantea,* (h) *Lolium perenne,* (i) *Lolium multiflorum,* (j) *Festuca rubra,* (k) *Festuca ovina,* (l) *Festuca arundinacea,* (m) *Cynodon dactylon,* (n) *Cynodon dactylonXtransvaalensis,* (o) *Zoysia tenuifolia,* (p) *Zoysia japonicaXtenuifolia,* (q) *Zoysia matrella,* (r) *Stenotaphrum secundatum,* (s) *Eremochloa ophiuroides,* and (t) *Paspalum notatum* illustrated to show some important anatomical and morphological features.

The bentgrasses (*Agrostis* species), as seen in Figure 3.3d, e, f, and g, have strongly ridged adaxial leaf surfaces with large vascular bundles beneath each ridge. Under moisture stress, these leaves roll with the ridges coming together tightly; thus, the bentgrasses have a rolled vernation. Ridged adaxial surfaces are also evident in the ryegrasses (*Lolium* species) as seen in Figure 3.3h and i. Perennial ryegrass, especially, has a

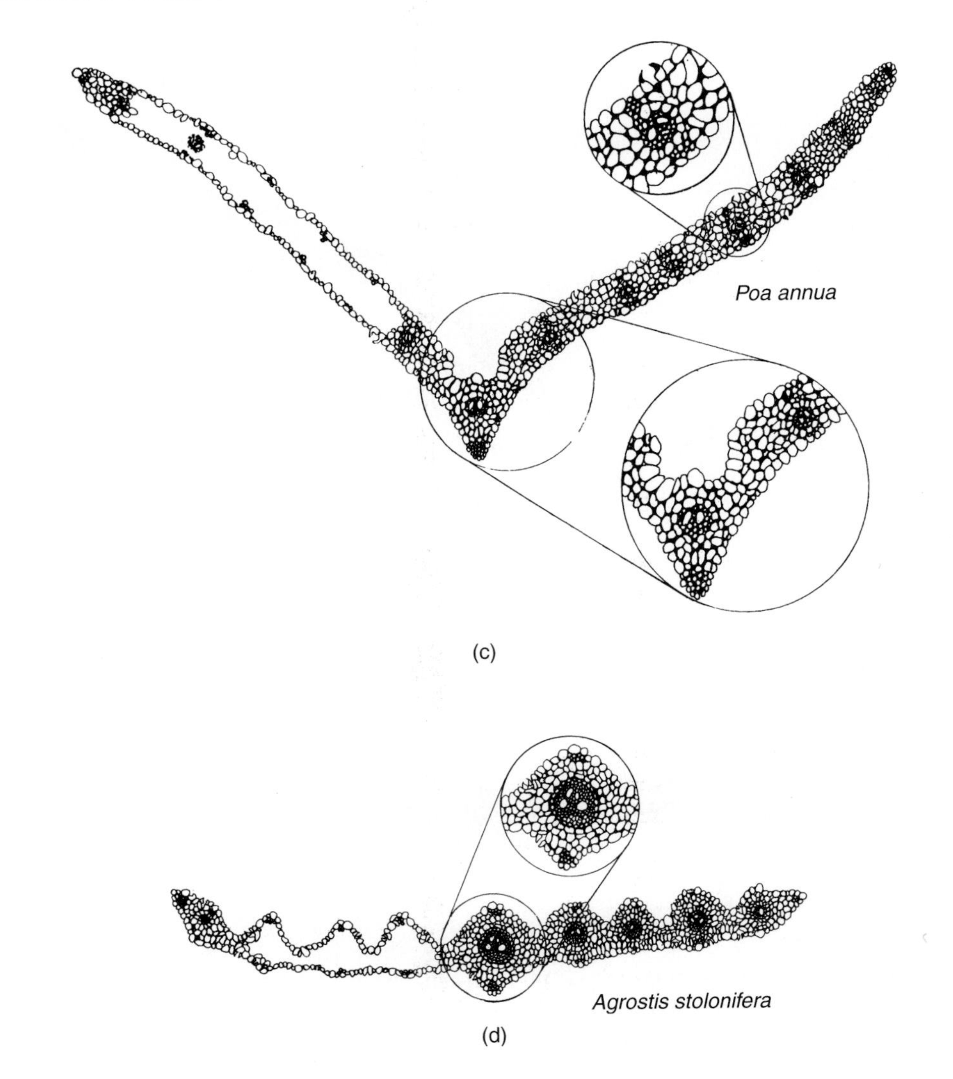

(c)

(d)

(e)

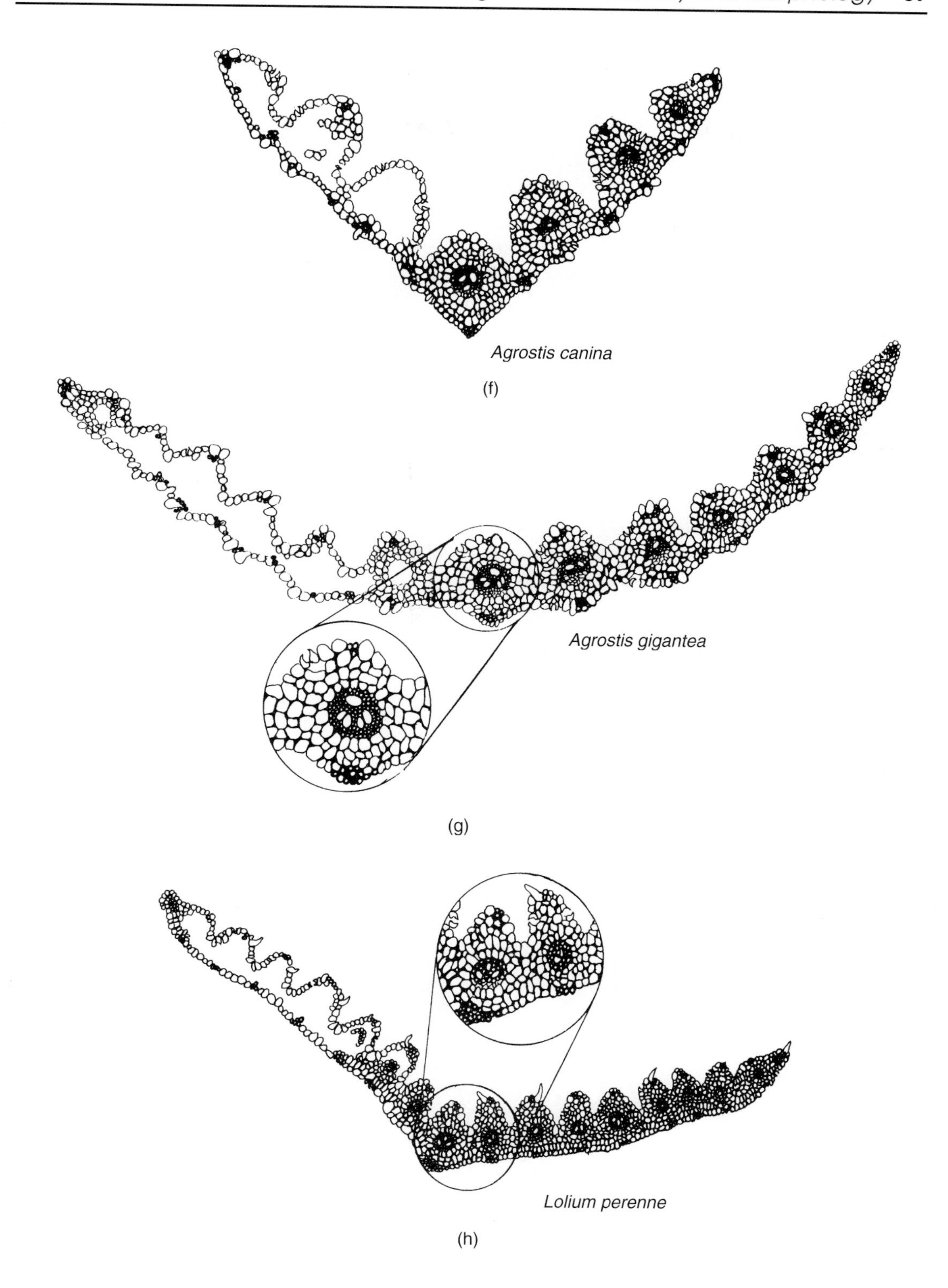

(f)

(g)

(h)

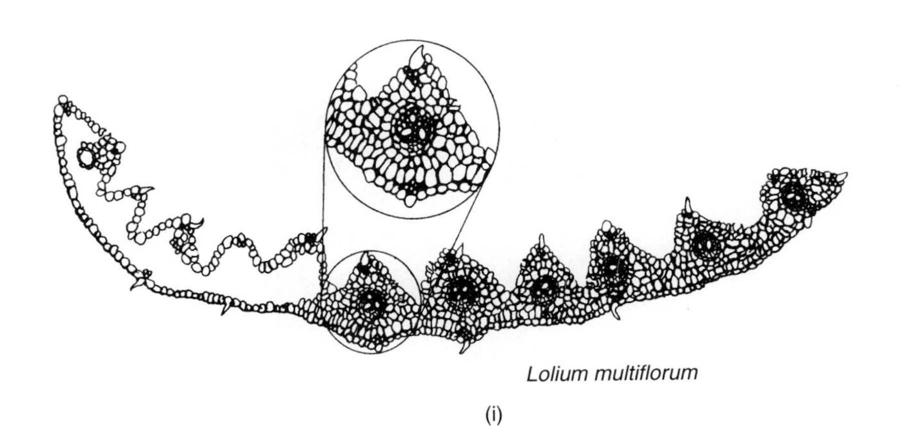

Lolium multiflorum

(i)

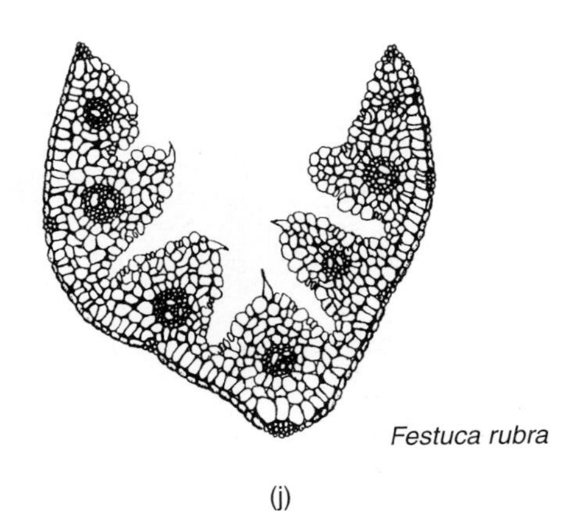

Festuca rubra

(j)

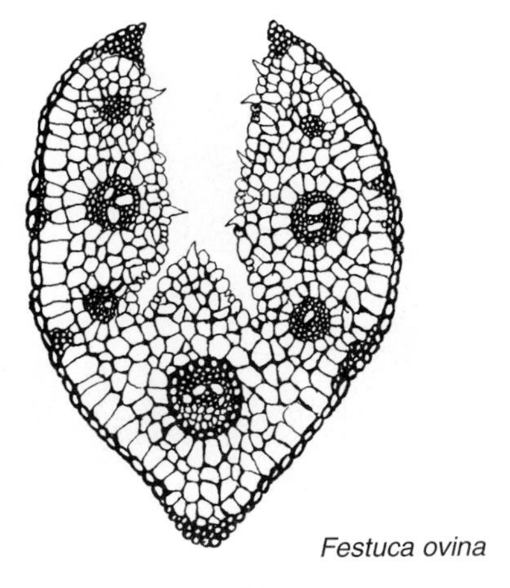

Festuca ovina

(k)

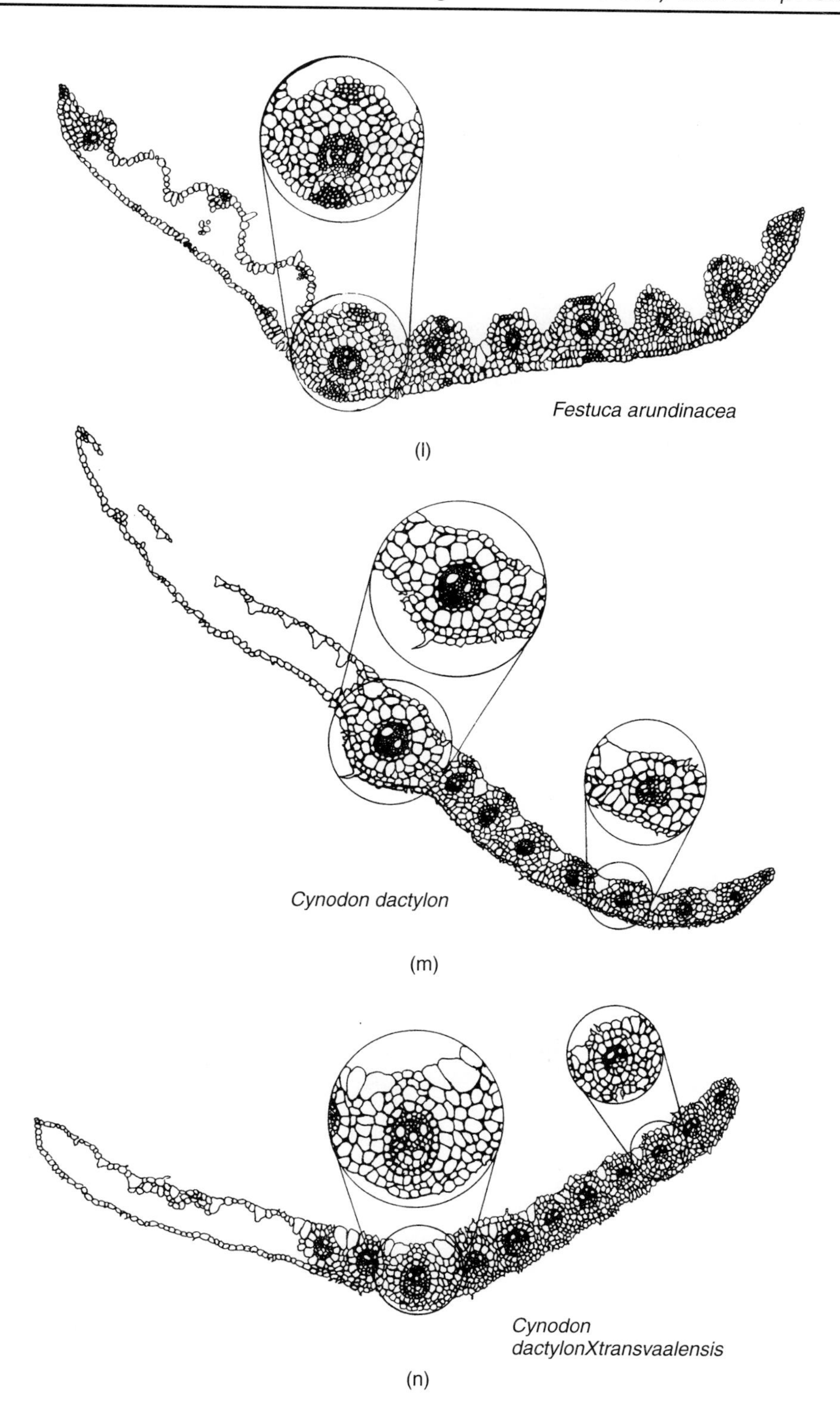
Festuca arundinacea
(l)
Cynodon dactylon
(m)
Cynodon dactylonXtransvaalensis
(n)

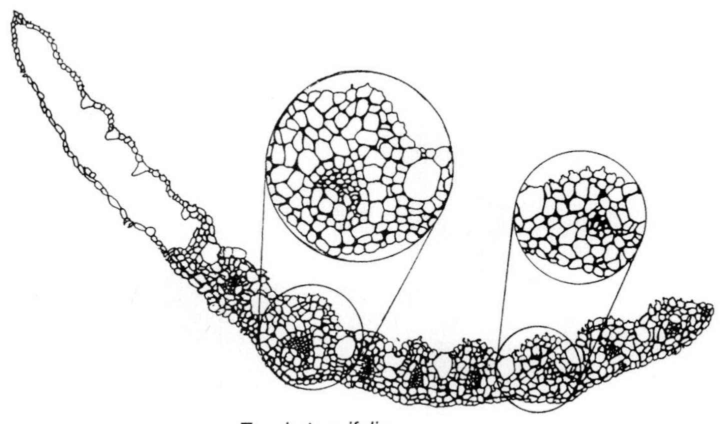

Zoysia tenuifolia

(o)

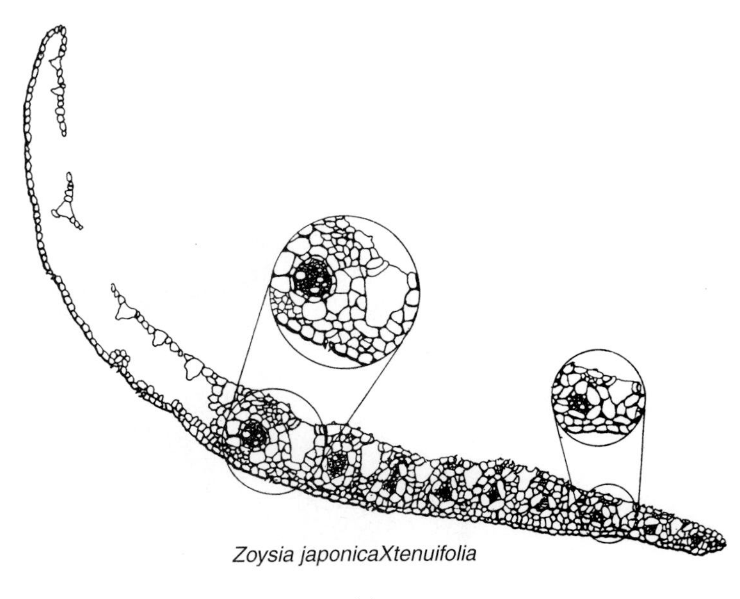

Zoysia japonicaXtenuifolia

(p)

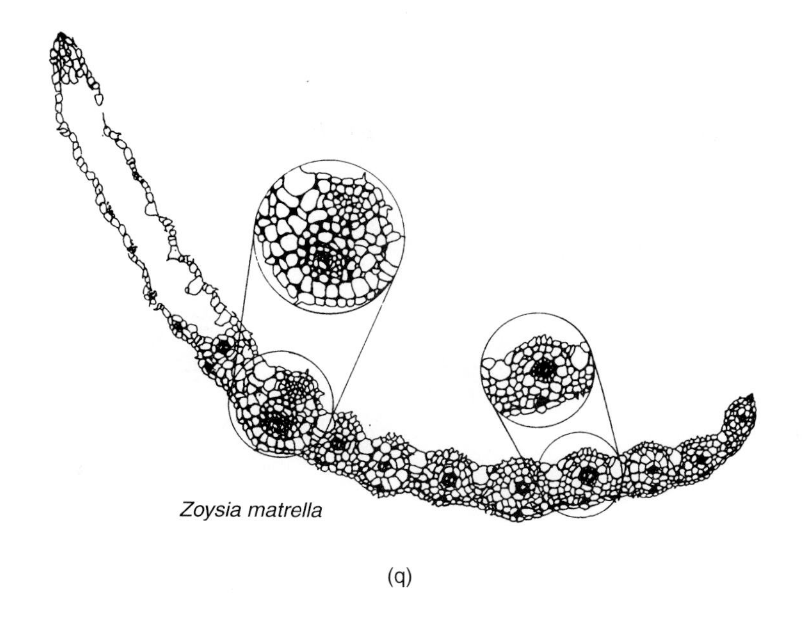

(q)

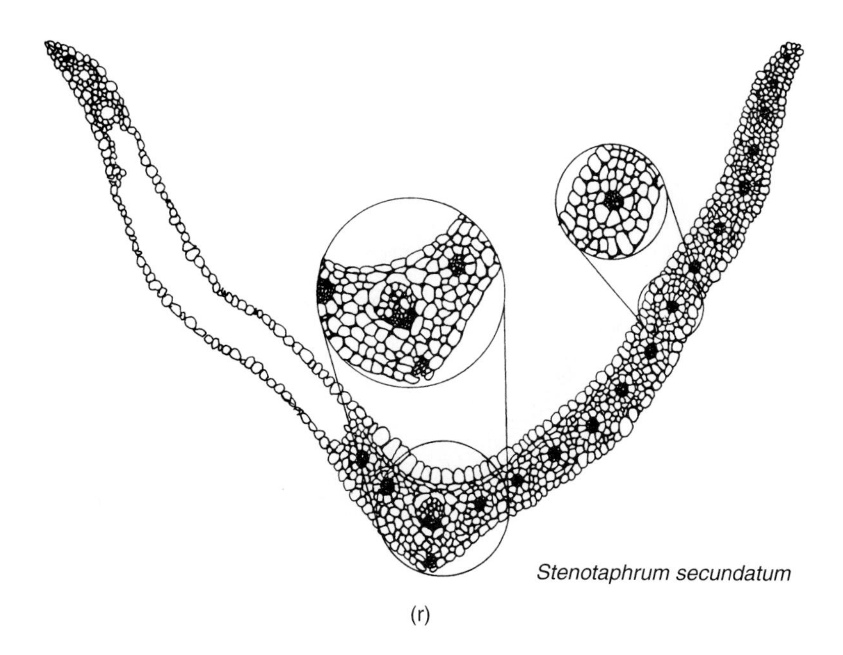

(r)

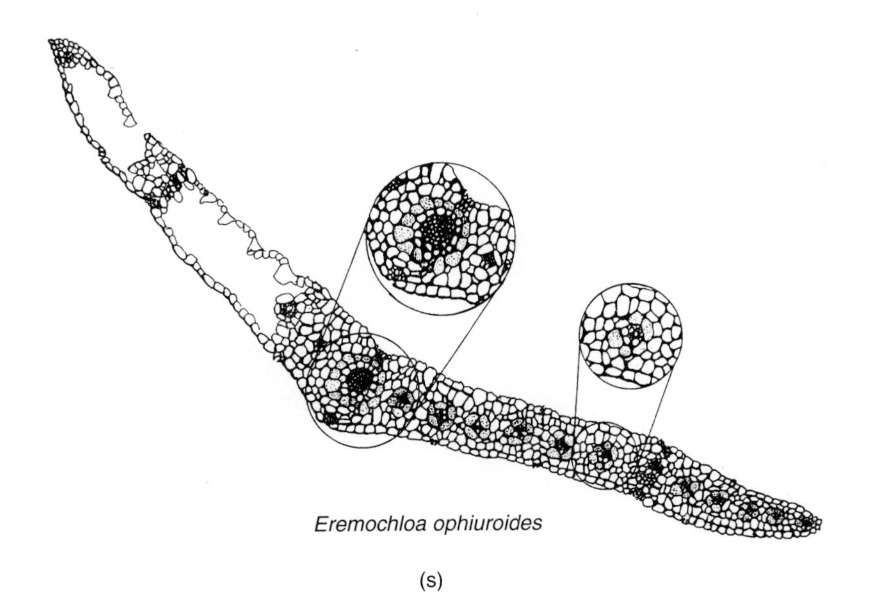

Eremochloa ophiuroides

(s)

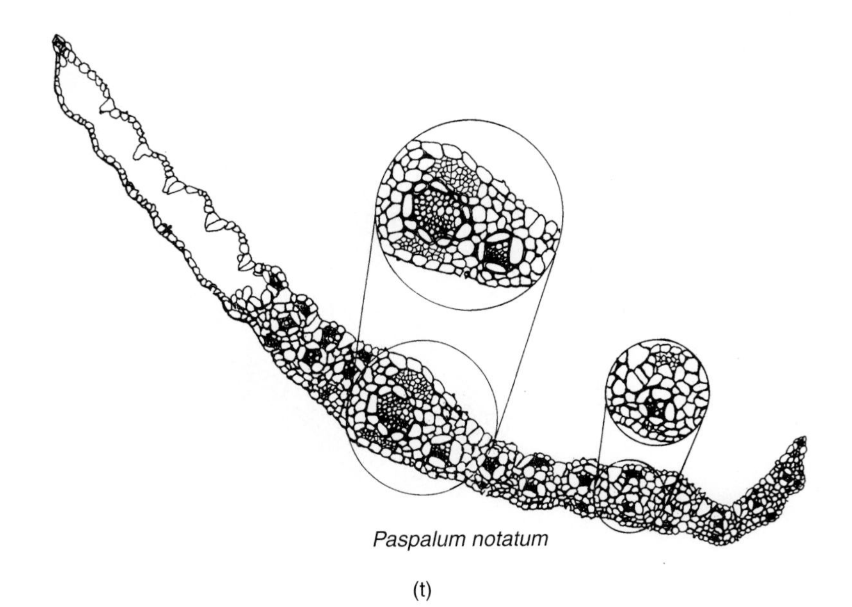

Paspalum notatum

(t)

strongly keeled abaxial leaf because of the cells protruding beneath the midrib, and a folded vernation; annual ryegrass has a rolled vernation. As shown in Figure 3.3j and k, the fine fescues (*Festuca rubra* and *ovina*) also have ridged adaxial leaf surfaces. Under moisture stress, the leaves tightly fold to form an almost needle-like shape. Tall fescue (*Festuca arundinacea*), as shown in Figure 3.3l, has a stiff blade that is deeply ridged on the adaxial surface and flat on the abaxial side; it has a rolled vernation.

Eragrostoid and Panicoid Species

The bermudagrasses (*Cynodon* species), as seen in Figure 3.3m and n, have relatively smooth, sometimes hairy, leaf surfaces and a folded vernation. The zoysiagrasses (*Zoysia* species), as shown in Figure 3.3o, p, and q, also have relatively smooth leaf surfaces; however, they all have a rolled vernation. Some zoysias have sparse, long hairs that are usually confined to the adaxial leaf surface. As seen in Figure 3.3r, St. Augustinegrass (*Stenotaphrum secundatum*) has a large, relatively flat, smooth leaf and a folded vernation. Centipedegrass (*Eremochloa ophiuroides*), shown in Figure 3.3s, is similar in that it also has a flat, smooth leaf and a folded vernation. Finally, as seen in Figure 3.3t, bahiagrass (*Paspalum notatum*) has a relatively smooth, flat leaf, and a rolled vernation.

CRITERIA FOR SELECTING TURFGRASS SPECIES

Turfgrasses are selected for planting based on a variety of considerations. The anticipated persistence and quality of a turf and its rate of establishment are of primary importance. Rapid establishment is desired for soil stabilization, minimized postplanting care, and overall utility of a site. A comparison of the six most popular cool-season and warm-season turfgrasses shows the relative ranking of these grasses in establishment vigor (Table 3.2).

Representative germination rates for Kentucky bluegrass and perennial ryegrass are eighteen and seven days, respectively. Thus, where perennial ryegrass is used as a nursegrass with Kentucky bluegrass, its purpose is to promote quick cover while allowing Kentucky bluegrass to eventually dominate the turfgrass community. Warm-season turfgrasses are not usually planted in mixtures; therefore, the relatively slow establishment rate of zoysiagrass must be tolerated or an alternative species selected.

Leaf texture and shoot density are features of turfgrasses that frequently dictate species selection (see Tables 3.3 and 3.4). In northern Europe, colonial bentgrass and fine fescues are usually preferred over the coarser ryegrasses and bluegrasses for developing the "classic English lawn."

At locations in the United States where tall fescue is uniquely adapted, other less-adapted, finer-textured turfgrasses are frequently selected even though a higher cultural intensity is required to sustain them during midsummer stress periods. The extremely high densities of creeping bentgrass and hybrid bermudagrass are necessary for optimum

Table 3.2. Establishment Vigor of Popular Turfgrasses

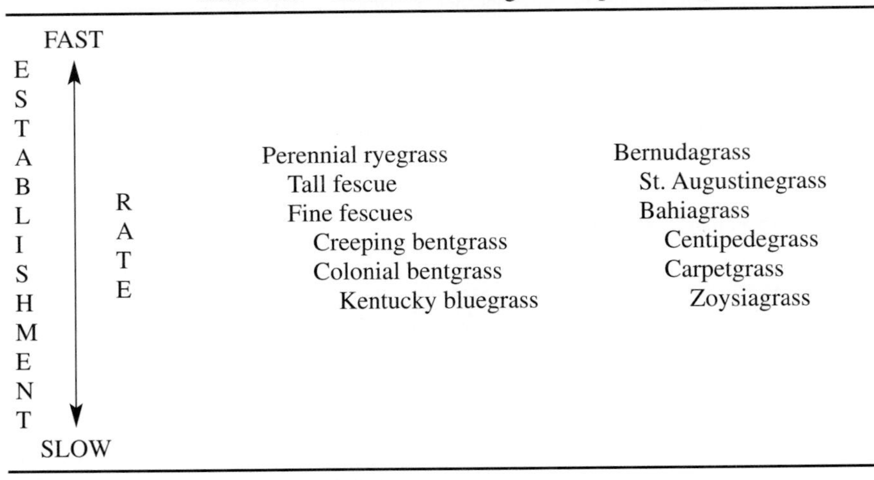

ESTABLISHMENT RATE		
FAST	Perennial ryegrass	Bernudagrass
	Tall fescue	St. Augustinegrass
	Fine fescues	Bahiagrass
	Creeping bentgrass	Centipedegrass
	Colonial bentgrass	Carpetgrass
SLOW	Kentucky bluegrass	Zoysiagrass

Table 3.3. Leaf Texture of Popular Turfgrasses

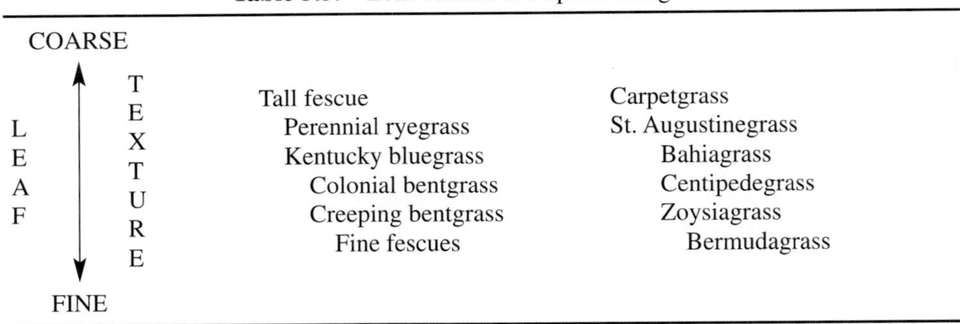

LEAF TEXTURE		
COARSE	Tall fescue	Carpetgrass
	Perennial ryegrass	St. Augustinegrass
	Kentucky bluegrass	Bahiagrass
	Colonial bentgrass	Centipedegrass
	Creeping bentgrass	Zoysiagrass
FINE	Fine fescues	Bermudagrass

Table 3.4. Shoot Density of Popular Turfgrasses

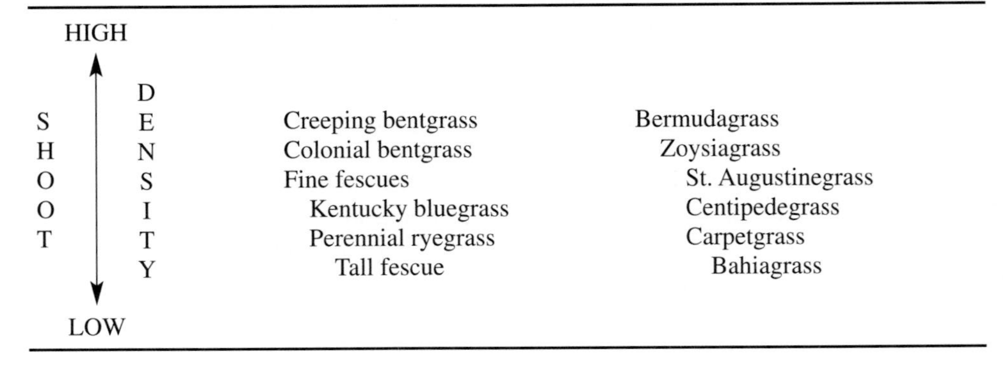

SHOOT DENSITY		
HIGH	Creeping bentgrass	Bermudagrass
	Colonial bentgrass	Zoysiagrass
	Fine fescues	St. Augustinegrass
	Kentucky bluegrass	Centipedegrass
	Perennial ryegrass	Carpetgrass
LOW	Tall fescue	Bahiagrass

playing conditions on golf turf, while the relatively low density of tall fescue or bahiagrass roadside turf is quite acceptable, especially to the observer traveling at fifty-five miles per hour.

Although cool- and warm-season turfgrasses are generally restricted to their respective climatic zones, species within these groups vary widely in their tolerance of temperature and moisture extremes. Creeping bentgrass has outstanding cold tolerance among cool-season grasses, while tall fescue and perennial ryegrass may be severely thinned under the same conditions (Table 3.5).

Bermudagrass is intermediate among warm-season turfgrasses in cold tolerance, and one species of zoysiagrass (*Z. japonica*) can survive subarctic conditions. The heat tolerance of warm-season grasses is generally greater than that of the cool-season species but, within the cool-season group, tall fescue will withstand considerably more heat stress than perennial ryegrass (Table 3.6).

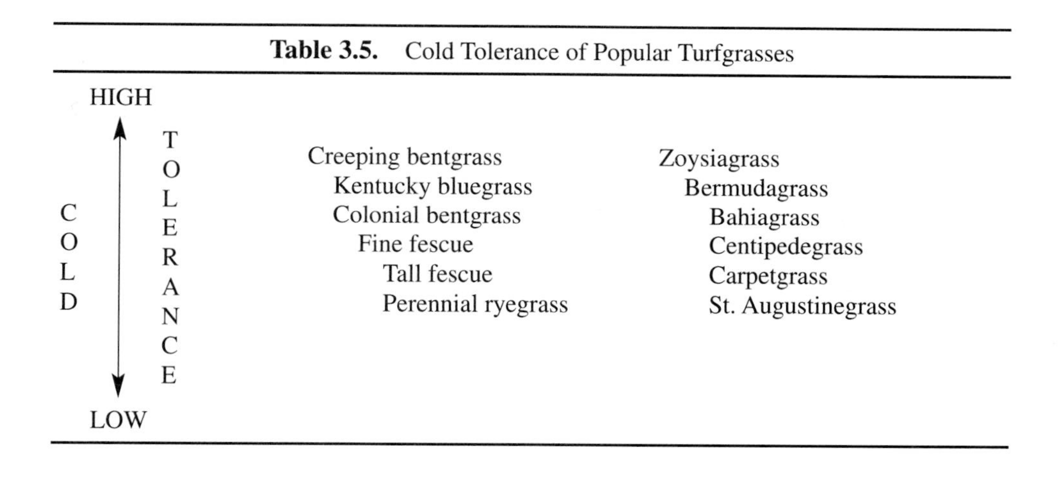

Table 3.5. Cold Tolerance of Popular Turfgrasses

COLD TOLERANCE		
HIGH	Creeping bentgrass	Zoysiagrass
	Kentucky bluegrass	Bermudagrass
	Colonial bentgrass	Bahiagrass
	Fine fescue	Centipedegrass
	Tall fescue	Carpetgrass
LOW	Perennial ryegrass	St. Augustinegrass

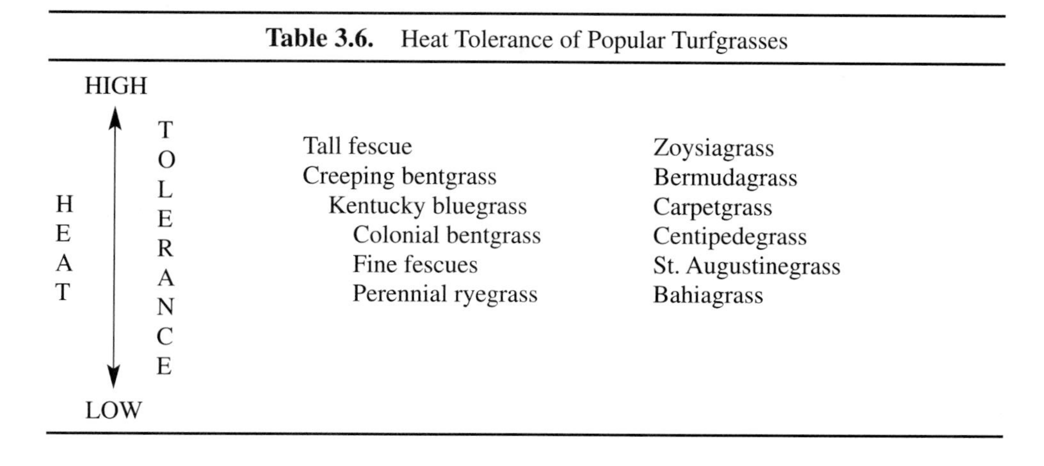

Table 3.6. Heat Tolerance of Popular Turfgrasses

HEAT TOLERANCE		
HIGH	Tall fescue	Zoysiagrass
	Creeping bentgrass	Bermudagrass
	Kentucky bluegrass	Carpetgrass
	Colonial bentgrass	Centipedegrass
	Fine fescues	St. Augustinegrass
LOW	Perennial ryegrass	Bahiagrass

Table 3.7. Drought Tolerance of Popular Turfgrasses

DROUGHT TOLERANCE		
HIGH ↑		
	Fine fescues	Bermudagrass
	Tall fescue	Zoysiagrass
	Kentucky bluegrass	Bahiagrass
	Perennial ryegrass	St. Augustinegrass
	Colonial bentgrass	Centipedegrass
	Creeping bentgrass	Carpetgrass
LOW ↓		

Some warm-season turfs are annually overseeded with one or more cool-season species in order to provide green color and a nondormant playing surface during cool months. The transition back to a warm-season turf is encouraged by warm temperatures and competitive growth of the warm-season grasses.

Drought tolerance is an extremely important criterion for selecting a turfgrass where irrigation cannot be provided as needed during extended periods of inadequate rainfall (see Table 3.7). Within the zones of cool-season turfgrass adaptation, fine fescues provide excellent drought tolerance in cool regions, while tall fescue is preferred in warmer regions. The bentgrasses are highly dependent on a frequent irrigation regime during dry periods, especially when temperatures are high. Warm-season turfgrasses may lose color during an extended drought, but they will usually survive within their zones of adaptation.

One of the most frequent causes of turfgrass deterioration is shade. Many lawns that were established with Kentucky bluegrass sod became weed infested or simply died under trees or along the shaded side of buildings. Although Kentucky bluegrass/fine fescue seed mixtures are not as widely used as in previous years, the shade-adapted fine fescues are still important grasses for use where insufficient sunlight limits the adaptation of Kentucky bluegrass. Among the warm-season grasses, bermudagrass is virtually shade intolerant, while St. Augustinegrass (with the exception of certain cultivars, most notably Floratam) has generally good shade adaptation. Since these grasses are not texturally compatible, St. Augustinegrass is usually the preferred turfgrass in its zone of adaptation for home lawns and other turfs where shade is a significant feature in the environment (see Table 3.8).

Soil pH can be adjusted by soil incorporation or topical application of lime or other amendments to provide a suitable soil reaction for turfgrass growth. However, this may not be feasible on extensive turfgrass sites sustained under a low cultural intensity. Therefore, soil acidity may be an important determinant in the selection of a turfgrass for planting. Fescues and bentgrasses are more tolerant of acid soil conditions than are bluegrasses and

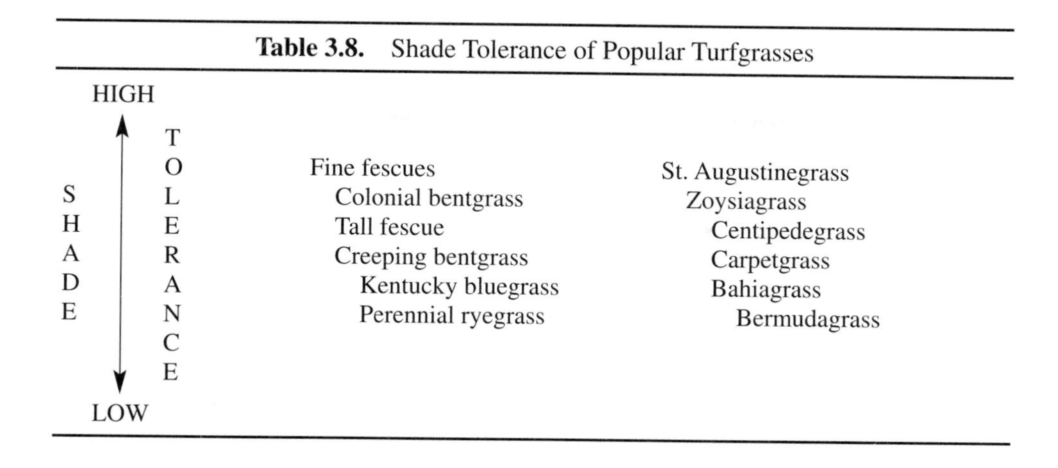

Table 3.8. Shade Tolerance of Popular Turfgrasses

SHADE TOLERANCE (HIGH → LOW)	Cool-season	Warm-season
HIGH	Fine fescues	St. Augustinegrass
	Colonial bentgrass	Zoysiagrass
	Tall fescue	Centipedegrass
	Creeping bentgrass	Carpetgrass
	Kentucky bluegrass	Bahiagrass
LOW	Perennial ryegrass	Bermudagrass

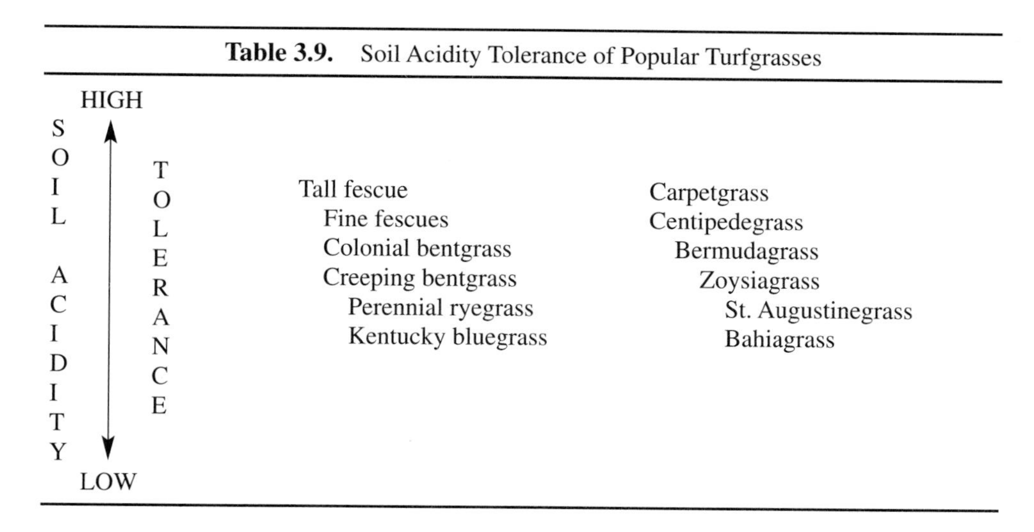

Table 3.9. Soil Acidity Tolerance of Popular Turfgrasses

SOIL ACIDITY TOLERANCE (HIGH → LOW)	Cool-season	Warm-season
HIGH	Tall fescue	Carpetgrass
	Fine fescues	Centipedegrass
	Colonial bentgrass	Bermudagrass
	Creeping bentgrass	Zoysiagrass
	Perennial ryegrass	St. Augustinegrass
LOW	Kentucky bluegrass	Bahiagrass

ryegrasses (Table 3.9). Most warm-season species are fairly well adapted to acid soil conditions; however, carpetgrass and centipedegrass are especially well adapted to acid soils.

Turfgrasses are frequently planted on flood plains and on slopes adjacent to water bodies where temporary submergence occasionally occurs. This condition frequently results in the loss of Kentucky bluegrass and zoysiagrass turfs. However, creeping bentgrass is quite tolerant of flooding, and tall fescue is valuable for use in drainage ditches along roadsides. Among the warm-season grasses, bermudagrass and bahiagrass are outstanding in their submersion tolerance (see Table 3.10).

Along highways and sidewalks, calcium and sodium chlorides are sometimes used to melt snow and ice, and subsequent runoff of these salts may result in the death of

Table 3.10. Submersion Tolerance of Popular Turfgrasses

SUBMERSION TOLERANCE		
HIGH		
↑	Creeping bentgrass	Bermudagrass
	Tall fescue	Bahiagrass
	Colonial bentgrass	St. Augustinegrass
	Kentucky bluegrass	Carpetgrass
	Perennial ryegrass	Zoysiagrass
↓	Fine fescues	Centipedegrass
LOW		

Table 3.11. Salinity Tolerance of Popular Turfgrasses

SALINITY TOLERANCE		
HIGH		
↑	Creeping bentgrass	Bermudagrass
	Tall fescue	Zoysiagrass
	Perennial ryegrass	St. Augustinegrass
	Fine fescues	Bahiagrass
	Kentucky bluegrass	Carpetgrass
↓	Colonial bentgrass	Centipedegrass
LOW		

adjacent turfgrasses. Some sources of irrigation water are high in salts, resulting in saline soil conditions in the irrigated turf. In dry climates, salts are brought to the soil surface in conjunction with upward water movement in response to evaporation and transpiration. Turfgrass tolerance to soil salinity varies with species and cultural factors (Table 3.11). Creeping bentgrass and tall fescue are cool-season grasses with good salinity tolerance. Among the warm-season species, seashore paspalum, bermudagrass, zoysiagrass, and St. Augustinegrass are quite tolerant of saline soil conditions.

Turfgrasses that are adapted to a high intensity of culture may be unsatisfactory on sites where a low cultural intensity is practiced (see Table 3.12). Conversely, turfgrasses adapted to a low intensity of culture may quickly deteriorate under a high cultural intensity due to environmental stress, disease, weed invasion, or other causes. For example, tall fescue will form a satisfactory turf when mowed at 2 inches or higher but will deteriorate under mowing heights of 1 inch or less. Creeping bentgrass is tolerant of mowing

Table 3.12. Mowing Height Adaptation of Popular Turfgrasses

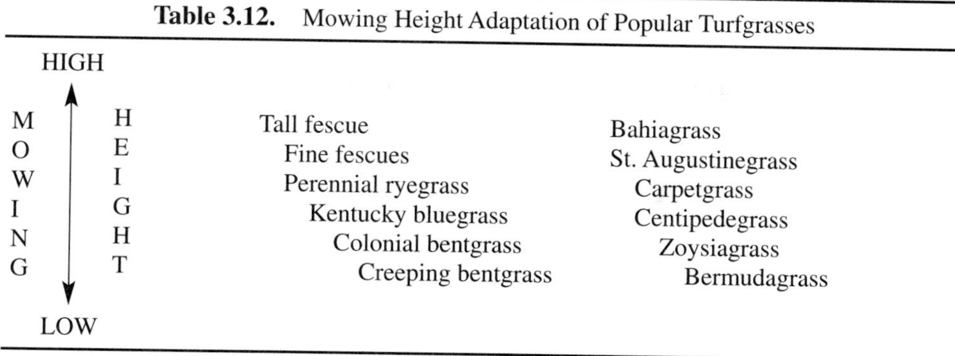

HIGH ↑ MOWING HEIGHT ↓ LOW

Cool-season	Warm-season
Tall fescue	Bahiagrass
Fine fescues	St. Augustinegrass
Perennial ryegrass	Carpetgrass
Kentucky bluegrass	Centipedegrass
Colonial bentgrass	Zoysiagrass
Creeping bentgrass	Bermudagrass

Table 3.13. Mowing Quality Range of Popular Turfgrasses

GOOD ↑ MOWING QUALITY ↓ POOR

Cool-season	Warm-season
Kentucky bluegrass	St. Augustinegrass
Colonial bentgrass	Bermudagrass
Creeping bentgrass	Centipedegrass
Tall fescue	Carpetgrass
Fine fescues	Zoysiagrass
Perennial ryegrass	Bahiagrass

heights below 1/4 inch; however, it becomes puffy and unsightly when mowed above 1 inch. The warm-season analogues of tall fescue and creeping bentgrass are bahiagrass and bermudagrass, respectively. Mowing quality varies widely among turfgrass species (Table 3.13); leaves of perennial ryegrass and bahiagrass have tough vascular bundles that resist cutting, while Kentucky bluegrass and St. Augustinegrass usually cut cleanly as long as the mowing blades are sharp. Creeping bentgrass and bermudagrass have a fairly high fertility requirement for satisfactory growth and density (see Table 3.14). In contrast, fine fescues and bahiagrass are well adapted to low fertility, and excessive fertilization may result in serious weed, disease, and insect problems.

Disease proneness, in particular, must be considered in selecting a turfgrass species for use on a specific site (Table 3.15). Grasses that have a high potential for disease incidence may require frequent treatment with various fungicides. Closely mowed creeping bentgrass and bermudagrass turfs may require fungicide spraying to prevent the occurrence of dollar spot, brown patch, and other diseases. Tall fescue and centipedegrass are relatively disease free.

Some turfgrass species, including creeping bentgrass and bermudagrass, can develop excessive accumulations of thatch, which, in turn, can predispose the turf to

Table 3.14. Fertility Requirement Range of Popular Turfgrasses

FERTILITY REQUIREMENT		
HIGH		
	Creeping bentgrass	Bermudagrass
	Colonial bentgrass	St. Augustinegrass
	Kentucky bluegrass	Zoysiagrass
	Perennial ryegrass	Centipedegrass
	Tall fescue	Carpetgrass
	Fine fescues	Bahiagrass
LOW		

Table 3.15. Disease Potential of Popular Turfgrasses

DISEASE POTENTIAL		
HIGH		
	Creeping bentgrass	St. Augustinegras
	Colonial bentgrass	Bermudagrass
	Fine fescues	Zoysiagrass
	Kentucky bluegrass	Carpetgrass
	Perennial ryegrass	Bahiagrass
	Tall fescue	Centipedegrass
LOW		

rapid deterioration from numerous causes (see Table 3.16). Thatch is seldom a problem in tall fescue and bahiagrass turfs.

Intensively trafficked turfs, including athletic fields, playgrounds, and golf course greens, tees, and fairways, should be composed of turfgrasses that are tolerant of wear from traffic and that have the capacity to recover quickly from injury. Although tall fescue is the most wear-resistant of the cool-season turfgrasses (see Table 3.17), it may not be suitable for many sports turfs because of its poor adaptation to high intensities of culture (for example, close mowing and high fertility).

Kentucky bluegrass and perennial ryegrass are fairly wear resistant, and the recuperative potential of Kentucky bluegrass is good. Furthermore, these grasses are well adapted to moderate to high cultural intensities. On putting greens, the very close mow-

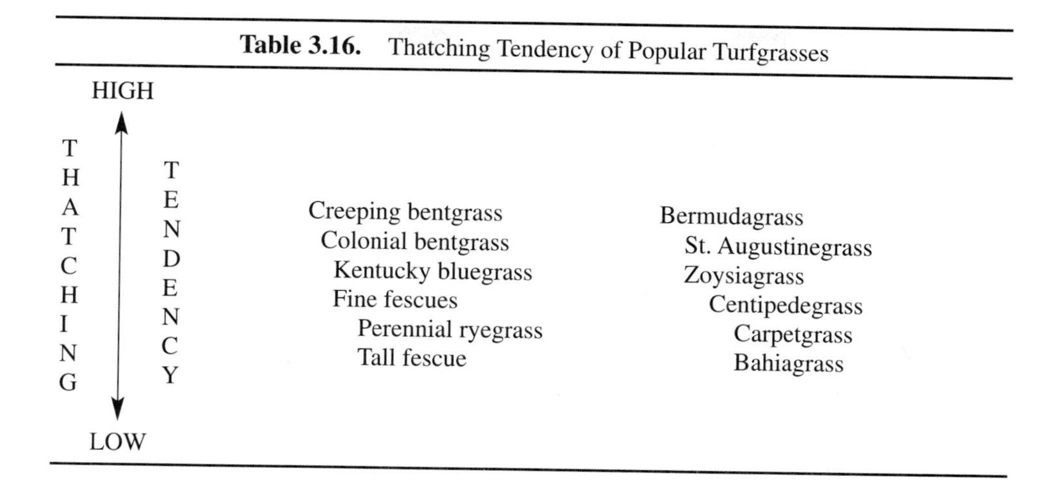

Table 3.16. Thatching Tendency of Popular Turfgrasses

THATCHING TENDENCY (HIGH → LOW)		
HIGH	Creeping bentgrass	Bermudagrass
	Colonial bentgrass	St. Augustinegrass
	Kentucky bluegrass	Zoysiagrass
	Fine fescues	Centipedegrass
	Perennial ryegrass	Carpetgrass
LOW	Tall fescue	Bahiagrass

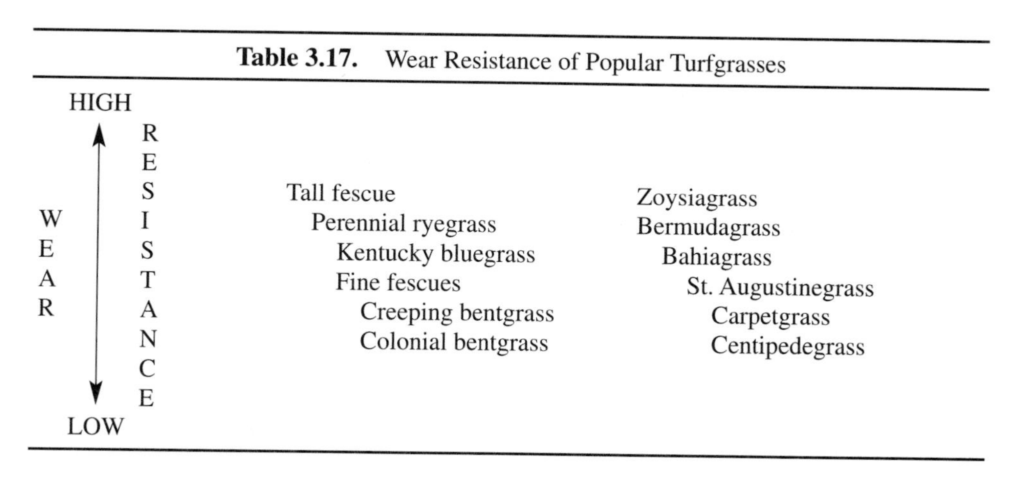

Table 3.17. Wear Resistance of Popular Turfgrasses

WEAR RESISTANCE (HIGH → LOW)		
HIGH	Tall fescue	Zoysiagrass
	Perennial ryegrass	Bermudagrass
	Kentucky bluegrass	Bahiagrass
	Fine fescues	St. Augustinegrass
	Creeping bentgrass	Carpetgrass
LOW	Colonial bentgrass	Centipedegrass

ing heights that are employed preclude the use of all but the bentgrasses where cool-season turfgrasses are adapted.

Creeping bentgrass ranks only fair in wear resistance, but has excellent recuperative ability (Table 3.18). Among the warm-season species, bermudagrass is quite wear resistant and has excellent recuperative capacity. In contrast, the very wear-resistant zoysiagrasses are very slow to recover from injury. In some climatic zones, zoysiagrass is suitable for use on golf course tees provided the tees are large enough to permit adequate recovery of worn areas.

The ranking of turfgrass species for these criteria may change with the development of new cultivars that differ from the norm in their environmental adaptation. For example, Nugget and A-34 Kentucky bluegrasses have persisted quite well in some moderately shaded environments. Studies with numerous Kentucky bluegrass cultivars in Illinois

Table 3.18. Recuperative Capacity of Popular Turfgrasses

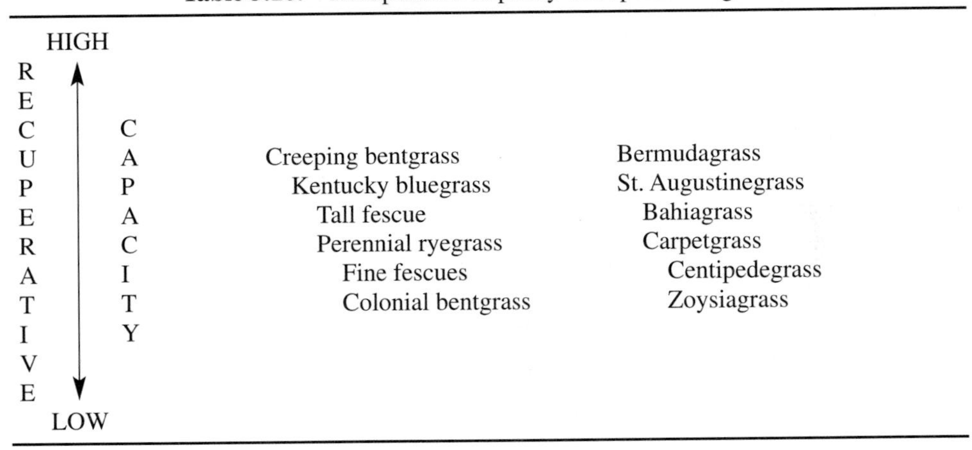

RECUPERATIVE CAPACITY		
HIGH	Creeping bentgrass	Bermudagrass
	Kentucky bluegrass	St. Augustinegrass
	Tall fescue	Bahiagrass
	Perennial ryegrass	Carpetgrass
	Fine fescues	Centipedegrass
LOW	Colonial bentgrass	Zoysiagrass

have shown that Touchdown and Brunswick can persist as very dense turfs under a 3/4-inch mowing height. Thus, the results of turfgrass breeding efforts of recent years are already changing the performance characteristics and adaptation limits of the primary turfgrass species.

QUESTIONS

1. Differentiate among the three grass **subfamilies** containing common turfgrass species.
2. List and characterize the eleven **climatic types** and indicate the turfgrass species adapted to each.
3. Characterize the following **festucoid** turfgrass genera: *Festuca, Poa, Lolium,* and *Agrostis.*
4. Characterize the following **eragrostoid** turfgrass genera: *Cynodon* and *Zoysia.*
5. Characterize the following **panicoid** turfgrass genera: *Paspalum, Eremochloa,* and *Stenotaphrum.*
6. Rank at least five cool-season and five warm-season turfgrasses with respect to their **heat** and **cold tolerances.**

ANNUAL BLUEGRASS

ROUGH BLUEGRASS

KENTUCKY BLUEGRASS

PERENNIAL RYEGRASS

ANNUAL RYEGRASS

CREEPING BENTGRASS

VELVET BENTGRASS

REDTOP

COLONIAL BENTGRASS

TALL FESCUE

RED FESCUES

HARD FESCUE

BERMUDAGRASS

BUFFALOGRASS

ZOYSIAGRASS

KIKUYUGRASS

BAHIAGRASS

ST. AUGUSTINEGRASS

CENTIPEDEGRASS

SEASHORE PASPALUM

CHAPTER 4

The Turfgrass Environment

The turfgrass community exists in intimate association with its environment. All components of the natural and artificially induced environment affect the persistence and quality of a turf. The environment consists of a highly integrated and dynamic array of forces that, in the aggregate, determine the adaptation and growth of plant species. For academic treatment, it is useful to consider the environment in three distinct, but interacting, dimensions: the atmosphere above and immediately surrounding the turfgrass aerial shoots; the edaphic environment (thatch-soil), including roots and other belowground plant organs; and the biotic component of the environment, encompassing cultural practices, pest organisms, and use of the turf by people. The turfgrass community and its environment form the turfgrass ecosystem.

ATMOSPHERIC ENVIRONMENT

The atmospheric conditions affecting turf result from seasonal and daily fluctuations in the weather. These conditions are measurable as temperature, moisture, light, and wind. Each results from the influence of the sun on the earth's atmosphere. The sun provides light to brighten an otherwise dark universe. Temperature is a reflection of the sun's warming effects on the earth's surface and, subsequently, the atmosphere above it. Moisture conditions result from evaporation of water from oceans, lakes, and other bodies, and the subsequent influences of pressure systems, temperature, geographic features, and winds. Finally, wind occurs as a result of differential heating of the earth's surface, rotation of the earth about its axis, and the formation and movement of pressure systems.

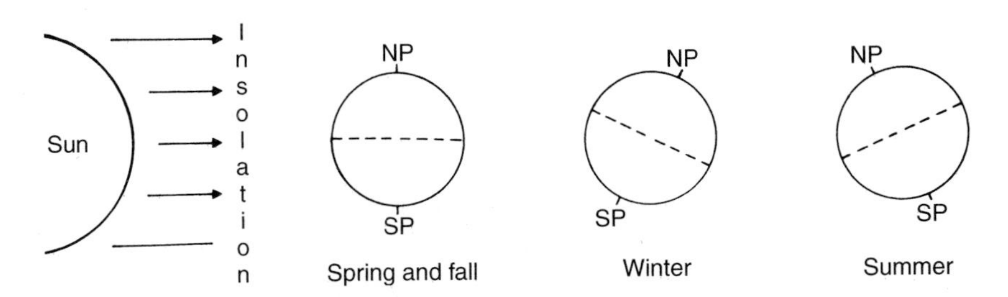

Figure 4.1. Revolution of the earth around the sun and the resultant shift in the equator's position relative to insolation account for seasonal changes in weather.

The orientation of the equator relative to incoming solar radiation (insolation) changes as the earth resolves around the sun. This accounts for seasonal changes in weather (Figure 4.1). In spring and fall, insolation is most direct at the equator. During winter in the northern hemisphere, insolation is most direct below the equator, and, during summer, the opposite is true. Therefore, in the northern hemisphere, winters tend to be more severe and summers milder as one proceeds poleward; summer heat tends toward higher intensity and winters are milder as one moves closer to the equator. These trends are substantially modified by variations in topographical relief and the proximity of large bodies of water. As pointed out in Chapter 3, sites at high elevations in the tropics may be quite suitable for the growth of cool-season turfgrasses, since average temperatures decrease appreciably with relatively small increases in altitude. Because water bodies tend to modify diurnal and seasonal temperature fluctuations, oceanic sites are milder in summer and winter compared to continental areas at the same latitude.

Light

Turfgrasses absorb and convert to chemical energy, through photosynthesis, only about 1 to 2% of the incident radiation. Most of the absorbed energy is reradiated at longer wavelengths with the release of heat that significantly affects atmospheric temperature. At the turfgrass leaf surfaces, solar radiation may also be reflected or transmitted through the leaf for possible absorption by other leaves (Figure 4.2). The amount of reflected radiation varies among plants and is affected significantly by moisture conditions. Glossy or wet leaf surfaces are more reflective than dry, dull leaves. The amount of radiation absorbed by the leaves varies from about 50 to nearly 80% of insolation, depending on leaf orientation; the more horizontally oriented leaves are more efficient absorbers of solar radiation (Figure 4.3).

The amount of light received by turfgrasses is influenced by many factors in the environment. Clouds, buildings, trees, and other features can reduce light intensity through shading. Turfgrass response to variations in the intensity of light ranges from a

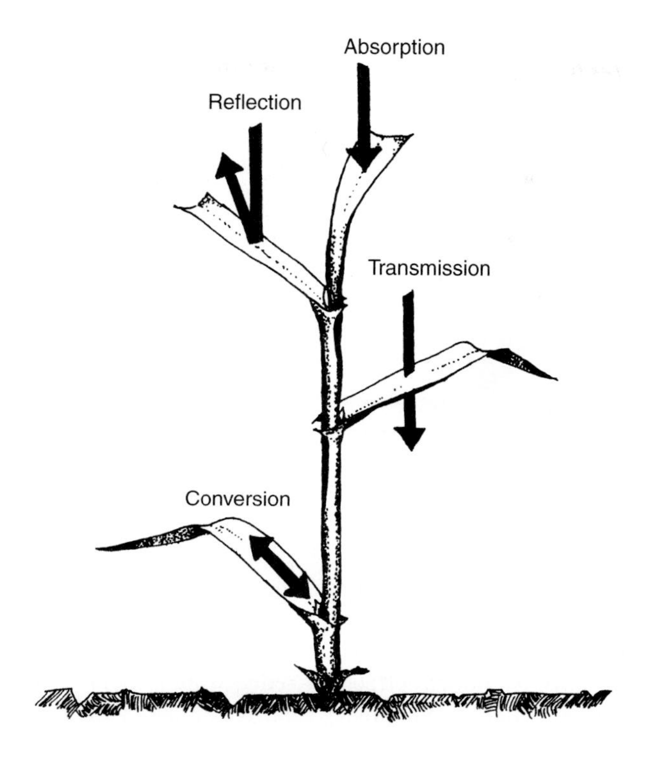

Figure 4.2. Solar radiation may be absorbed, transmitted, or reflected by turfgrass leaves. A small amount of absorbed solar energy is converted to chemical energy by photosynthesis.

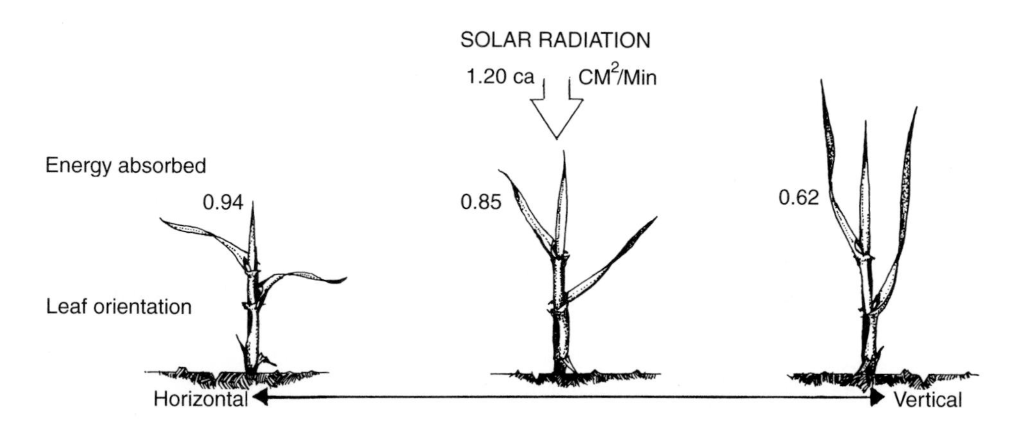

Figure 4.3. Effect of leaf orientation on the efficiency with which solar radiation is absorbed by turfgrasses.

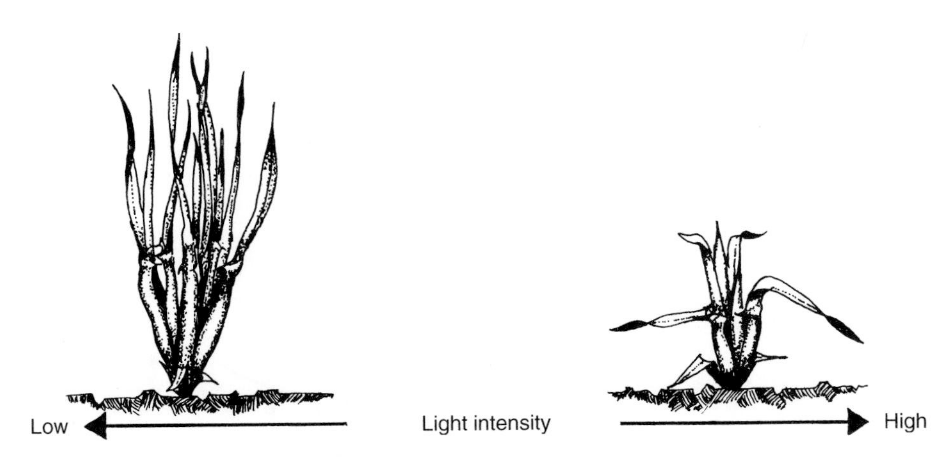

Figure 4.4. Comparison of turfgrass leaf orientations under low- and high-light intensities.

relatively horizontal orientation of the leaf blades at high intensities to a failure of the turf to persist under extremely low light. With moderate shading, turfgrass leaf blades typically assume a more upright orientation as if they were "reaching up" for more light (Figure 4.4). Additional effects of reduced light include thinner, longer leaves; reduced density and tillering; shallower rooting; thinner cuticles; and lower reserve carbohydrates within the plants. A "shade" turf is thus more delicate and less tolerant of wear, disease, and environmental stresses.

Sufficient light is necessary for sustaining a level of photosynthetic activity adequate for the plant's respiratory requirement and for new growth. Since older plant tissues eventually die, growth must proceed at a rate sufficient to at least replace those tissues lost to senescence. In a heavily trafficked turf, the loss of plant tissue to mechanical damage necessitates additional growth to sustain turfgrass quality. Thus, several dimensions of essential photosynthetic activity are recognized in turfgrasses: an adequate level to reach the *compensation point* at which photosynthesis equals respiration, an additional increment to sustain the plant's respiratory requirement during dark hours when no photosynthesis is taking place, and extra carbohydrate production to promote a growth rate sufficient to offset the losses of plant tissues from natural senescence and mechanical damage. If the level of photosynthetic activity is inadequate to provide for the necessary rate of growth, deterioration of the turf is inevitable unless some adjustments can be made. Some possible courses of action include increasing light penetration by pruning or removing interfering plants, reducing traffic intensity to minimize mechanical damage, or establishing shade-adapted plant species or cultivars in place of the existing turfgrass community.

Adjustments in a turfgrass cultural program can improve the appearance and persistence of a shaded turf. Raising the mowing height can partially compensate for the more upright leaf orientation that results from reduced light. The rate of nitrogen fertilization should be reduced from that of turf in full sun to reduce the potential for carbohydrate depletion. Trees should be fertilized separately by employing deep-root feeding devices or applying fertilizer to holes produced on 2-foot centers. Irrigation should be conducted infrequently but with sufficient intensity to encourage deep rooting of the turf and to provide for the moisture requirements of trees. Finally, fungicides may be needed to control disease-causing organisms that impose an additional stress on shaded turf.

The ultimate shading problem is called *light exclusion.* This results where heavy clippings from mowing, or other obstacles to light penetration, occur directly on the turf. Injury from light exclusion can take place within hours if temperatures are high. Objects that exclude light should not be allowed to remain on the turf for any longer than absolutely necessary, and heavy clippings should be broken up or removed.

Where trees are responsible for a shaded condition, other factors further complicate the problem. Trees may severely compete with turfgrasses for available moisture and nutrients in the soil, and the roots of some trees exude chemicals that may be toxic to turfgrasses. Tree leaves not only intercept light and thus reduce the amount of light energy available for the underlying turf, but they may also deplete the light of its photosynthetically active wavelengths. Therefore, the light penetrating to the turfgrass surface may be of little value in sustaining the turf. Finally, the accumulation of tree leaves on turf can result in injury from light exclusion. Therefore, in tree-shaded sites, special measures may be necessary to ensure the survival and desired quality of the turf. These measures include root pruning of trees to reduce competition, trimming lower tree limbs and pruning the tree's crown to allow better light penetration, and immediately removing fallen leaves.

The effects of shading on the turfgrass microenvironment include moderation of diurnal and seasonal temperature fluctuations, restricted air movement, and increased relative humidity. Depending on the intensity of these factors, the results may be beneficial or highly detrimental to the turf. Partially shaded turfs often appear healthier than nonshaded turfs during summer stress periods because of the apparent differences in heat and drought stresses. Where air movement is severely restricted and high relative humidities occur, however, the turf may be much more susceptible to disease. This is especially evident on low, poorly drained sites.

One final important aspect of light is photoperiod, or day length. At intermediate latitudes short photoperiods occur in the early and late portions of the growing season, and long photoperiods occur during mid-season. In equatorial regions photoperiods are relatively constant during the year, while in arctic regions continuous light or dark periods may persist for months. Photoperiodic conditions influence flowering (Chapter 2) and vegetative growth and development of turfgrasses. Turfgrasses growing under short day lengths typically exhibit increased density and tillering, shorter leaves, smaller shoots (including rhizomes and stolons), and a more prostrate growth habit compared to plants under long day lengths (Figure 4.5).

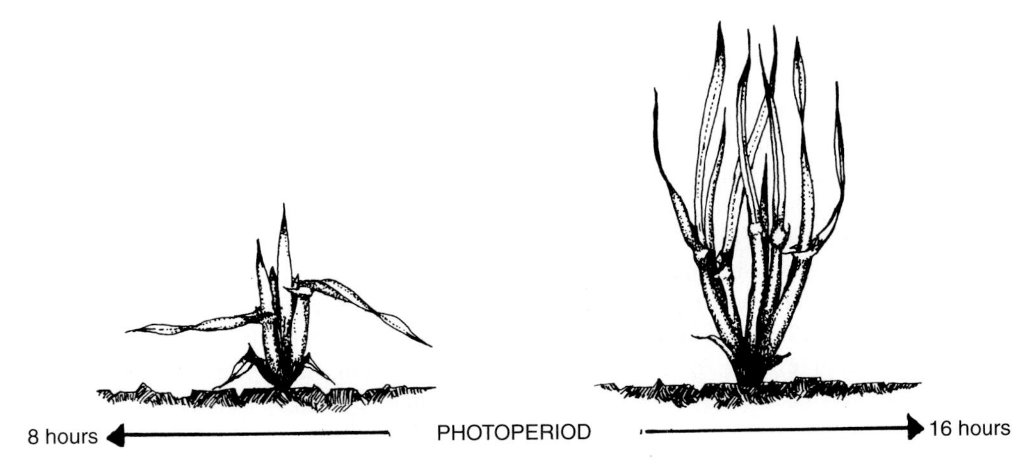

Figure 4.5. Comparison of cool-season turfgrass plants under short versus long photoperiods.

Temperature

Temperature is the most important environmental factor affecting the adaptation of turfgrasses to a particular geographic region. It is a measurable expression of heat energy from solar radiation. Many factors affect the net accumulation of heat energy by specific objects in the landscape. Much of the absorbed heat energy can be transferred from one environmental feature to another by various processes, including evaporation, radiation, conduction, convection, and advection.

Evaporation is the process by which water is changed from a liquid to a gaseous state with the concurrent conversion of sensible heat to latent heat. The sensible heat of an object or of the atmosphere is measurable as temperature; latent heat is absorbed by water, causing a change from the liquid to the gaseous state with no temperature change. Thus, water acts as an energy sink to effectively moderate temperature fluctuations through changes in its physical state (Figure 4.6). Transpiration by plants is essentially an evaporative process by which internal plant moisture (liquid) is converted to water vapor (gas) and released to the atmosphere through the stomates. This conversion requires 570 calories per gram of water, and the plant is, thus, cooled through transpiration.

Much of the solar radiation absorbed by plants is *reradiated* at longer (infrared) wavelengths to the atmosphere. This process transfers heat from the plant to the ambient air with a resultant increase in air temperature.

Heat energy can also be transferred through *conduction* by objects or molecules that are in contact with each other. Heat is transferred by adjacent soil particles from the surface downward. Air molecules in contact with a warmed object can conduct some heat to the surrounding air; however, air is a relatively poor conductor, and conduction probably contributes little to the cooling of plants.

Convection occurs when plumes of heated air rise from plant surfaces. This process is important in the transfer of heat from plants and other warmed objects to the

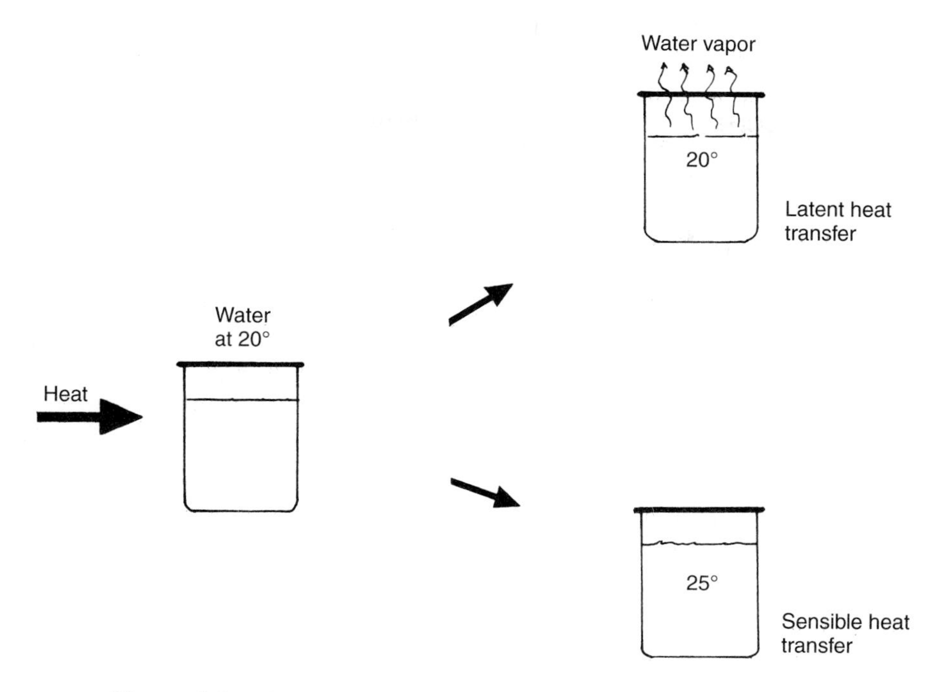

Figure 4.6. Comparison of latent and sensible heat transfer processes.

atmosphere. A similar process associated with air movement is *advection.* Air passing slowly over a warmed surface picks up heat that is transferred to an adjacent site located downwind. Small strips of turfs located close to efficient energy-absorbing surfaces (paved surfaces, gravel beds, and the like) encounter greater heat stress in summer than do large turfgrass sites. Similarly, advective cooling can occur on turfs located near large bodies of water when the air above the cooler water moves inland.

Acting together, energy transfer processes can sustain turfgrass temperatures near those of the ambient air. A ranking of these processes in the order of their relative importance would be radiation > convection > transpiration > wind. In turfgrass culture, the only processes over which there is usually some control are transpiration and wind. Adequate transpirational cooling can be promoted by ensuring an adequate supply of plant-available moisture through irrigation and other cultural practices. Air movement across some turfgrass sites can be improved by removing trees and other obstacles. Such measures could be decisive in promoting a desired level of turfgrass quality.

Turfgrass Adaptation to Temperature Conditions

The growth of each turfgrass species and cultivar is characterized by three cardinal temperatures that establish the temperature range within which growth takes place: maximum

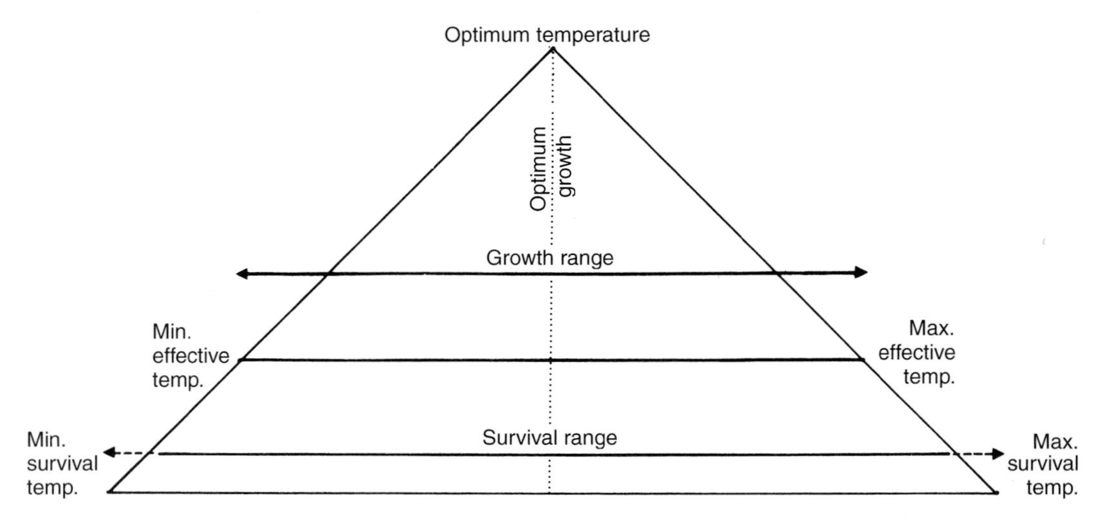

Figure 4.7. Cardinal temperatures influencing turfgrass growth and survival.

effective temperature, minimum effective temperature, and the temperature at which optimum growth occurs (Figure 4.7). In addition, two other cardinal temperatures (maximum survival and minimum survival) establish the temperature extremes beyond which the turfgrass cannot survive. The position of these latter temperature points varies depending on the exposure period. Turfgrasses may survive at very high or very low temperatures if the period of exposure is short; however, death may occur during a period of prolonged exposure to those same temperatures.

In the natural environment, temperature is continually changing. The highest temperature typically occurs at midday, while temperatures are usually lowest just prior to sunrise. Furthermore, the temperature of the soil surface (bare soil) generally fluctuates more widely than that of the air 5 feet above ground—the height at which temperatures are measured by weather-reporting services (Figure 4.8). This occurs because the temperature of a surface (soil, turf, and the like) results from the net accumulation of heat energy from solar radiation. The temperature of the atmosphere, however, is the result of the transfer of heat from an absorbing surface to the air above it. Since these processes occur in sequence and usually at different rates, soil surface temperature tends to be higher than that of the atmosphere during the day and slightly lower at night when no solar radiation is being received. Turfed surfaces do not accumulate heat as rapidly as most nonturfed surfaces because of transpirational cooling and other differences in their energy transfer characteristics. Thus, the temperatures of a turf and the air above it are generally cooler than those of many other terrestrial features. This is most evident on athletic fields constructed with artificial surfacing materials or around residences where large expanses of gravel or other materials have been substituted for lawn turf. The temperatures above these surfaces can be appreciably higher than those occurring above turfgrass plantings.

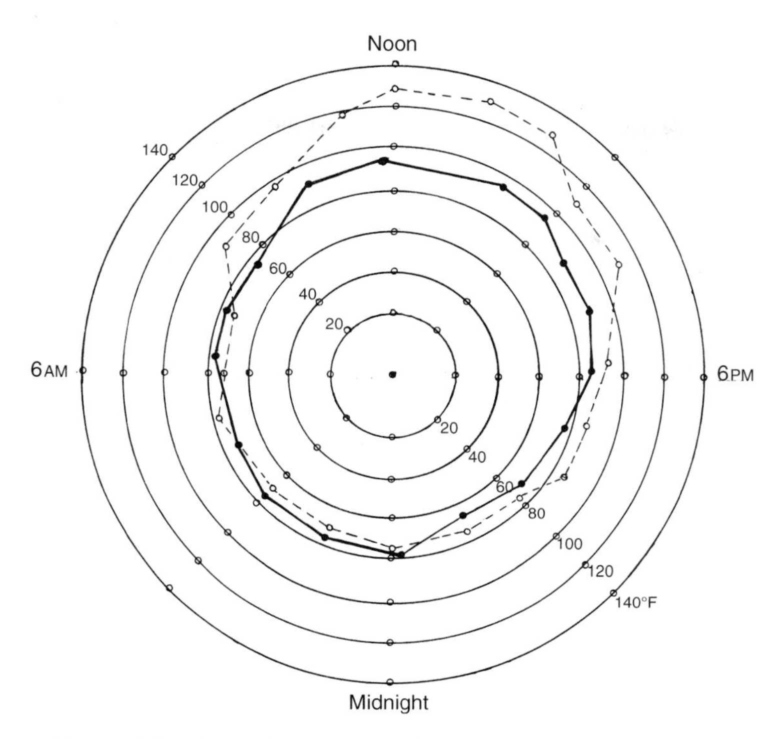

Figure 4.8. Diurnal temperature fluctuations of the soil surface.

Growth measurements of turfgrasses in controlled-environment chambers set at constant temperatures cannot be directly related to the growth responses of these plants in the field. Experiments with Kentucky bluegrass and bermudagrass at constant and alternating day and night temperatures showed that shoot growth was promoted by fluctuating diurnal temperatures. The response of plants to rhythmic variations in temperature is termed *thermoperiodism.* The presumed basis for the favorable thermoperiodic growth of plants is that optimum rates of photosynthesis and growth occur at different temperatures. Photosynthetic production of carbohydrates occurs during daylight hours, while plant growth proceeds at an optimum rate at night when temperatures are cooler.

Seasonal fluctuations in temperature substantially influence the growth of turfgrasses in two important respects: they determine the duration of optimum, or at least favorable, growing conditions, and they establish the limits of adaptation for each turfgrass. Thus, many warm-season species may grow well during the summer months in regions with temperate climates but not survive the winter. Conversely, cool-season grasses may grow vigorously during late fall to midspring in subtropical climatic zones, only to die or be overtaken by warm-season grasses during the summer. Mulching of bermudagrass turf prior to winter may improve its survival in locations where it is marginally adapted. Likewise, carefully controlled irrigation of annual bluegrass turf can aid summer survival. Correct cultural practices can thus extend the limits of adaptation of specific grasses to some extent.

Besides diurnal and seasonal fluctuations, temperature varies with latitude, altitude, and topography. Increases in latitude and altitude are generally associated with cooler temperatures. Topographical features, however, exert such substantial influences on the climate of a region that predictions of the adaptation and growth of turfgrasses based strictly on latitude and altitude are unreliable. Important topographical factors that affect the turfgrass environment include the proximity and size of water bodies; the size, shape, and orientation of elevated land relative to prevailing winds; and the size, position, and density of plants and other features in the landscape. Turfgrass sites located near large bodies of water typically have less severe diurnal and seasonal temperature fluctuations than do inland sites at the same latitude. Large mountain ranges, such as the Rocky Mountains in North America and the Andes Mountains in South America, result in relatively humid conditions over vast areas of their windward sides and relatively dry conditions on their leeward sides. Even small variations in relief can substantially affect the temperature and moisture conditions for turfgrasses on different sides of elevated sites. South-facing slopes receive more direct solar radiation than do north-facing slopes in the northern hemisphere, resulting in the greatest heat accumulation at the southern and southwestern exposures. This effect can be so dramatic at some locations that, along highways, the south-facing slopes have predominantly bermudagrass turf, while Kentucky bluegrass or tall fescue thrives on the opposing north-facing slopes. Reflection of heat from the sides of buildings can also significantly affect growing conditions for turfgrasses planted around them. Summer heat is most severe along the south side, while on the north side, where heat stress is minimal, shading may threaten turfgrass survival.

Moisture

Water is the most important requirement for turfgrass growth and survival. Turfgrasses are composed of living containers (cells) of water within which all metabolic processes take place. Since the water content of actively growing turfgrasses approaches 90% of total mass, a small reduction in the moisture content of a plant can dramatically reduce growth and appearance, and even cause death.

The various functions of water in the plant include maintaining cell turgidity for structure and growth; transporting nutrients and organic compounds throughout the plant; comprising much of the living protoplasm in the plant cells; constituting a raw material for various chemical processes, including photosynthesis; and, through transpiration, buffering the plant against wide temperature fluctuations. This last function accounts for the greatest utilization of water by plants.

Transpiration involves the absorption, transport, and release of water to the atmosphere by plants (Figure 4.9). The evaporation of water films surrounding leaf cells requires heat energy (570 calories per gram of water); thus, transpiration is an important

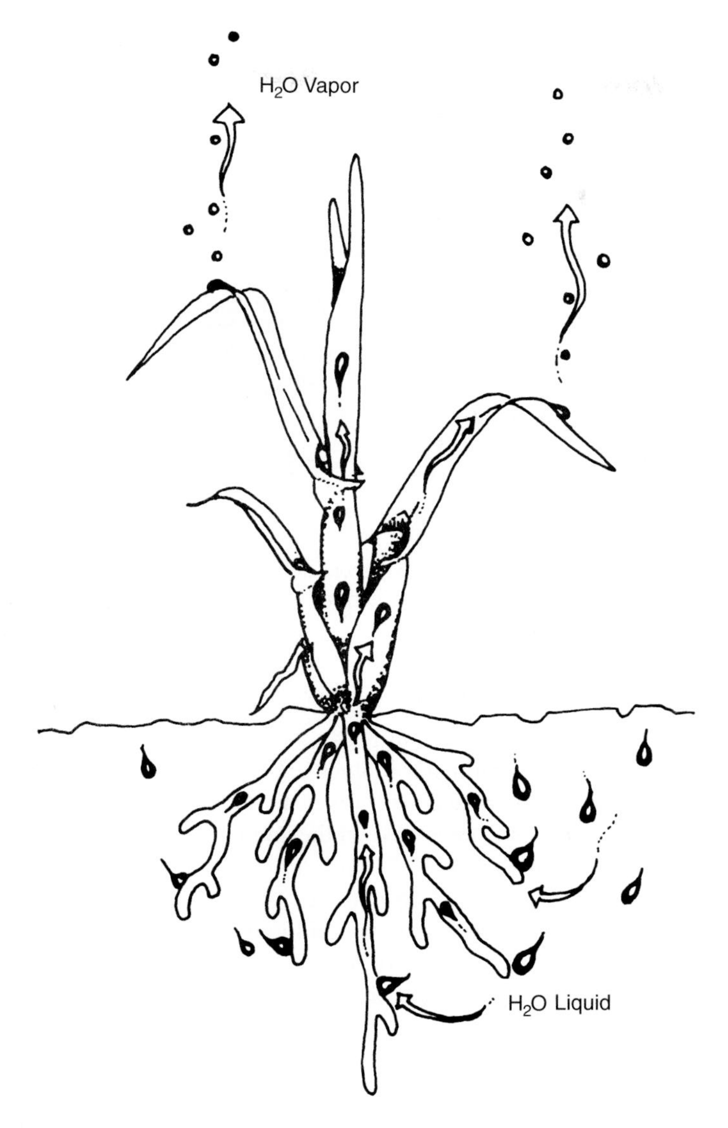

Figure 4.9. Illustration of transpirational cooling of a turfgrass plant.

means of maintaining the plant within a tolerable temperature range. The exit of water vapor from the plant occurs primarily through the stomates, small openings distributed throughout the leaf epidermis (Figure 4.10). As it exits the plant, much of the water vapor surrounds the leaf to form a layer of moist air called the *boundary layer.* The thickness of the boundary layer is determined by numerous factors, including transpiration rate, relative humidity of the ambient air, and wind velocity.

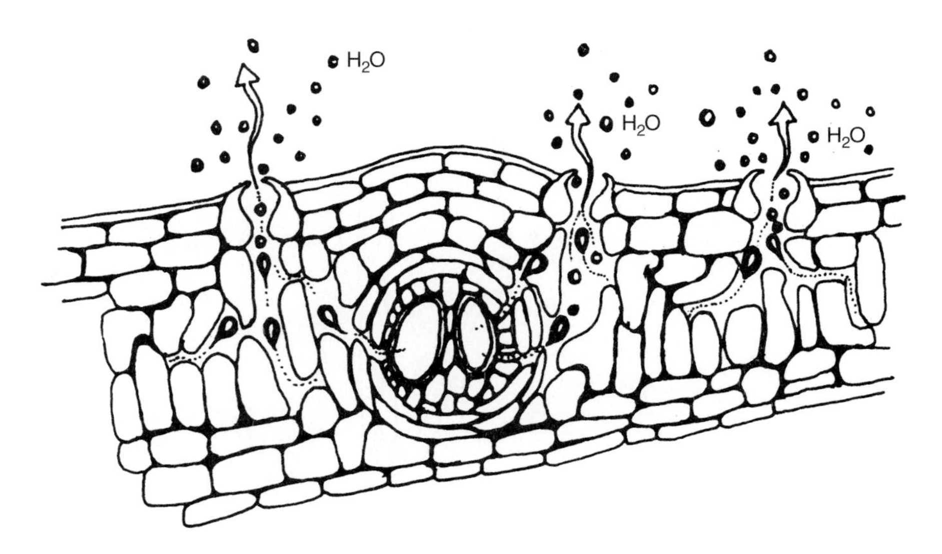

Figure 4.10. Illustration of leaf cross-section showing evaporation of water from mesophyll cell surfaces and the movement of water vapor through stomatal openings in the leaf.

With rapid transpiration, high relative humidity, and no wind, the boundary layer may reach up to 1 centimeter in thickness. As wind velocity increases, the moist air surrounding the leaf is mixed with dryer air to reduce the boundary layer to as little as a few millimeters in thickness. When the ambient air is at 40% relative humidity (RH) and 86°F, its vapor pressure is 12.74 mm of mercury (Hg), while that of the saturated boundary layer (100% RH) is 31.85 mm Hg. Thus, the large vapor pressure gradient of 19.11 mm Hg (31.85 – 12.74) results in rapid movement of water vapor away from the boundary layer and in the direction of lower vapor pressure. On a very humid day (80% RH) with the same air temperature, the vapor pressure of the ambient air would be 25.48 mm Hg and the vapor pressure gradient only 6.37 mm Hg; therefore, the tendency for water vapor to move away from the boundary layer would be considerably reduced. The boundary layer is important in that it reduces the vapor pressure gradient between the intercellular spaces within the leaf and the air immediately outside the leaf. The result is a much reduced rate of transpiration and, therefore, a smaller amount of water consumed by the plant.

Under some conditions, plants may lose water faster than it is absorbed by the roots. This creates an internal moisture deficit that may be favorable or highly unfavorable to the plant, depending on the magnitude and duration of the moisture imbalance. Carefully controlled moisture deficits can result in more extensive root growth and other morphological alterations that generally improve the tolerance of the turfgrass to environmental stresses. When severe moisture deficits are allowed to develop, however, the plants wilt and may eventually die or become dormant. The capacity of plants to survive periods of moisture stress is called *drought resistance.*

Drought Resistance

Drought-resistance mechanisms in plants are classified as escape, avoidance, and tolerance. *Drought escape* is of little practical importance to turfgrass species, but it is evident in some annual species that may invade the turfgrass community (i.e., weeds) and that respond to drought stress by producing large quantities of seed to ensure subsequent generations of the population. *Drought avoidance* mechanisms include those that function in maintaining high plant-water potentials under conditions in which moisture availability is low. Maintenance of high plant-water potentials may be affected by reduced transpirational water loss through reductions in leaf area and increases in stomatal, cuticular, or canopy resistance. Increased root density and depth contribute to the plant's water-uptake capacity. *Drought tolerance* mechanisms include those that function in maintaining turgor pressure at low water potentials and others that contribute to the survival of dehydrated plants.

The role of fungal endophytes (*Neotyphodium* and *Epichloe* **spp.**) in drought tolerance and avoidance mechanisms in cool-season turfgrasses is not clear, but one likely explanation is the prevention of injury and associated water loss from parasitic nematodes, shoot-feeding insects, and some disease-inciting agents. Another possible explanation is an endophyte-induced alteration in the production of phytohormones that enhance cell wall extensibility and cell expansion rate. Biologically active alkaloids occurring in turfgrass-endophyte associations may favorably influence the physiology and growth of infected turfgrasses by encouraging the formation of larger, more competitive plants.

Forms of Atmospheric Moisture

The oceans covering approximately three-fourths of the earth's surface are the basic sources of atmospheric moisture. Water evaporates from bodies of water and, as water vapor, becomes a component of the atmosphere. When the air is cooled, water vapor condenses around small particles (condensation nuclei) to form clouds. Precipitation from clouds transfers moisture to terrestrial areas, where it supports life. Subsequent return of moisture to the atmosphere or to bodies of water completes the hydrologic cycle.

Forms of moisture that are important in turfgrass culture include precipitation, irrigation, water vapor, and dew. The distribution of precipitation over land masses greatly affects the quantity and type of vegetation that can be sustained. In some areas even the most drought-hardy turfgrasses may not persist without supplemental irrigation. Under different climatic conditions, turfgrasses may thrive strictly from naturally occurring precipitation. A detailed discussion of turfgrass irrigation is provided in Chapter 5.

Water vapor is a relatively small component of air, but it is an extremely important factor in the turfgrass environment. Precipitation, transpiration, and temperature are influenced by atmospheric water vapor. These, in turn, control the distribution and growth of turfgrasses and other plant species. The water vapor content, or relative humidity, of the

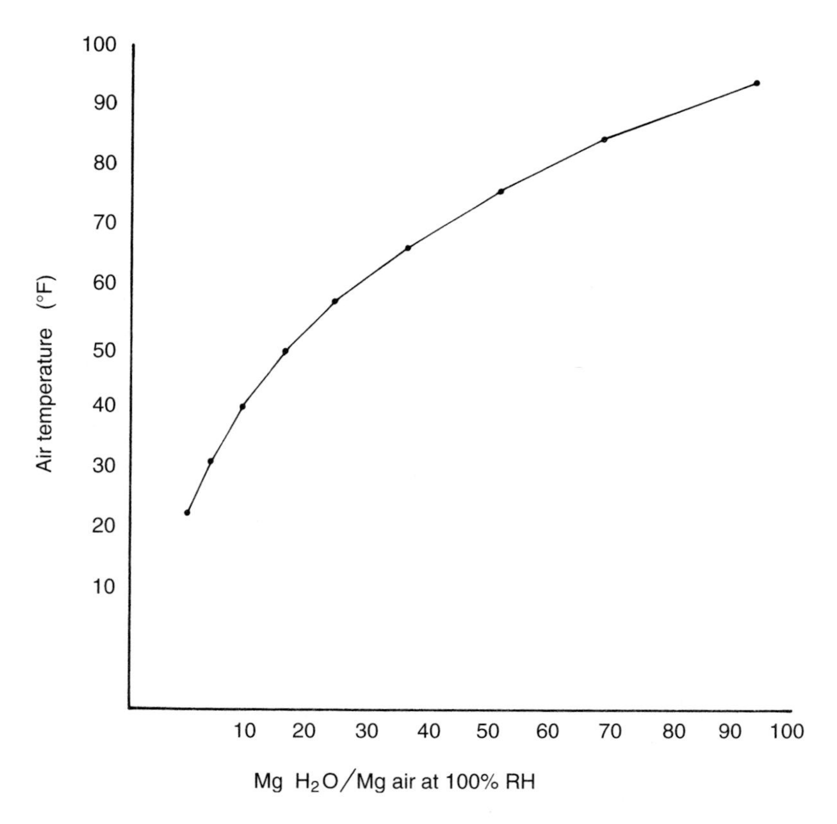

Figure 4.11. Relationship between temperature and the amount of water vapor held by saturated air.

air surrounding turfgrass shoots influences the incidence of disease through its effect on the growth and survival of pathogenic organisms (see Chapter 7). The amount of water vapor that can be held by air is directly proportional to the temperature of the air (Figure 4.11). Increasing the air temperature from 62° to 87°F allows for a doubling of the water vapor content of saturated air. Conversely, a substantial drop in the temperature of moist air can result in condensation of moisture around particles or on plant surfaces.

Dew is the accumulation of visible moisture on plant leaf surfaces observed in the early morning hours. It is the result of several processes occurring independently or together. One process, called *guttation,* involves the diffusion of plant moisture through openings (hydathodes) at the ends of uncut leaves (Figure 4.12). Guttation occurs when water pressure builds up in the roots during periods of minimal transpiration and rapid water absorption. When leaves are cut, resulting in the removal of hydathodes, plant moisture may still diffuse through openings at the wound sites. This is called *exudation.* Guttation and exudation fluids contain various minerals and simple organic compounds collected from within the plant. As droplets of these fluids form on the leaves, these

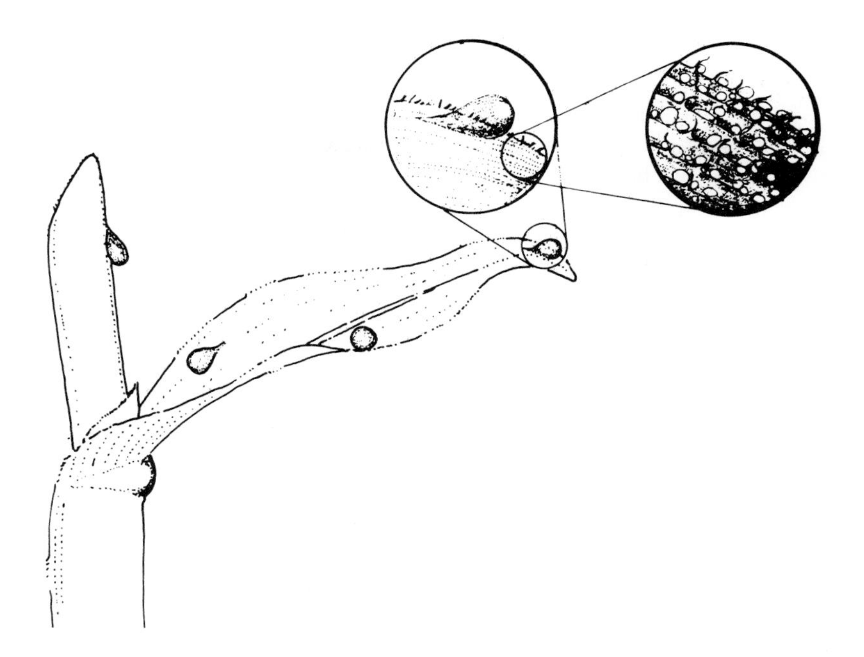

Figure 4.12. Guttation water (large droplets) and condensation water (small droplets on grass leaf surfaces).

dissolved materials accumulate on the leaf surfaces and may significantly enhance the growth of fungal pathogens and increase disease incidence. Some burning of the leaf tips may also result as the fluids evaporate and leave high concentrations of salts on the leaves.

Condensation, another dew-forming process, may occur on leaf surfaces when radiation cooling of the leaves reduces leaf temperature to below that of the surrounding air. The moist air immediately adjacent to the leaf is cooled to its dew-point temperature (temperature at which air is saturated) and, with further cooling, water vapor condenses on the leaves. Condensation is the opposite of evaporation; latent heat is converted to sensible heat during condensation, and the drop in leaf temperature from radiation cooling is reduced. Dew formation typically occurs during the evening hours, especially on clear nights when radiation cooling is proceeding rapidly. Under favorable conditions, so much dew may form that water drips from the leaves onto the soil. The amount of moisture gained from these processes, however, is usually not very substantial.

During cold seasons when night temperatures drop to below freezing, frost is formed in place of dew. Traffic should be avoided on frosted turfs until the frost has disappeared; otherwise, damage to the leaf tissue may occur and be evident until sufficient new growth has taken place.

Dew may be of some benefit to the turfgrass community. It delays the onset of transpiration during the early morning hours so that soil moisture is conserved. In summer, the rise in leaf temperature is retarded until later in the day when the dew has evaporated.

The application of various pesticides is facilitated by the presence of dew. Sprayer operators are less likely to miss strips of turf where dew clearly marks the unsprayed areas.

An undesirable effect of dew is the enhanced disease development that results, especially on greens and some other closely mowed turfs. A traditional disease control practice has been to remove dew by poling, dragging, or syringing with water during the early morning hours.

During winter months, snow and ice may accumulate on turf and remain for extended periods. Snow cover protects the turf from winter desiccation and traffic-induced injury; however, several winter disease problems may be favored in snow-covered turfs and, for this reason, greens are usually treated with preventive fungicides prior to snowfall. Freezing rain may quickly form ice layers over a turf during winter. There is little evidence to suggest that temporary ice cover can directly result in serious injury to turf; however, when the ice thaws on sites with poor surface drainage, submerged turfgrass crowns may absorb enough water to reduce their cold-hardiness. A rapid freeze could then result in some direct low-temperature injury. Partial removal of a thick ice cover may be advisable on greens to reduce the potential for this type of injury.

Wind

Wind is air in motion. Its direction and velocity can have important effects on turf through mixing action and transport of debris. Concentrations of atmospheric gases, water, and other materials in the immediate vicinity of turfgrasses can be dispersed through the mixing action of wind (Figure 4.13). In effect, parcels of "clean" air from above the turf are carried into the shoot zone to displace parcels of "dirty" air. Thus, water vapor and other gases are diluted with air containing lower concentrations of these substances. Likewise, temperature differentials between air from the shoot zone and air from higher altitudes are also reduced. Evapotranspiration from the turf is accelerated by wind because of the reduced boundary layer surrounding leaf surfaces. Finally, winds carry soil particles and other debris that can cause direct abrasion of the foliage or other effects associated with deposition of materials onto the turf. Deposition of small amounts of soil is usually not harmful and may actually be beneficial as a topdressing. However, large quantities of deposited soil can completely cover the turf and entirely exclude light from photosynthetically active tissues. Layers of different soil materials in the turf profile resulting from deposition by wind can interfere with soil-water movement. This is of greatest concern where fine-textured material is deposited on top of a relatively coarse-textured soil. Such layering can result in reduced infiltration and poor drainage from the surface layer, as discussed later in this chapter.

Wind Patterns

Air circulation around the earth is initiated by differential heating of the earth's surface from solar radiation. Since the most direct solar radiation occurs in the vicinity of the equator, the heated air expands upward and flows toward the poles aloft. Descending air

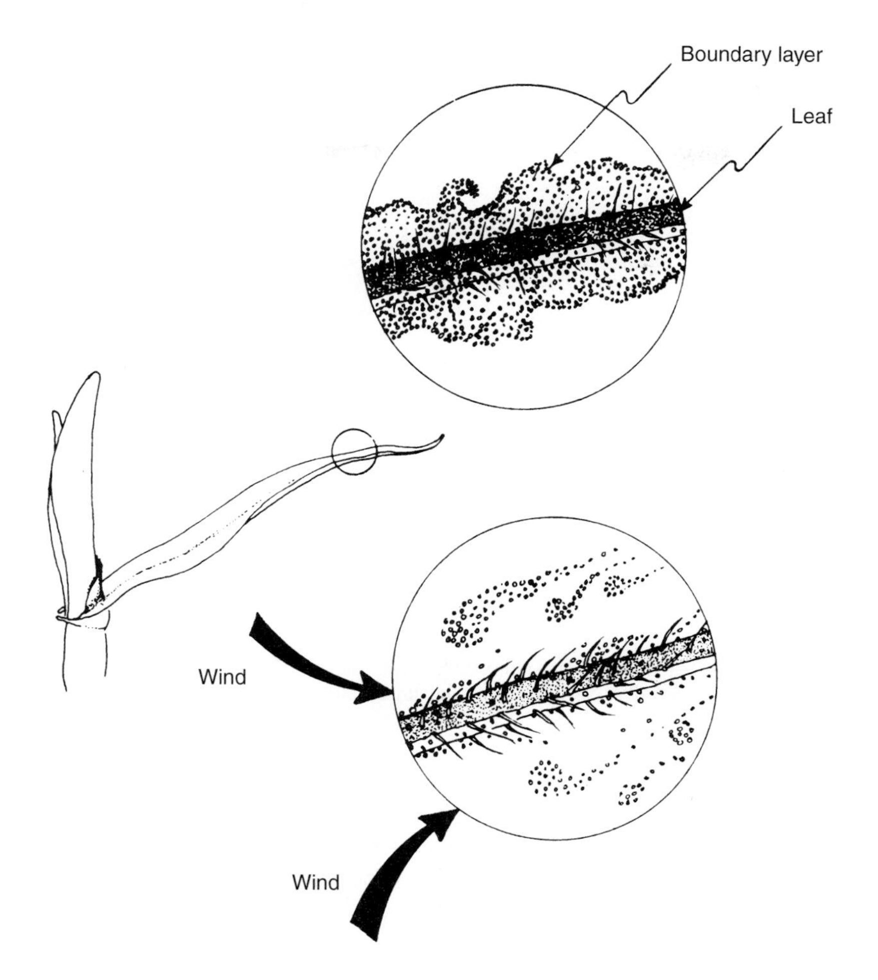

Figure 4.13. Wind-induced dispersal of water molecules constituting the boundary layer surrounding turfgrass leaves.

at about 30° latitude and at the poles produces high-pressure systems that migrate over water and land masses. At intermediate latitudes where cold and warm air masses converge, the warmer air is lifted and cooled. If sufficient moisture is present in the air, clouds form and produce frontal weather.

As high-pressure systems move across the ground, air also moves around and away from the centers of high pressure due to the rotation of the earth on its axis (Figure 4.14). The direction of this airflow is clockwise in the northern hemisphere and counterclockwise in the southern hemisphere. Atmospheric pressure is lowest where two high-pressure systems (air masses) meet; this results in the formation of a low-pressure system in which air converges and rises. Air movement around low-pressure systems is counterclockwise in the northern hemisphere and clockwise in the southern hemisphere. In the United

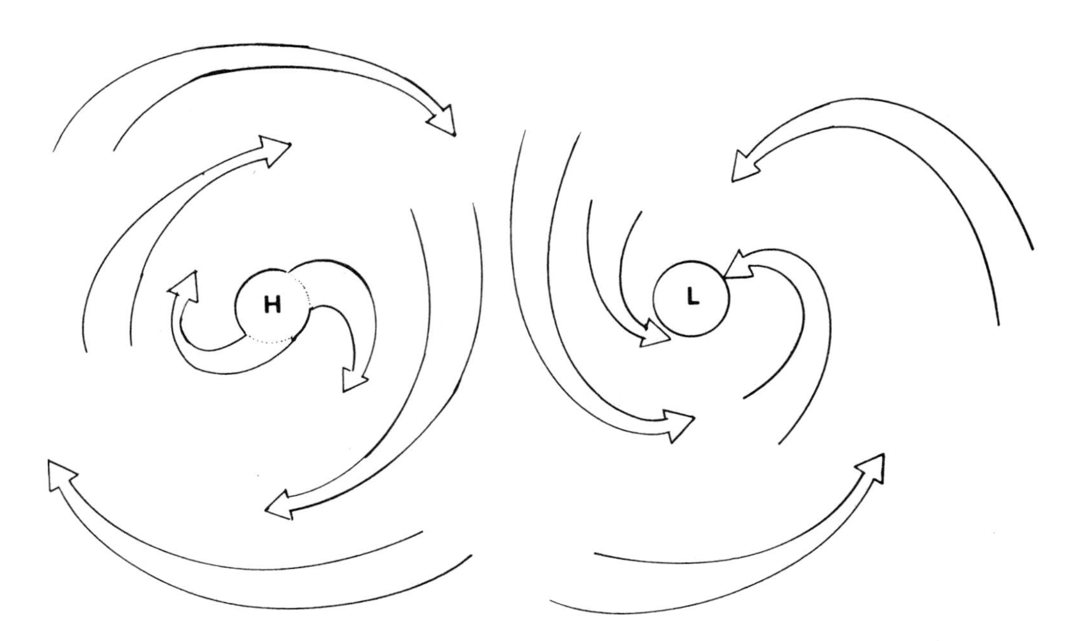

Figure 4.14. Airflow patterns associated with high- and low-pressure systems in the northern hemisphere (airflow patterns are reversed in the southern hemisphere).

States, air masses generally move from west to east. Their moisture and temperature are influenced by the nature of the surfaces over which they travel. Canadian highs moving into the U.S. interior have relatively dry, cold air, while air from the Gulf of Mexico carries considerable moisture to states located to the north and east. Pacific highs carry moisture to the western coast and deposit most of it west of the Rocky Mountains. The continental United States is rich in climatic diversity due, in part, to the differential origin and movement of air masses across the continent.

Air movement over a specific turfgrass site is largely due to general weather patterns; however, local topography can have considerable influence on the nature of winds. In coastal areas, sea breezes result from differential warming of land and sea surfaces during the day and the consequent pressure force that accelerates the air from sea to land. At night, when the land is colder than the sea, a gentle flow from land to sea, called a *land breeze,* develops. Mountain and valley winds result where air adjacent to mountain slopes warms faster than the air at some horizontal distance from the slope, causing circulation similar to sea breezes; air flows toward and up the mountain slope during the day and downward at night. Drainage winds occur from the gravitation of cold air off high ground in intermountain areas. This air seeps down the slopes and gathers in valleys to produce a gentle or moderate cool or cold breeze. Where mountains are located close to a coast, drainage winds can be severe. An example is the Santa Ana wind, which flows down the Santa Ana Canyon of southern California and spreads over the lowlands toward the coast. Chinook winds are strong, dry, warm winds that develop along the lee slope of mountain ranges. In the United States, lowlands just east of the Rocky Mountains are subjected to chinook winds from westerly airflow across the mountains.

Wind Effects on Turf

Disruption by winds of the boundary layer surrounding turfgrass leaves and wind-induced soil deposition have already been discussed. Wind can be very beneficial to turfgrasses or highly detrimental, depending on its intensity. Air flowing at a few miles per hour can accelerate heat transfer to substantially cool a turf during hot weather. The importance of wind in drying turfgrass foliage and, thus, reducing disease incidence is most evident on sites where dense trees and shrubs obstruct air movement. The incidence of brown patch, *Pythium,* and *Helminthosporium* diseases is usually much higher on these sites. Exposed sites at relatively high elevations, however, may be subjected to severe winds that promote rapid drying of the turf. Such sites may require more irrigation during the summer. In winter, severe winds across turfgrass sites without snow cover can cause substantial desiccation and loss of turf. Protective windbreaks, including trees, mounds, and other landscape features, can effectively reduce the potential for desiccation injury to turf.

Wind is also important in disseminating weed seeds and vegetative propagules, fungal spores, salt sprays, and nearly all other foreign substances and organisms that can adversely affect turf. Atmospheric pollutants from industrial and other sources are also carried by wind. Sulfur dioxide, fluorides, ozone, and some nitrogen-containing gases can be transported in sufficient concentrations to directly injure turfgrasses or to weaken them so that they are less resistant to other environmental stresses.

EDAPHIC ENVIRONMENT

The science that deals with the influence of soil and other media on the growth of plants is called *edaphology.* The edaphic environment of a turfgrass community may be composed of synthetic materials, native soil, organic residues, or any combination thereof. Where combined, these media may form a uniform mixture or occur as distinct layers within the turf soil profile. Naturally occurring soil profiles have more or less distinct layers, or horizons, resulting from various activities termed soil-forming processes (Figure 4.15). The surface soil (A horizon) is a zone that has received leachates from organic materials situated at its surface and from which various substances have been leached by percolating water. It is typically higher in organic matter and more favorably structured than the lower soil horizons. Below is the subsoil, or B horizon, where accumulation of leachates has occurred. Subsoil is a far less suitable medium for supporting plant growth than is soil from the A horizon. The C horizon is a zone of partially weathered parent material resulting from the deposition of transported materials or from decomposition of underlying bedrock. Thousands of years of weathering, leaching, and organic matter accumulation are necessary to transform parent materials into soil.

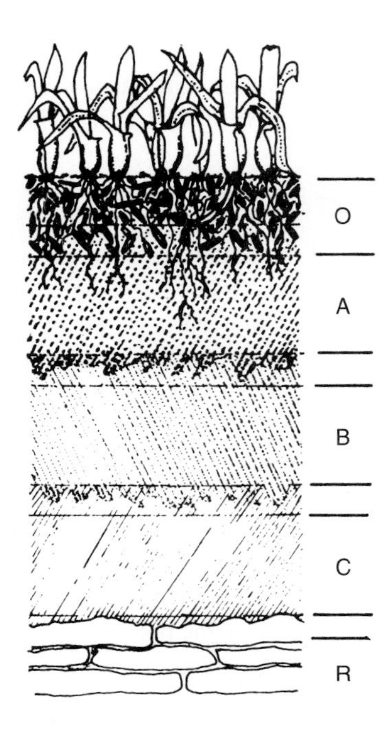

Figure 4.15. Undisturbed soil profile under turf showing surface organic residue (O), topsoil (A), subsoil (B), substratum (C), and bedrock, if present (R).

On turfgrass sites, the zonation of the soil profile may be obscured by mixing or transfer of materials from different horizons, especially where deep excavations of soil have been made during construction operations. Also, the planned incorporation of various amendments may have substantially altered the characteristics of a naturally formed soil. Because soil formation requires many thousands of years, productive soils must be regarded as limited environmental resources requiring careful and intelligent management.

Soil is important as a plant growth medium in terms of its fertility, water relations, gas exchange, and physical support of plant roots. For turfgrass plantings, the soil and plant community should also provide a firm, but resilient, surface that resists compaction from traffic and use.

If a block of turf, including soil several inches deep, were examined, three distinct phases of physical composition would be evident: solid, liquid, and air. Soil, organic matter in different stages of decomposition, and roots and lateral shoots form a solid phase. Organic matter may occur as indistinguishable additions to the soil mineral matter or as organic strands positioned within or above the surface soil. The undecomposed or partially decomposed layer of organic matter situated above the soil surface is called *thatch.* When integrated with the surface soil, it forms a thatchlike derivative called *mat.* A

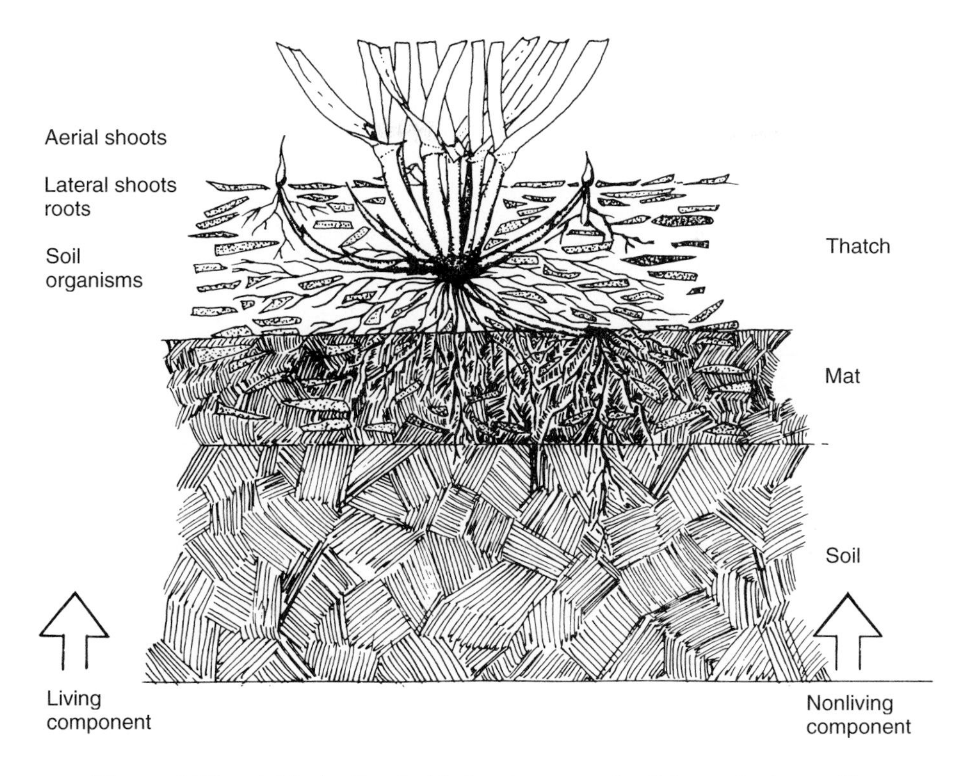

Figure 4.16. A block of turf with thatch, mat, and soil layers in the profile.

specific block of turf may contain one or more of these edaphic features (thatch, mat, and soil), depending on naturally occurring phenomena, cultural practices, and turfgrass genotype (Figure 4.16). Thatch and mat formation will be discussed further in a later section of this chapter.

Pore spaces of varying size and shape permeate the soil and organic materials. If the turf block were completely saturated, all pores would be filled with water and dissolved substances (liquid phase). If it were allowed to dry, all but the smallest pores would be filled with air (air phase).

The relative proportions of the three compositional phases within the total volume of the block largely determine the suitability of the soil as a growth medium for turfgrasses. Generally, a composition of one-half solids, one-fourth liquid, and one-fourth air is considered an optimum proportion for a plant growth medium. The capacity of a growth medium to sustain this proportion is a function of numerous physical, chemical, and biological factors within the edaphic environment. Therefore, turfgrass edaphology involves the physics, chemistry, and biology of soil media.

Table 4.1. Textural Classification of Soil Particles (USDA)

Separate	Diameter size (mm)
Very coarse sand	2.00–1.00
Coarse sand	1.00–0.50
Medium sand	0.50–0.25
Fine sand	0.25–0.10
Very fine sand	0.10–0.05
Silt	0.05–0.002
Clay	<0.002

Soil Physics

The physical properties of a soil, including texture and structure, directly affect its aeration, moisture, and temperature, and indirectly affect fertility and the activity of soil organisms.

Texture

Soil texture is determined by the size of soil particles and their relative proportion. Soil particles range in size from less than 1 micrometer (Mm) to 2 millimeters (2 mm or 2000 Mm) in diameter. Larger particles may occur in soil, but these are considered separately as gravel or rocks. Depending on their size, soil particles are grouped into various soil separates, including clay, silt, and sand. Further divisions are recognized to separate sand into five textural groups. Therefore, there are a total of seven soil separates according to the USDA system of classification (Table 4.1). The proportion of sand, silt, and clay determines the textural class. There are twelve textural classes, as illustrated in the textural triangle (Figure 4.17).

Associated with texture is the amount of surface area that exists within a given weight of soil. One gram of very coarse sand has approximately 11 cm^2 of surface, while the same weight of clay may have a total surface area of 8 million cm^2. Clay is important in chemical reactions involving the adsorption and exchange of plant nutrients. The pore sizes within a predominantly clay soil are so small, however, that much of the water contained in them is generally unavailable to plants. Silt pores are larger and retain higher amounts of plant-available moisture, while sand pores are so large that they contribute little to water retention. Sand is important, however, in promoting soil aeration and drainage.

Structure

Soil structure refers to the arrangement of soil particles. Clay forms aggregates in which individual particles are held together in various configurations. An individual aggregate

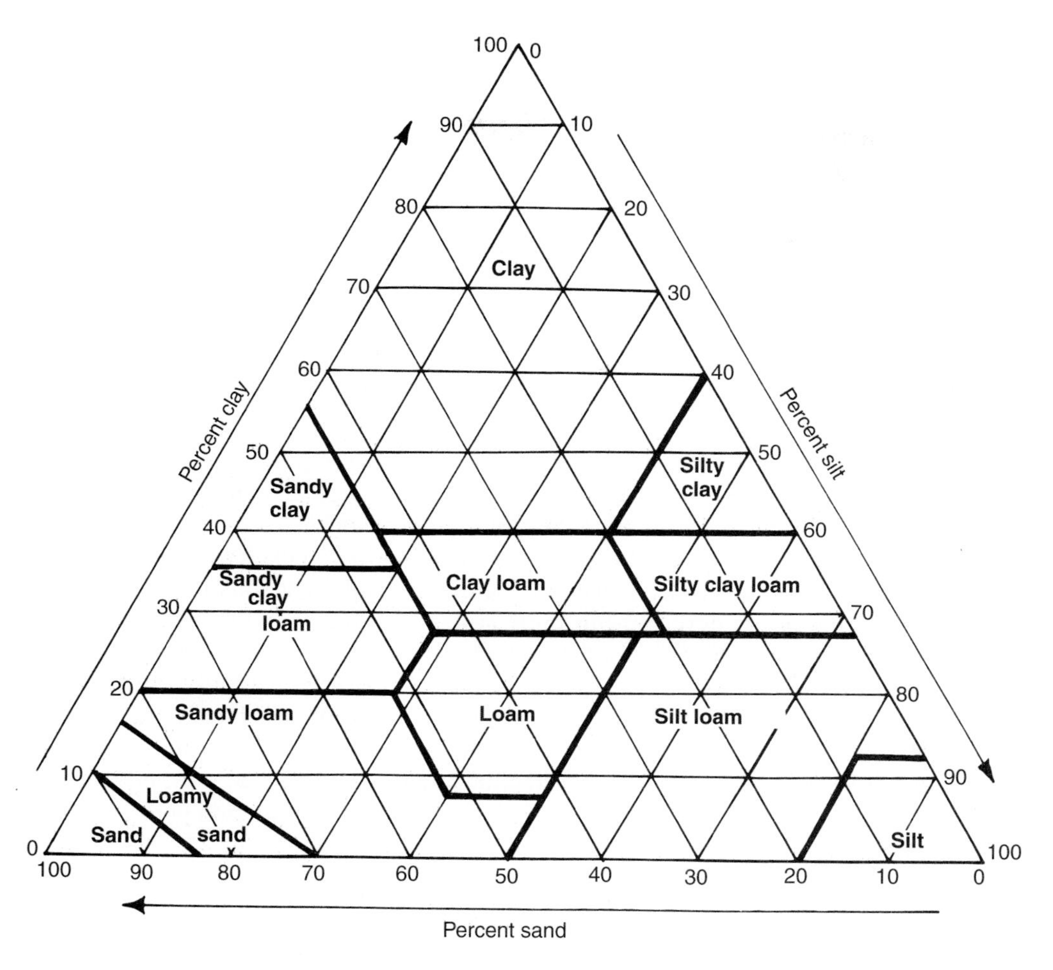

Figure 4.17. Textural triangle with soil textural classes reflecting relative percentages of sand, silt, and clay.

may be as large as, or larger than, a sand particle. Thus, aggregation provides a medium in which pore sizes covering a broad range exist. If soil aggregates are sufficiently stable, a well-aggregated clay soil can serve as an excellent medium for plant growth. However, aggregates differ in their structural stability depending on the specific clay minerals present and the strength of the binding agents holding particles together. Also, traffic, splashing rain or irrigation, and cultivation tend to destroy soil structure. Usually, the best soils for plant growth are those that contain a reasonable proportion of various soil separates and possess a favorable soil structure that has been promoted and sustained through proper cultural practices.

The first step in aggregate formation is the bringing together of soil particles by various forces, including freezing and thawing, wetting and drying, root growth, and the activities of soil organisms. Once joined, polyvalent cations (Ca^{2+}, Mg^{2+}, Al^{3+}, and others) can bridge adjacent soil particles by forming electrostatic bonds. Various cementing agents resulting from decomposing organic matter or hydroxylation of cations can also bind soil particles together. Without structure, a fine-textured soil would be a dense, plastic mass when wet, and bricklike when dry. If crushed, a dry structureless clay will be flourlike in appearance.

Density

Density is the mass of substance per unit volume, usually expressed as grams per cubic centimeter (g/cc). Water has a density of 1 g/cc (at 4°C and 760 mm Hg). Two density measurements, particle density and bulk density, are commonly used for soils. Particle density (PD) is the density of dry, solid soil particles. It averages 2.65 g/cc. Bulk density (BD) is the dry weight of an undisturbed volume of soil. A well-structured, fine- or medium-textured soil will have an abundance of large and small pores, and its bulk density will be low compared to the same soil in a compacted state. If a soil could be so compacted that virtually all pore spaces were removed, bulk density would equal particle density. Comparisons of the bulk densities of different soils may not provide reliable indices for determining their suitability for growing plants. Sands have predominantly large pores; yet their bulk densities are high because of the relatively large mass of solids making up these media. For example, a bulk density of 1.5 g/cc may indicate a rather compacted loam soil with insufficient large pores for drainage. However, in a coarse sand of the same bulk density, aeration would not be a limiting factor in plant growth, but water retention might be. Within soil type, bulk density values are valuable for assessing the physical condition of a soil. Soil porosity can also be determined from bulk density:

$$\text{percent porosity} = 100 - (\text{BD/PD} \times 100)$$

A loam soil with a bulk density of 1.3 would, therefore, have 49% pore spaces:

$$100 - (1.30/2.65 \times 100) = 40\%$$

The percent pore space is only a measure of total porosity and does not directly indicate the distribution of different-sized pores. Within a particular soil type, however, the percent porosity can provide a reasonable assessment of pore size distribution if a sufficient bank of information exists to compare plant growth response to different soil porosities.

Moisture

The importance of soil structure and density lies in their influence on the number and size of pores and the consequent movement of water and air within the soil. Water drains

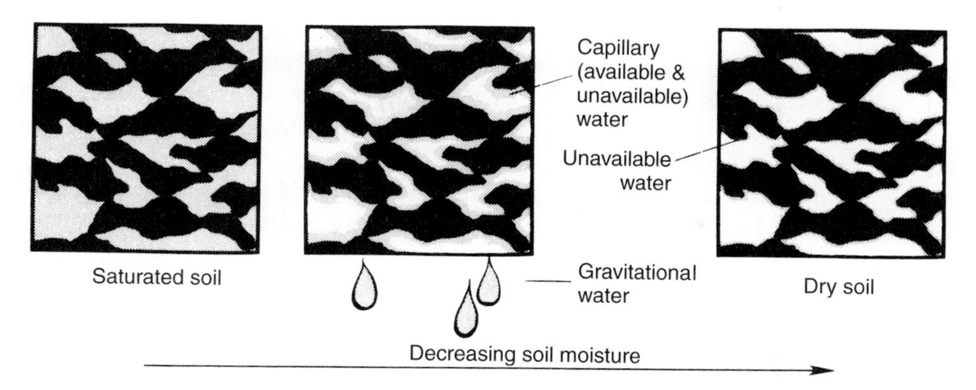

Figure 4.18. Illustration of different types of soil water. Gravitational water drains from the large (aeration) pores, while capillary water is held as films on soil surfaces.

rapidly from large (aeration) pores due to gravitational force. This is called *gravitational water* (Figure 4.18). The remaining water is retained as thin films on the surfaces of soil particles and as wedges where two or more particles come together. To absorb this water, plants must overcome forces of adhesion and cohesion that hold water in the soil. Adhesion is the attraction between soil surfaces and water; cohesion is the attraction between water molecules. The portion of retained water that plant roots can absorb is called *available water.* When water films are reduced in thickness, the attraction of the soil for water becomes greater, and eventually plants can no longer secure enough water to satisfy their needs. The tightly held water that is essentially unavailable to plants is simply called *unavailable water.*

A well-structured soil will release enough water in response to gravity so that aeration porosity is adequate to sustain healthy plants. The water content of the soil following drainage of the aeration pores is called *field capacity,* and the air content is called *aeration capacity.* These are not precise values, but are useful as indications of soil structure for a given soil. In a poorly structured soil, removed water may not be replaced by air because of soil shrinkage. Such soils form massive clods with large cracks instead of friable granules.

The thickness of a water film at which plants can no longer absorb sufficient water to sustain growth is about the same for all soil types. However, the water content, measured in grams of water per gram of dry soil, at which this occurs varies among soils because of large differences in total surface area. When plants growing in a particular soil wilt irreversibly, the soil is said to be at its *permanent wilting point,* and the amount of water remaining in the soil is called the *permanent wilting percentage.* This amount ranges from 1 to 2% for sandy soils to 25 to 30% for fine-textured (clayey) soils. Therefore, a measure of soil water content cannot serve as a guide to a plant's soil-water requirement, but a good relationship exists between a plant's water needs and the work required to remove a unit of water from the soil. Water potential (Ψ_ω) is an expression of the energy status of soil water relative to pure, free water. Soil-absorbed water is not capable of doing as much work as pure, free water; thus, its energy, or water potential, is lower. Differences in water potential between two locations in a soil, called the *water potential gradient,* provide a force that causes water to flow from locations of higher to lower water potential.

Soil water conductivity is an expression of the ease with which the soil conducts water. In compacted soils, conductivity is low because of high resistance to flow. Well-structured soils conduct water more rapidly. Where a plant is pulling water from the soil immediately surrounding its roots, replenishment of this root-zone soil moisture depends on (1) the water potential gradient between the root-zone soil and a location in the soil where water is available and (2) the conductivity of the soil.

In a saturated soil, water potential near the soil surface approaches zero. As the soil drains, water potential becomes progressively lower (more negative). Water movement in saturated and unsaturated soils should be considered as two distinct processes. Saturated flow occurs when all or most of the pores are filled with water. It takes place through large pores and, therefore, is most rapid in coarse-textured soils. The principal force acting on the water is gravity, and the direction of flow is primarily downward. If the number of large pores decreases suddenly at a given soil depth, downward movement is restricted, and water may accumulate above the interface where the two media meet, resulting in a temporary water table. This occurs in sand or thatch overlying loam soil or with a compacted subsoil.

Unsaturated water flow occurs in soils in which the large pores are not filled with water. The rate of unsaturated flow depends on the thickness of water films surrounding soil particles; thicker water films allow faster flow rates than thinner films due to the differences in water potential. Thus, water moves faster in moist soils than in dry soils. Unsaturated flow proceeds in any direction, irrespective of gravitational force. The "wick action" or capillary flow of water from lower to upper soil locations is actually unsaturated flow.

Where the continuity of water films is disrupted, as at the interface between a fine-textured soil and an underlying coarse-textured soil, unsaturated flow is slowed or may stop altogether (Figure 4.19). This can result in an accumulation of water, called a *perched water table,* above the interface. Water will not move across the interface until the water potential in the above soil builds to a level sufficient to overcome the attraction between the water and the fine-textured soil. When sufficient water potential has built up from continued downward flow toward the interface, water will enter the coarse-textured soil and be conducted away. Where a coarse-textured soil is underlain by a fine-textured soil, downward flow through the soil profile will slow to a rate determined by the hydraulic conductivity of the fine-textured soil. If that rate is slower than the rate at which water is entering the coarse-textured soil, a *temporary water table* may form as long as water continues to enter the soil at the same rate and persists until the fine-textured soil absorbs all of the free water situated above it.

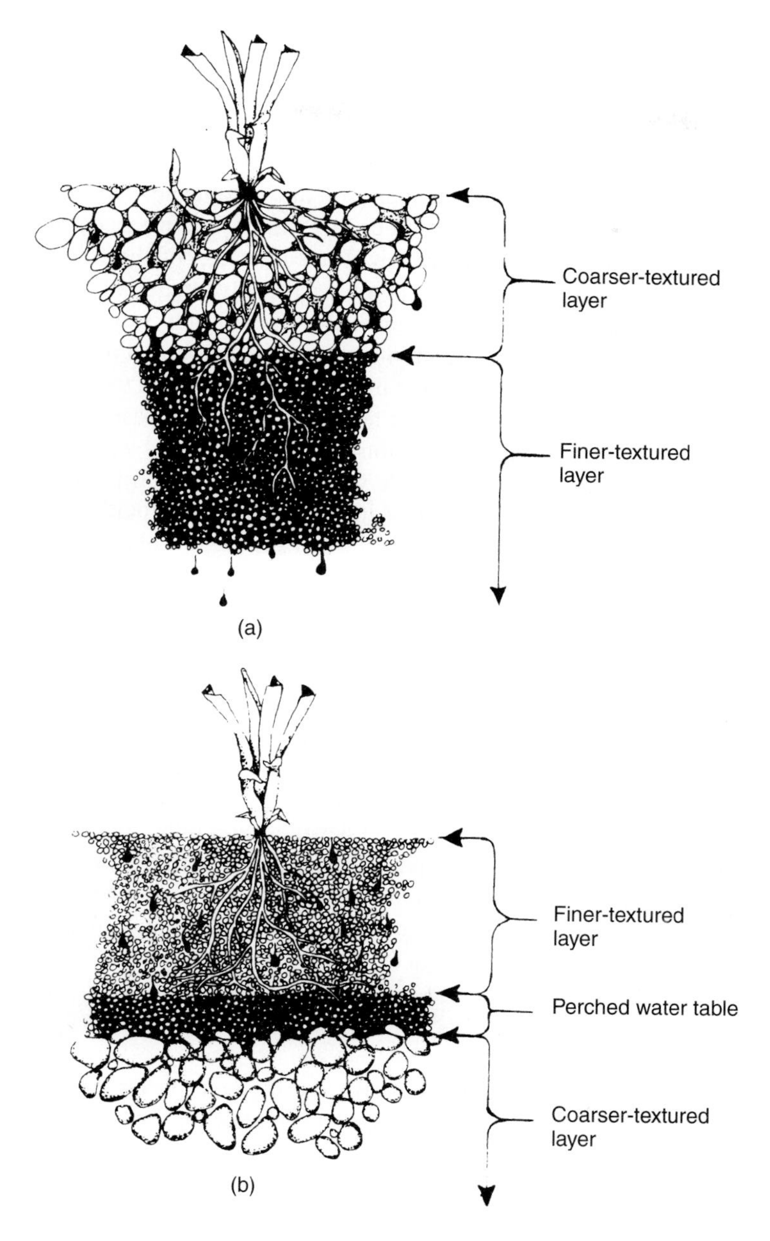

Figure 4.19. Layered turf-soil profiles in which a coarser-textured layer (thatch, sand) overlies a finer-textured soil (a), and in which a finer-textured soil overlies a layer of coarser texture (b), resulting in a perched water table.

This principle can be applied to turf soils in which layers exist either by design or by error. Consider a fine-textured soil that has been modified by incorporation of sand or other coarse amendments. If the incorporation is not uniform but results in subsurface layers of coarse material, water flow within the profile can be disrupted. On the other hand, where a layer of coarse sand has been intentionally positioned beneath finer-textured sand with some loam soil and organic amendments, a perched water table will develop following irrigation or rainfall. In this instance, the perched water table is desired to compensate for the low water retention of the fine sand. A USGA green is constructed in this fashion to combine the advantages of compaction resistance and moisture retention in the root zone. Results are only satisfactory, however, where the design includes a critical depth of the surface medium; a too-shallow surface layer will not drain properly, while a layer that is too deep will be too dry at the surface. This can be illustrated using a rectangular household sponge measuring 5 by 3 by 1 inches (Figure 4.20). When the sponge is saturated and positioned with its 5-inch and 3-inch sides in the horizontal plane, very little water drains out. Turning the sponge 90° to position the 3-inch side vertically results in more drainage. Rotating the sponge so that its 5-inch side is vertical results in still further drainage. This demonstration shows the relationship between height of the water column and water retention within the pore volume of the sponge. Cutting the sponge horizontally into three equal sections while it is still in the 5-inch vertical orientation and squeezing each section to remove internal moisture would reveal that the uppermost section was driest and the lowermost section was wettest.

The sponge analogy facilitates understanding of some soil water phenomena. In a USGA green, the sponge represents the surface soil medium, and the air, or "free space," beneath the sponge represents the coarse sand or gravel layer underlying the surface medium. A workable greens design depends on a rather precise application of principles of soil physics. The settled depth (depth after soil settling) of the surface medium must be correct, and the physical composition of the medium must be such that air and moisture within the turfgrass root zone are sufficient to sustain growth. The USGA green and other designs are discussed further in Chapter 9.

Aeration

The process by which soil air is replaced by atmospheric air is called soil aeration. Soil air has higher concentrations of carbon dioxide and water vapor but less oxygen than atmospheric air due to the consumption of oxygen and production of carbon dioxide by soil organisms. The magnitude of difference depends on the rate of gaseous exchange between the atmosphere and soil. Aeration is brought about by processes of diffusion and mass flow.

Diffusion is the movement of gases through air-filled pores from regions of higher to lower concentration of the gas and is proportional to air-filled porosity. Diffusion is

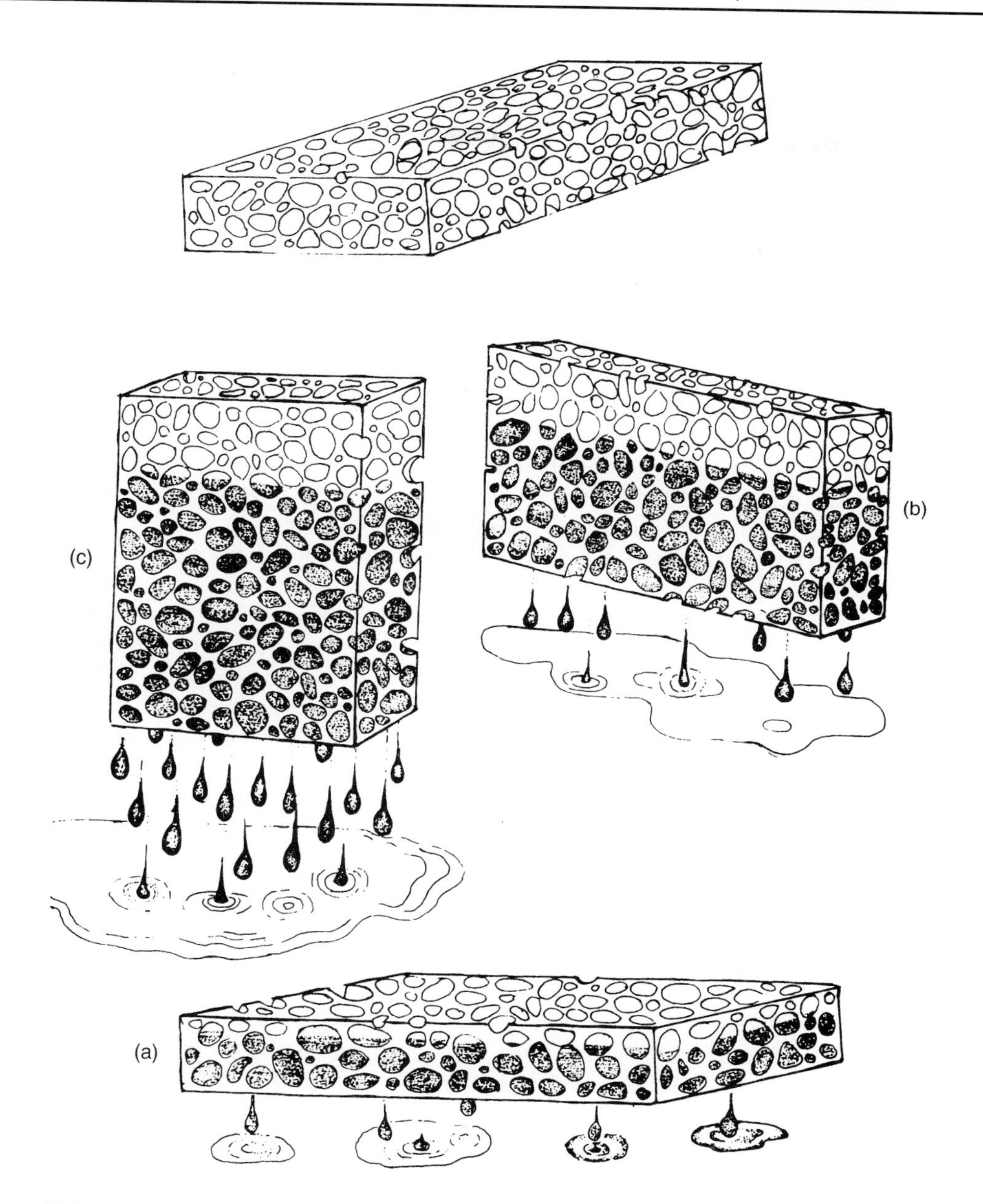

Figure 4.20. Water drainage from a saturated sponge (a) with the thin side oriented vertically, (b) after being rotated 90°, and (c) after its longest side is oriented vertically.

low in compacted soils because of reduced pore size and number, and the discontinuity of soil pores. Similarly, diffusion is low in wet soils because of the absence or reduction of air-filled pores, and the extremely low diffusion rate of air in water. Mass flow occurs as a result of the following:

1. Expansion and contraction of soil gases due to temperature and barometric pressure changes.
2. Soil air removal through precipitation and irrigation, and replacement as water is removed by drainage, plant use, and evaporation.
3. Wind action causing air to be forced into the soil at some locations and pulled out at others. Relative to diffusion, mass flow is considered to have a minor influence on soil aeration.

Poorly aerated soils are often deficient in oxygen. Oxygen is used by plant roots and soil organisms for respiration, which produces carbon dioxide. Without an adequate exchange of gases between the atmosphere and soil, oxygen levels decline and carbon dioxide levels increase in soil air. This can result in reduced absorption of nutrients and water by plant roots, since they must have sufficient oxygen for respiration to generate the energy necessary for this process. Microbial decomposition of organic matter is also inhibited in oxygen-deficient soils, as is the bacterial oxidation of ammonia to nitrate nitrogen. Denitrification, the conversion of nitrate to N_2 and N_2O gases, occurs in persistently wet soils, resulting in a loss of soil nitrogen to the atmosphere.

Turfgrass communities growing in compacted or persistently wet soils are often invaded by various weed species. This reflects, in part, the differential adaptation of plants to poorly aerated soils. Some weed species that typically grow under these conditions may have the capacity to transmit foliar-absorbed oxygen to their roots to satisfy respiratory requirements. Thus, specific weeds may have an advantage over many turfgrasses through their ability to survive persistently wet conditions.

Temperature

Many physical, chemical, and biological events that take place in soil are strongly temperature dependent. Soil temperature is, in turn, affected by (1) atmospheric conditions (air temperature, moisture, wind, and solar radiation), (2) thermal absorption and conductivity of the soil, and (3) plant cover. Atmospheric influences on soil temperature have been discussed in an earlier section of this chapter.

Thermal absorption is a function of the color, moisture level, and organic matter content of the soil. Generally, darker soils high in organic matter are more efficient in absorbing heat from the atmosphere. Heat absorption occurs faster in drier soils, since the specific heat (the amount of energy necessary to raise the temperature of 1 gram of a substance by 1°C) for water, dry mineral soil, and dry humus is 1.0, 0.2, and 0.4, respectively. Therefore, as the water content of a given soil increases, the amount of energy from solar radiation or atmospheric air required to raise its temperature increases proportionately.

Changes in soil temperature are influenced by the air-moisture-solid balance in the soil. Sandy soils warm and cool at a faster rate than clayey soils due to generally higher aeration porosity and lower retained moisture. Similarly, compacted soils undergo slower warming and cooling than well-structured soils. Temperature fluctuations during the winter months in cooler climates can be beneficial in promoting desirable soil structure. Alternate freezing and thawing of soil, and wetting and drying cycles, tend to rearrange compacted soil particles in such a way that aeration porosity is increased. Other temperature-dependent processes, including the formation and expansion of ice crystals and the shrinking and swelling of organic matter, also result in improved soil structure.

Most reactions within the soil occur more rapidly at higher temperatures. Microbial activity, which is so important for nitrogen transformations, organic matter decomposition, and other processes, is highly temperature dependent.

Adaptation of turfgrass species is considerably influenced by soil temperature. Root growth of Kentucky bluegrass is slowed at soil temperatures above 75°F, while 85°F is favorable to bermudagrasses.

Soil Chemistry

Soil chemistry deals with the chemical reactions that occur on colloidal surfaces in the soil. A knowledge of soil chemistry is essential for understanding nutrient availability to plants, phytotoxic effects from soil constituents, and the relationship between fertility and soil physical and biological properties.

Soil Colloids

Soil colloids include small soil particles measuring 0.2 mm or less in diameter. Clay and humus make up the colloidal component of a soil. Most clay colloids are secondary crystalline minerals formed from such primary minerals as quartz, feldspars, micas, hornblende, and augite. Clay colloids are made up of planes of oxygen atoms (O, OH) with silicon (Si^{4+}) and aluminum (Al^{3+}) atoms holding the oxygens together by ionic bonding (Figure 4.21). Several planes of oxygen atoms with intervening silicon and aluminum planes make up each crystal layer within a clay particle, which has many layers stacked like a deck of cards. Silicon and aluminum atoms making up the cationic planes within the crystal layer may be substituted with cations of lower valence. For example, Al^{3+} may be substituted for Si^{4+}, Mg^{2+}, and Fe^{2+}, or Zn^{2+} for Al^{3+}. This is called isomorphous substitution and results in unsatisfied negative charges in adjacent oxygen atoms. The net effect is the colloid's negative surface charges and consequent capacity for attracting cations (*cation exchange capacity*). Other sources of negative charges include unsatisfied edge-of-clay oxygens, ionized hydrogen from hydroxyl groups, and ionized hydrogen from organic materials.

Not all clays are alike. Most have a definite, repeating arrangement of component elements (*crystalline structure*). Others are not formed from well-oriented crystals and,

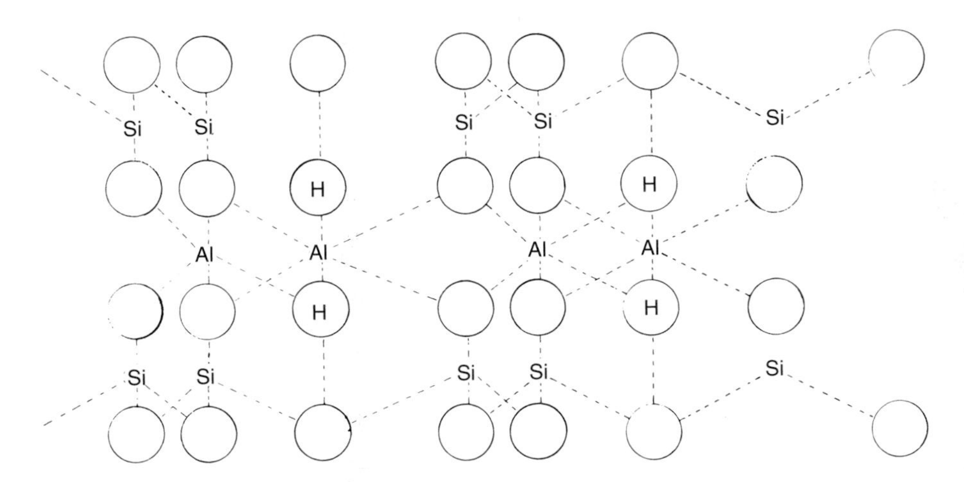

Figure 4.21. Clay particle configuration with many layers, each consisting of several planes of oxygen atoms with intervening silicon and aluminum ions (two-to-one-type clay).

therefore, are said to have an amorphous structure. In temperate climates, the silicate clays are predominant. Sesquioxide clays, which are hydrous oxide clays of iron and aluminum (rather than silicon), are found mostly in tropical climates. Within the silicate clays, composition of the crystal layer varies depending on the number of oxygen and cationic (Si^{4+}, Al^{3+}) planes. A one-to-one-type clay has one plane of silicon and one plane of aluminum cations, along with two outer planes and a shared inner plane of oxygens. A two-to-one-type clay has two planes of silicon and one of aluminum with two outer planes and two shared planes of oxygens, as in Figure 4.21. Kaolinite is an example of a one-to-one-type silicate clay. It has almost no isomorphous substitution; its cation exchange capacity is low; and because adjacent crystal layers are strongly attracted to each other by hydrogen bonds, water does not penetrate between layers and almost no swelling of the colloids occurs. Montmorillonite is a two-to-one-type silicate clay. It has a relatively high cation exchange capacity due to considerable isomorphous substitution, and adjacent crystal layers allow water to penetrate between them. This causes substantial swelling of the clay particle and subsequent shrinking upon drying. Illite is a two-to-one-type silicate clay similar to montmorillonite except that potassium ions hold adjacent crystal layers together, thus limiting swelling.

Sesquioxide clays have a crystal layer composed of a single plane of iron and aluminum cations with planes of hydroxyl ions on either side. These clays do not swell, are not sticky, and behave more like fine sands than silicate clays.

Humus is a semistable end product of decomposing plant and animal residues. Its organic components continue to decompose, but very slowly compared to the original organic material. Humus colloids are amorphous organic particles consisting of proteins, lignins, and complex sugars. On a dry-weight basis, humus has a cation exchange

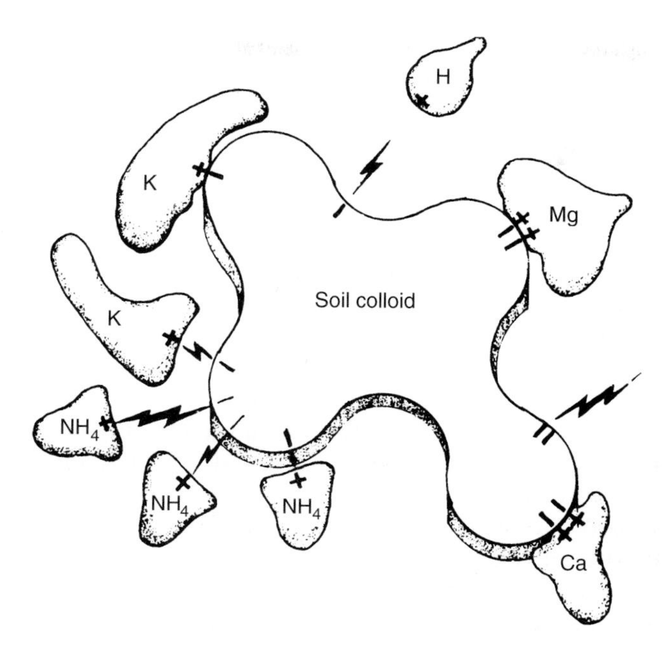

Figure 4.22. Electrostatic attraction between negatively charged soil colloid and positively charged ions.

capacity that is many times greater than that of clay colloids. Negative charges are due to ionization of hydrogen from oxygen-containing groups including the following:

$$\text{carboxyl groups } (\text{R}\overset{\text{O}}{\overset{\|}{\text{C}}}\text{OH} \rightarrow \text{R}\overset{\text{O}}{\overset{\|}{\text{C}}}\text{O}^- + \text{H}^+) \text{ in acids}$$

and

$$\text{hydroxyl groups } (\text{ROH} \rightarrow \text{RO}^- + \text{H}^+) \text{ in alcohols and phenols}$$

Cation Exchange

Positively charged ions (cations) are adsorbed onto colloids at sites with unneutralized negative charges. The adsorbing force is electrostatic attraction between the positively and negatively charged entities (Figure 4.22). The adsorbed cations resist removal by leaching water, but they can be replaced by other cations in solution due to mass action and the preferential adsorption of some cations over others. This exchange of one cation for another is called cation exchange. When such cations as NH_4^+ (ammonium), K^+

(potassium), and Ca^{2+} (calcium) are supplied to the soil by fertilizers and lime, these ions replace other cations already adsorbed to exchange sites on the surfaces of soil colloids and plant roots.

The distribution of different cations on exchange sites can dramatically affect soil physical properties. Saturation of the exchange sites by Na^+ (sodium) causes a dispersal of clay particles and a consequent loss of desirable soil structure. Sodium-saturated soils become severely compacted and highly impermeable to water. Replacement of adsorbed sodium cations by calcium can improve flocculation of clay particles, resulting in an overall improvement in soil physical properties.

Cation exchange capacity (CEC) is the amount of exchangeable cations per unit dry weight of soil. It is measured in milliequivalents (mEq) per 100 grams (g) of soil.

The amount of a specific cation in the soil (in pounds per acre) can be calculated from CEC values. First, determine the milliequivalent weight of the cation by dividing its molecular weight by its valence. Calcium (Ca^{2+}) has a molecular weight of 40 g/mole and a valence of 2 (two positive charges). Therefore, 40 g/mole $\times$ 2 = 20 g/equivalent or 20 mg/mEq. Then, multiply the milliequivalent weight (20 mg/mEq) by the milliequivalent value for calcium given as part of the total cation exchange capacity. If the calcium portion of CEC is determined to be 13 mEq/100 g soil, then 20 mg/mEq $\times$ 13 mEq/100 g = 260 mg/100 g or 260 mg/100,000 mg or 2600 ppm (parts per million) calcium. Finally, since the surface 6 2/3 inches of an acre of field soil (acre furrow slice) weighs approximately 2 million pounds, 2600 ppm Ca is equal to 5200 lb Ca/acre. The principal cations and their equivalent weights are shown in Table 4.2.

A soil with a cation exchange capacity of 20 mEq/100 g may have 65% of the exchange sites satisfied with calcium (13 mEq/100 g), 10% with magnesium (2 mEq/100 g), 5% with potassium (1 mEq/100 g), and the rest with sodium, ammonium, hydrogen, and other miscellaneous cations. The sum of all milliequivalents of all cations adsorbed onto exchange sites in each 100 g of soil is the cation exchange capacity.

The CEC of a soil depends on its texture, type of clay, percent organic matter, and, possibly, pH. Fine-textured soils typically contain more clay and, therefore, have higher CECs than coarse-textured soils. The specific type of clay is important, since montmorillonite ranges from 80 to 150 mEq/100 g, while kaolinite has CECs between

Table 4.2. Equivalent Weights of Principal Cations

Cation	Molecular Weight (g/mole)	Valence	Milliequivalent Weight (mg/mEq)
Calcium (Ca^{2+})	40	2	20
Magnesium (Mg^{2+})	24	2	12
Potassium (K^+)	39	1	39
Sodium (Na^+)	23	1	23
Ammonium (NH_4^+)	18	1	18
Hydrogen (H^+)	1	1	1

3 and 15 mEq/100 g. Organic matter varies widely in CEC, depending on origin and degree of decomposition; some types contribute significantly to soil CEC due to the multiple adsorptive sites on organic colloids for retaining cations. Since many of these adsorptive sites are the result of hydrogen ionization, increasing pH usually results in more ionization and, therefore, higher CEC.

Soil Reaction

Soil reaction is an indication of the acidity or alkalinity of a soil and is measured in units of pH. Soil pH is actually the negative logarithm of the hydrogen (H^+) ion concentration. The pH scale extends from 0 to 14; each whole-number unit reflects one magnitude of change in the hydrogen (or hydroxide OH^-) ion concentration. At a pH of 7, the H^+ and OH^- concentrations are both 10^{-7} (0.0000001) moles/liter, and the soil solution is said to be at neutrality. As the pH decreases to 6, the H^+ concentration increases to 10^{-6} (0.000001). The solution becomes more acidic by a factor of 10 (one magnitude) with each pH unit decrease. At pH units above 7, the OH^- concentration is greater than the H^+ concentration, and the solution is said to be alkaline. The relationship between H^+ and OH^- ions in the soil solution is expressed by the formula

$$[H^+][OH^-]/[HOH] = 10^{-14}$$

which states that the product of the H^+ and OH^- ion concentrations always equals 10^{-14}.

Soil pH can be measured with a pH meter, which actually measures the H^+ concentration in the soil solution. This is called *active* acidity. Because of the exchangeable H^+ and Al^{3+} on soil colloids, soils also have *potential* acidity. An equilibrium exists between active and potential acidity. If bases are added to neutralize active acidity, potential acidity is released from the exchange sites to maintain equilibrium. Thus, soils containing appreciable amounts of clay and organic matter resist pH change and are said to be buffered due to their relatively high cation exchange capacities and, therefore, high potential acidities.

The normal pH range in soil is 4.0 to 8.0. In areas where sufficient precipitation occurs to leach soluble basic salts, the soil pH tends to decrease. Irrigation may induce a similar effect, except where the irrigation water contains basic salts that accumulate in the soil. In dry regions where evapotranspiration exceeds precipitation, soil pH tends to be alkaline due to the accumulation of basic salts in the surface soil. Sodic soils (high in sodium) may reach a pH of 8.5 to 10.

Turfgrass species are adapted to a wide range of soil pHs; however, optimum growing conditions usually exist where the pH is neutral to slightly acid (7.0–6.0). Probably the most important detrimental effect of excessive acidity is the high solubility of aluminum and manganese, which can reach toxic concentrations in acid soils. Soil pH can be adjusted upward through periodic applications of lime, which result in better rooting, increased turfgrass vigor, improved availability of some plant nutrients, reduced availability of toxic elements, and more favorable microbial activity. Liming materials include quicklime [CaO], hydrated lime [$Ca(OH)_2$], calcitic limestone [$CaCO_3$], and dolomitic limestone [$CaCO_3$–$MgCO_3$]. Lime moves through the soil profile very slowly; therefore, applications are especially effective when made prior to establishing a new turf, so that the lime can be tilled in, or possibly in conjunction with core cultivation of established turfs. The amount of lime necessary for neutralizing excessive soil acidity varies with soil texture. Clayey soils may require nearly twice as much lime as sandy soils for comparable pH adjustments. Results from soil-testing laboratories usually include recommendations for lime application where needed. Depending on the type of lime used, the amount applied at any one time should not exceed 25 to 50 lb/1000 ft^2. If hydroxide [$Ca(OH)_2$] or oxide [CaO] forms are used, rates higher than 25 pounds may severely injure the turfgrass, especially when atmospheric temperatures are high.

Excessive soil alkalinity can result in deficiencies of several plant micronutrients. The problem often can be reduced by additions of acidifying fertilizers, such as elemental sulfur and sulfates of ammonium, iron, and aluminum. As with pH adjustments by liming, the amount of sulfur or sulfates required depends on the buffering (cation exchange capacity) inherent in the soil, free $CaCO_3$ in the soil, and dissolved salts in the irrigation water. The maximum rate of finely ground elemental sulfur that should be applied at any one time is 200 lbs/acre/year on lawns and fairways, 100 lbs/acre/year on greens, and 35 lbs/acre/year on annual bluegrass greens. Due to the high potential for sulfur-induced injury, applications should be confined to periods of moderate to low atmospheric temperature.

Salted Soils

Salted soils are classified on the basis of two criteria: soluble salt content and exchangeable sodium percentage. The three types of salted soils are saline, sodic, and saline-sodic.

Soluble salts are inorganic chemicals that have a water solubility greater than 2.4 g/liter (the solubility of gypsum). The amount of total soluble salts in a soil can be estimated by measuring the electrical conductivity (EC) of the soil solution. The procedure involves mixing water with a weighed soil sample to form a saturated paste, removing the water by suction filtration, and measuring the EC of the extract in decisiemens per meter (dS/m). If it is below 4 dS/m, the soil is nonsalty. The exchangeable sodium percentage (ESP) is the percentage of total exchangeable cations that are sodium.

Saline soils contain sufficient soluble salts to reduce plant growth. Characteristics of saline soils include EC $>$ 4 dS/m; ESP $<$ 15% of CEC; and pH $<$ 8.5. Saline soils are usually well structured and quite permeable. On the other hand, sodic soils contain sufficient Na^+ to raise the pH to 8.5 or higher, EC is less than 4 dS/m, and ESP is 15% or more of CEC. High concentrations of Na^+ cause dispersion of clay particles, resulting in a structureless, impermeable soil. Saline-sodic soils are high in both sodium (ESP $>$ 15% of CEC) and soluble salts (EC $>$ 4 dS/m).

Controlling problems associated with saline soils depends on reducing the concentration of soluble salts. Irrigation must be performed beyond plant moisture requirements in order to leach excess salts from the turfgrass root zone, and it must be frequent enough to prevent drying of the soil so that salts do not move upward in capillary flow. On sites where the intensity and frequency of irrigation cannot be sufficient to adequately reduce salt concentrations in the root zone, or where the irrigation water contains high concentrations of salts, turfgrass species that are tolerant of saline conditions must be used. Warm-season turfgrasses with good salt tolerance include bermudagrass, zoysiagrass, and St. Augustinegrass. The most tolerant of the cool-season turfgrasses is creeping bentgrass, followed by tall fescue and perennial ryegrass. A relative newcomer to this group is alkaligrass (*Puccinellia distans*), which appears to be uniquely adapted to saline conditions.

In sodic soils, exchangeable Na^+ must be replaced by other cations that do not disperse clay particles but, rather, will cause flocculation to occur. The principal cations used for this purpose include Ca^{2+} and H^+ generated from gypsum ($CaSO_4$), sulfur, or other materials. Gypsum usually causes the fastest response due to the direct exchange of Ca^{2+} for Na^+ on colloidal surfaces. This remedy is most effective when the material can be incorporated into the soil as opposed to being surface applied. The gypsum requirement (GR) is the amount of gypsum, in tons per acre, required to reclaim the soil and is calculated by the formula $GR = 1.72\ (Na_x)$, in which Na_x is the milliequivalents of sodium per 100 grams of soil (mEq/100 g) to be replaced. For example, if the exchangeable sodium has been measured to be 10 mEq/100 g (CEC = 28 mEq/100 g; therefore, ESP = 36%), and the desired exchangeable sodium is 3 mEq/100 g, then $GR = 1.72\ (10–3) = 12$ tons of gypsum per acre. Gypsum is usually surface applied at 1 ton per acre; up to 2 tons per acre per year are permissible. Elemental sulfur can be used to reduce sodium in soils with free lime present, while gypsum can be used in soils with or without free lime present. The efficiency of sulfur in reclaiming sodic soils is approximately 5.6 times that of gypsum; therefore, only about 2 tons of sulfur per acre would be required to achieve the same effect. However, since sulfur must be oxidized microbially to sulfuric acid before it can yield H^+ ions to replace exchangeable Na^+, reaction time would be much slower than with gypsum, which directly supplies Ca^{2+}.

Soil Biomass

Soil biomass consists of the sum total of all living organisms and their residues that become part of the soil profile. *Live biomass* is that portion of soil biomass that includes an array of microscopic and macroscopic organisms, plant roots, and subsurface shoots. *Residual biomass,* or dead organic matter, is made up of undecomposed and partially decomposed remnants of plant and animal material occurring within and above the soil. Given the dynamic nature of soil biomass, changes within and above the soil dramatically influence the suitability of the medium for sustaining turfgrasses.

Live Biomass

The soil is virtually alive with huge populations of microscopic plants called *microflora* (bacteria, fungi, actinomycetes, algae) and microscopic animals called *microfauna* (protozoa, nematodes). Occasionally, we find evidence of larger, soil-inhabiting animals such as earthworms, arthropods (insects, mites, centipedes, millipedes), gastropods (slugs, snails), and even larger burrowing animals including moles, gophers, and mice. These are referred to collectively as *macrofauna*. Live roots and other plant parts constitute the soil *macroflora*.

Soil organisms are largely beneficial in that they decompose organic materials, convert materials into plant-available forms, and improve soil structure. Some, however, are pests that can cause extensive damage to turf.

Of the soil microflora, bacteria are the most numerous. Bacteria occur in two principal categories: *autotrophic* bacteria, which derive their nutritive carbon from CO_2, and *heterotrophic* bacteria, which obtain carbon from organic matter. Autotrophs are very beneficial because of the processes by which they obtain energy. Oxidation of ammonia to nitrites and then to nitrates, as well as oxidation of sulfur, iron, manganese, and other elements, is an important process for providing plant-available nutrients. Conversion of carbon monoxide (CO) to methane (CH_4), a reduction reaction, or to CO_2, an oxidation reaction, is certainly important in highly populated areas where CO emissions would otherwise accumulate to phytotoxic concentrations in the atmosphere. Some autotrophs engage in less desirable activities, such as denitrification, or the reduction of nitrate (NO_3^-) to N_2 or N_2O gases. This may account for substantial losses of fertilizer nitrogen to the atmosphere, especially from warm, persistently wet soils.

Heterotrophs include most of the soil bacteria. Soils may contain more than 100 million bacteria per gram of dry soil. The optimum pH for most bacteria is near neutrality. Bacteria are the principal organisms for decomposing organic matter and, as such, provide for recycling of nutrients contained within plant and animal residues. Were it not for heterotrophic bacteria, turfgrass and other plant communities (indeed, all organisms) would be virtually inundated by their own debris. Some heterotrophs fix atmospheric nitrogen and, thus, enable plants to obtain nitrogen from the air. Certain nitrogen-fixing heterotrophs attack root hairs of host plants, resulting in the formation of protective nodules. A symbiotic union is, thus, formed in which the plant supplies organic substances and minerals to the bacteria, and the bacteria fix and supply nitrogen to their host plant. Fixation actually occurs as the bacteria utilize atmospheric nitrogen for synthesizing body proteins. Since the life span of a single bacterium is only a few hours, a portion of the bacterial population is continually dying, decomposing, and releasing ammonium and nitrate ions to host and adjacent plants. The principal turf-type host plants that support nitrogen-fixing bacteria are the clovers.

Compared with other microflora, fungi are present in the lowest numbers in soil, especially sands. Usually, their numbers range from 1,000 to 1 million per gram of dry soil. They are most competitive at a relatively low soil pH. Fungi occur in three primary categories depending on their nutritive processes: parasitic, saprophytic, and symbiotic fungi. Parasitic fungi that cause plant diseases will be discussed in Chapter 7. Saprophytes are fungi that function as decomposers of residual biomass. Some are especially important for breaking down cellulose and lignin, which are resistant to many other types of biodecomposition. Symbiotic fungi form an intimate association with some plant roots that aids in the absorption of specific nutrients. This association is referred to as *mycorrhizae*.

Actinomycetes tend to dominate in high-pH soils. Usually, they are second to bacteria in numbers per gram of dry soil. They also give moist soil its characteristic smell. Actinomycetes are widely known for their production of antibiotics. They are important in turf soils as decomposers of residual biomass, especially cellulose and other resistant forms.

Algae are chlorophyll-containing microflora that serve as a source of organic matter. Some types function in nitrogen fixation. In turf, algal mats sometimes develop following severe disease incidence. Upon drying, they form a hard crust that, unless removed or broken up, makes the surface nearly impervious to water.

The microfauna include protozoa and nematodes. Protozoa are primitive, single-celled animals that feed mainly on bacteria. Nematodes are threadlike worms that are widely distributed in agricultural soils. Predaceous nematodes prey on soil flora and fauna, including other nematodes. Parasitic nematodes are of greatest concern because of their ability to infest plant roots and severely damage turfgrasses. Their entrance into plants facilitates entry by other pathogens. Control of turfgrass-parasitic nematodes will be discussed in Chapter 7.

Prominent among macrofauna are the earthworms. Of the 7000 species that have been identified, three are most commonly encountered. These are the garden worm (*Helodrilus calinginosus*), the red worm (*Helodrilus foetidus*), and the night crawler (*Lumbricus terrestris*). Earthworms do not feed on live plants but are extremely effective in reducing accumulations of residual organic materials. The ingested organic material and soil are excreted as small, granular aggregates containing substantial amounts of plant nutrients. The deposited casts are sometimes objectionable in closely mowed turf; however, the beneficial effects of earthworms often outweigh any problems resulting from their activity. In studies conducted in Illinois, control of earthworm populations by pesticides resulted in thatch development in an otherwise thatch-free Kentucky bluegrass turf. Other effects included higher soil bulk densities, reduced water infiltration and hydraulic conductivity, lower soil organic matter concentrations, shallow rooting, higher wilting proneness, and greater disease severity. Although present throughout the growing season, the earthworms appeared to be more active at the surface during spring and fall.

Occasionally, masses of dead earthworms may be present at the turf surface. This usually occurs when the sun suddenly appears following rainfall. At other times the surface of the ground may be disrupted from earthworm activity. Under these circumstances, it is tempting to schedule a pesticide application for control. However, we should carefully weigh the long-term advantages and disadvantages before deciding that a chemical control measure will invariably result in better turf.

Several arthropods and gastropods serve as decomposers of organic debris. Others, particularly certain insects, cause serious damage to turf. The appearance of a few of these small animals does not necessarily dictate that some control measure should be undertaken. Many insects or other small animals occurring in turf do not cause injury, and some of the potentially injurious ones are of no importance unless their populations build up to high levels. Strategies for controlling insects and other pests will be discussed further in Chapter 7.

Depending on location, numerous larger animals may occasionally be troublesome because of their burrowing into the turf. As with other macrofauna, their activities may be beneficial. However, when serious damage occurs or is threatened, control measures may be appropriately undertaken (see Chapter 7).

The soil macroflora include turfgrasses and other plants that are anchored to the soil. Root, rhizome, and basal shoot growth are influenced by soil conditions and, in turn, influence the physical, chemical, and biological properties of the soil. Where a fairly contiguous plant community (turfgrass) covers the soil there is a moderation of soil temperature and moisture extremes. The protective cover provided by turfgrasses reduces soil compaction from traffic, splashing raindrops, and other physical forces. Growth of roots and other belowground plant organs stirs the soil and promotes better structure. Deposition of plant debris within and atop the soil provides residual biomass, which is the carbon and energy source for many beneficial soil organisms. Decomposition of organic residues brings about numerous benefits measurable as improved soil properties and plant growth.

Residual Biomass

Residual biomass includes the undecomposed and partially decomposed organic residues that occur within the soil and above the soil surface. Cultivated mineral soils usually contain 1 to 5% organic matter. Turf soils may have somewhat higher percentages, especially near the surface. Three distinct types of residual biomass are common in turf: thatch, mat, and the organic fraction usually associated with cultivated soils.

Thatch is the layer of organic residue located immediately above the soil surface. It may be largely undecomposed or at an advanced stage of decomposition, especially where it meets the soil. Although considerable attention has been devoted to processes of thatch formation, its development in turf is still not completely understood. A typical explanation is that thatch occurs when plant biomass production exceeds decomposition, resulting in an accumulation of surface debris followed by growth of the turfgrass plants within this debris. Obviously, factors that suppress the decomposition rate or promote excessive production could trigger this imbalance. Another possible explanation is that growth of adventitious roots and stolons above the soil surface provides a medium within which dead or dying plant residues accumulate and decompose more slowly than in soil. Regardless of the mechanism of formation, the living plant community apparently exerts a stabilizing influence on thatch, since it has been shown that death of the turfgrass community is often followed by destruction of the thatch layer. Limited work to characterize thatch as a turfgrass growth medium has shown that it is a well-aerated, compaction-resistant medium with poor nutrient- and water-holding properties. Therefore, turfgrasses with most of their roots confined to the thatch layer are more prone to injury from environmental stresses. Diseases and insect injury are generally more severe where thatch is a problem. Measures for controlling thatch will be discussed in Chapter 6.

Mat is a surface layer of organic residue mixed with soil. Often mat results where soil has been incorporated into a thatch layer. Mechanisms by which this can occur include some naturally occurring phenomena, such as earthworms depositing soil within the thatch or wind depositing soil onto thatchy turf, and some culturally induced transformations, such as topdressing with soil or reincorporation of soil following core cultivation. Mat is generally considered a desirable feature in turf, especially in those turfs receiving frequent or intensive traffic. A mat layer not only provides better surface resiliency but seems to stabilize turf against impacting forces encountered in athletic activities. Although a mat layer may not be easily seen in a turf profile, its presence can be qualitatively determined by carefully working a soil core (taken with a soil probe) with the fingers, beginning at the base of the core. At moderate moisture, the soil should be easily separated from the roots until the mat layer is reached; then, the soil resists separation due to the organic material interspersed throughout the surface soil.

In addition to thatch and mat, other types of organic residues occur in mineral soils. These result from incorporating organic amendments during establishment or from deposition of organic materials during soil formation and under cropping practices. Since the soil organic matter is constantly undergoing change, it must be replenished continuously to maintain soil productivity. The principal replenishing sources are plant roots, residues of other plant parts, and soil microorganisms. Decomposition of organic matter results in the use of some carbon, nitrogen, and other elements by microorganisms; release of carbon compounds, water, and other elements to the soil and to the atmosphere; and the formation of a semistable organic residue called *humus*.

Nitrogen is the nutrient that most often controls the rate of organic matter decomposition because it is used to synthesize proteins in new microbial populations. The ratio of carbon to nitrogen, called the C:N ratio, controls the decomposition rate and largely determines whether nitrogen will be released to the soil or pulled from it by microbial activity. Most organic materials have a carbon content of about 40%; however, the nitrogen content is variable. Bacteria have C:N ratios of 4:1 to 5:1, while fungi have a C:N ratio of 9:1. Plant residues having C:N ratios of 20:1 or narrower contain sufficient nitrogen to supply microorganisms and also release nitrogen for plant use. This is because, in the decomposition of organic matter, much carbon is released to the atmosphere as CO_2, while most of the nitrogen is reused repeatedly by microbial populations, thus narrowing the C:N ratio. As the organic matter is used up, microorganisms die, decompose, and release more CO_2, as well as plant-available nitrogen. When organic materials with large C:N ratios, such as sawdust (250:1) and straw (90:1), are added to the soil, nitrogen is pulled from adjacent soil. Since microorganisms can compete more effectively than plant roots for available nitrogen within the soil, plants may show symptoms of nitrogen deficiency unless additional nitrogen is supplied through fertilization.

The organic matter of greatest importance in soil is that which is undergoing rapid decomposition. This is called the *active organic matter*. The principal benefits derived from active organic matter include soil aggregation and stabilization of soil aggregates by organic gums, which cement clay particles together; mineralization (release) of plant nutrients; and production of humus, which provides additional cation exchange capacity. Decomposition is favored by conditions that favor microbial activity, including adequate aeration and moisture, slightly acid to neutral pH, available nutrients, suitable temperatures (optimum is 95°F), and a favorable C:N ratio.

THE BIOTIC ENVIRONMENT

The biotic component of the turfgrass environment includes, primarily, the use and culture of the turf by people. Other biota are sometimes included, but, except for humans, most of these have been discussed under Soil Biomass. Human activities frequently constitute the most formidable of adverse influences against turfgrass persistence and quality. Conditions that reflect the type and intensity of these activities are wear and breakage of the turf, and displacement and compaction of the soil.

Wear is the physical deterioration of a turfgrass community resulting from excessive traffic. What constitutes "excessive" traffic will vary depending on turfgrass genotype, the physiological condition of the plants, and edaphic and climatic conditions. A rough bluegrass turf in a wet, shaded environment will be much more susceptible to wear than a well-cared-for bermudagrass turf on a sunny, sandy site. However, since bermudagrass would not survive where rough bluegrass is well adapted as a permanent turfgrass, the comparison is somewhat misleading. Additional comparisons of turfgrass species for wear tolerance are given in Chapter 3.

Direct pressure applied to the turf tends to crush shoots, especially when the plants are stiff with frost or wilted. When the pressure is accompanied by lateral forces, wear is greater. Since wear tolerance varies so much from site to site, and at different times on the same site, a traffic threshold level must be determined for each location and time of year. Then measures can be taken to control traffic intensity so that the threshold level is not exceeded. In many situations, wear can be minimized by making some simple adjustments. On golf courses, traffic patterns on greens, approaches, exit lanes, and tees can be made less concentrated by coordinating placement of pins and tee markers. Shifting sand-trap positions to provide wider traffic lanes and improving drainage can dramatically reduce wear. On athletic fields used for practice, intensive traffic can be shifted to different field locations to reduce wear and promote recovery of worn turf. Strategic placement of ornamental plantings, sidewalks, fences, and other landscape features is often effective in uniformly distributing, or concentrating, traffic for the purpose of promoting better turf. With a little imagination, small provisions that will result in less wear can be made on almost any site.

In succeeding chapters cultural practices are discussed in detail. Much of this discussion is oriented toward achieving healthy, vigorous turfgrass that is highly tolerant of environmental stresses, including traffic. Some rather basic cultural considerations for improving wearability include raising the mowing height to provide larger, more vigorous shoots; reducing plant succulence by adjusting fertilization and irrigation practices; and cultivating to promote better shoot and root growth. Some conditions require that traffic be avoided entirely. Frosted leaves, winter-dormant turfgrass, and wet soil conditions predispose the turf to serious injury from even light traffic. Moderately shaded turfgrass may be so weak that it is intolerant of all but minimal traffic.

Turf breakage results from those activities that disrupt the structural integrity of the turf. A golf ball landing on a green can tear the turf's surface, especially where plants are shallow rooted, resulting in "ball marks." If the damage is not repaired by carefully reshaping the turf's surface, mowing can scalp the upraised plants and soil from the turf causing even more severe injury. Golf shots, as well as a variety of other physical forces, may rip out sections of sod, causing "divots" in the turf. The size of the divots is influenced by the specific nature of the physical force, turfgrass species, cultural practices, and the overall condition of the turf. Repair operations may include replacing torn sections of sod or filling in the divots with soil (and perhaps seed) to promote recovery.

Soil displacement and associated rutting occurs where foot or vehicular traffic substantially changes the shape of the turf's surface. The susceptibility of a turf to soil displacement and rutting is a function of soil moisture, texture, and strength; the extent to which turfgrass roots and lateral shoots enhance the shear strength of the turf; and the intensity and duration of the traffic event. As soil displacement and rutting are most likely to occur on wet, fine-textured soils, traffic should be carefully controlled whenever and wherever these conditions exist.

Compaction is a physical condition of soil resulting from the compression of soil constituents into a relatively dense mass. A moderate amount of compaction may be desirable in turf soils to provide a firm surface and favorable contact between plant roots and soil particles. Severe compaction, however, is usually associated with inadequate soil aeration and poor drainage due to the loss of aeration porosity. Soil compaction can occur in the surface layer or at some depth within the soil profile. Subsurface compaction layers are called *pans;* the two principal types are genetic pans and induced pans. Genetic pans occur naturally due to very high clay content present during soil formation (claypan) or cementation of soil particles by organic matter or inorganic materials (hardpan). Induced pans result from pressure applied by cultivation or tillage operations. In field crop culture, a *plow sole* can develop at plow depth due to the alteration of soil physical structure from horizontal operation of the plow. With core cultivation of turf, induced pans can occur at the maximum depth of penetration of the tines and, possibly, along the sides of the holes resulting from cultivation.

A problem that is analogous to surface and subsurface compaction layers, but which occurs primarily in sandy soil profiles, is called "black layer." It is especially common where a sand topdressing program has been employed for turfs growing on finer sands or other fine-textured media. It appears as a coal black band ranging in thickness from a few millimeters to several inches. The characteristic foul-smelling odor of the band is due to the production of sulfide gases by bacteria under anaerobic conditions. The band itself is actually a precipitate of metal sulfides, including FeS, MnS, and MgS, that coats soil particles and clogs pore spaces. Some investigators believe that the existence of black layer in turf soils reflects the use of sulfur or sulfur derivatives for reducing the pH of calcareous sands. While the use of sulfur-containing materials may contribute to the formation of black layer, it is possible that the sulfur released from the biodegradation of soil organic matter may be sufficient when other conditions are favorable.

Surface compaction of turf soils occurs primarily in response to traffic. Particles of soil are pressed together with resultant increases in bulk density, soil strength, and water-holding capacity, and decreases in aeration porosity, water infiltration, and percolation. Increased bulk density results from the reduction in soil volume following compaction. The change in volume is attributed to a loss in total porosity; some pores are lost or reduced in size as clay particles become flattened in a plane perpendicular to the compacting force. The large (aeration) pores that are so important in drainage and soil aeration are most susceptible to loss during compaction. The accompanying shift in pore-size distribution (to generally smaller pores) accounts for increased moisture retention. Capillary pores assume a higher proportion of total porosity as aeration porosity is reduced. Increased soil strength, measurable as resistance to penetration, partially accounts for the poor root growth commonly observed in compacted soils. Inadequate oxygen concentration and relatively high concentrations of CO_2 and other potentially toxic gases are additional reasons for poor rooting in compacted soils. Low water infiltration and percolation rates typically accompany shifts in pore-size distribution toward smaller pores.

Physiological responses of plants to soil compaction include not only reduced root growth but reduced shoot growth, reduced water and nutrient uptake, reduced tolerance to heat and drought stresses, and increased susceptibility to scald and direct low-temperature injury. The generally weakened condition of plants growing in compacted soils coupled with prolonged periods of high humidity in the plants' microenvironment account for higher disease incidence as well.

Ecological responses of turfgrass communities subjected to soil compaction are frequently observed as dramatic shifts to compaction-tolerant weed populations such as knotweed, annual bluegrass, goosegrass, and white clover.

Soil compaction influences turfgrass cultural requirements. More frequent fertilization and irrigation are required to compensate for restricted rooting and generally poor growth. Cultivation practices are necessary to improve infiltration, drainage, and plant growth. Installation of drainage tiles, slit trenches, and catch basins may be essential to dispose of excessive surface moisture. Pesticides are required to control the higher incidence of weeds and diseases that frequently accompany soil compaction. Furthermore, mistakes in the application of pesticides that result in damage to the turf are of greater consequence where, because of compaction, suitable conditions for turfgrass recovery do not exist.

Methods for preventing soil compaction or reducing its effects will be discussed further in Chapters 6 and 8.

QUESTIONS

1. Differentiate among reflection, absorption, transmission, and conversion (to chemical energy through photosynthesis) of **light** by a turfgrass plant.
2. How is a turfgrass plant likely to respond to reduced light intensity?
3. How should cultural practices be adjusted to sustain a turfgrass community under low light intensities?
4. Explain the processes by which turfgrasses dissipate **heat** energy.
5. What factors influence turfgrass aerial shoot and root **temperatures**?
6. Explain the significance of **moisture** in turfgrass growth and survival.
7. What influence is **wind** likely to have on turfgrass growth and survival?
8. What is turfgrass **edaphology**?

9. Characterize a **block of soil** (supporting the growth of turfgrass plants) with respect to the distribution of solids, water, and air.
10. List the different forms of **organic residues** that are likely to be found in association with turf and explain their significance in turfgrass growth and quality.
11. Differentiate between **soil texture** and **structure.**
12. Characterize the different types of **soil moisture** that occur in a saturated soil and explain their significance with respect to turfgrass growth and quality.
13. What are **soil colloids** and in what ways are they likely to contribute to turfgrass nutrition?
14. What is meant by **soil reaction** and what is its significance with respect to turfgrass growth?
15. Characterize the **biotic environment** influencing a turfgrass community.

CHAPTER 5

Primary Cultural Practices

Mowing, irrigation, and fertilization are the primary turfgrass cultural operations needed to sustain turfgrass quality. These practices are highly interrelated. Changes in mowing height usually require adjustments in the frequency and intensity of fertilization and irrigation. On many established turfs additional cultural practices are rarely needed if primary culture is performed satisfactorily.

MOWING

Mowing is the most basic of all turfgrass cultural practices in that it influences most other cultural operations. It involves the periodic removal of a portion of turfgrass shoot growth. Depending on the cultural intensity required on a given site, mowing may be performed to maintain top growth within specific limits, control undesired vegetation that is intolerant of mowing, sustain an ornamental or recreational turf, produce a true putting surface, or develop a sod crop. From a purely botanical standpoint, mowing is detrimental to turfgrasses. It causes a temporary cessation of root growth, reduces carbohydrate production and storage, creates ports of entry for disease-causing organisms, temporarily increases water loss from cut leaf ends, and reduces water absorption by the roots. Yet turfgrasses persist because of their adaptation to regular defoliation. Turfgrass species are believed to have evolved under the selection pressure of grazing animals. Thus mowing practices are contemporary analogues to grazing, and the contention that mowing is an entirely artificial practice is true only in form, not in substance.

Mowing Equipment

The first "artificial" or mechanical mower was the scythe. It required great skill and considerable labor to reduce a grass field to a uniform-appearing turf by scything. In 1830 Edwin Budding, a textile engineer, developed the first reel mower. His device was an adaptation of the rotary shear used to cut the nap on carpets. In the early 1900s similar units were developed. They were pushed by hand, pulled by horses, or powered by steam until the advent of gasoline-powered mowers. Traditionally, allocations for mowing have constituted the largest portion of budgets for maintaining turfgrass sites. Today, however, the sophistication, efficiency, and ease of operation of modern mowing equipment enable the turfgrass manager to assign a higher percentage of the operating budget to other cultural operations.

Types of Mowers

Various types of equipment can be used to mow vegetation. The principal types used for mowing turfgrass are the reel, rotary, and flail. Basically, the *reel* mower consists of a rotating reel cylinder equipped with blades and a stationary bedknife (Figure 5.1). The reel blades guide leaves toward the bedknife, where they are cut by a shearing-type action. Mowing quality is, in part, a function of the sharpness of the cutting edges and proper adjustment of the bedknife against the reel blades. Dull cutting surfaces or poor adjustment result in tearing and bruising of the leaves. The mutilated leaves turn gray, then brown, at the tip and may be stunted as a result of being cut with dull blades. This can occur with any turfgrass, but it is especially pronounced with ryegrasses. The tough vascular bundles in ryegrass leaves resist mowing and result in frayed leaf ends even when a sharpened, properly adjusted mower is used. A comparison of turfgrass leaves mowed with sharp and dull mowers is illustrated in Figure 5.2.

Mowing quality is also influenced by the relationship between mowing height (MH) and the *clip of the reel* (CR), which is defined as the *forward distance traveled between successive clips*. The CR is determined by the number of blades on the reel, the rotational velocity of the reel blades, and the forward operating speed of the mower. Usually, the most uniform cut occurs when CR = MH. If the CR is appreciably longer than the MH, the turf surface takes on a wavy appearance, called *marcelling*, as illustrated in Figure 5.3. Marcelling usually occurs when the forward speed of a mower is excessive. With other factors being constant, doubling the forward operating speed will also double the CR. Reels on greens mowers are smaller in diameter and have more blades than reels used for mowing lawns and other higher cut turfs. The larger number of closely spaced blades on the reel makes a greens mower more suitable for mowing at very short heights.

If the CR is appreciably shorter than the MH, the efficiency with which the reel blades direct leaves to the bedknife is reduced and a ragged, nonuniform cut may result. This is due to the air movement generated by the reel, which prevents proper gathering of grass leaves by the reel blades. Thus a greens mower adjusted up to provide a 0.75-inch height of cut would be unsuitable for mowing lawn or fairway turf.

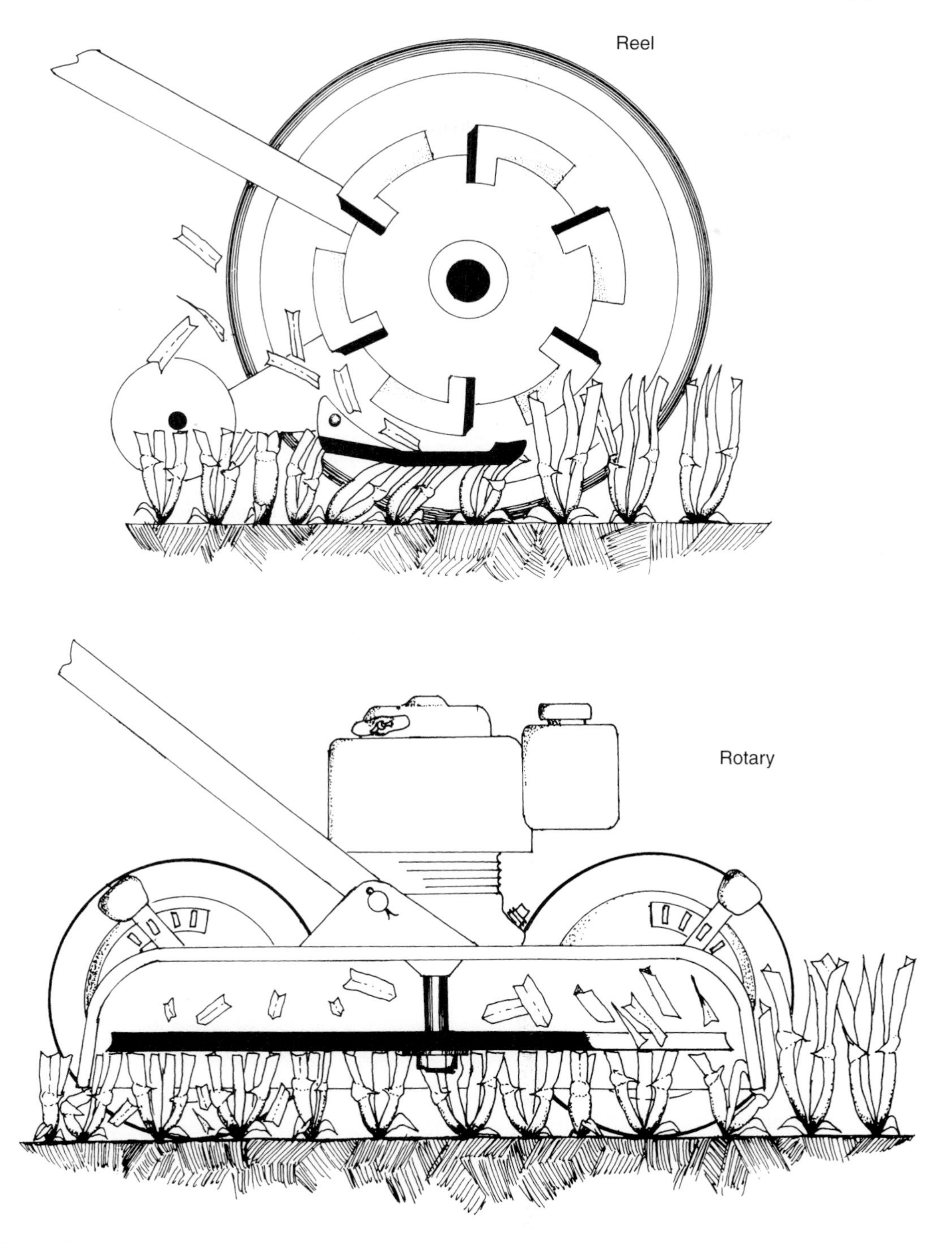

Figure 5.1. Comparison of reel (top) and rotary (bottom) mowing units.

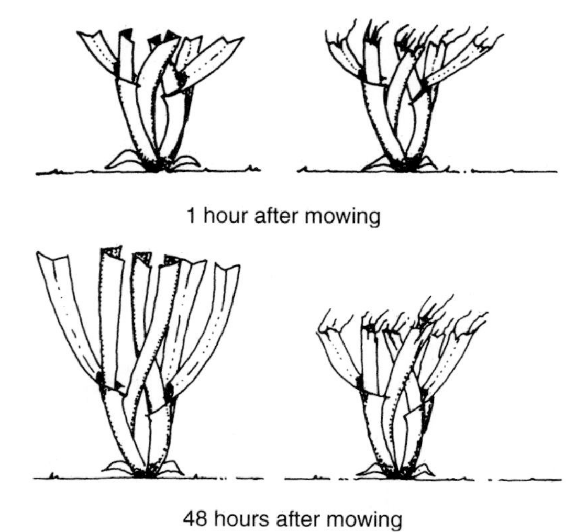

Figure 5.2. Comparison of turfgrass leaf growth following cutting with sharp (left) and dull (right) mower blades.

The inverse of CR is the *frequency of clip* (FC). This is defined as the number of clips per unit distance (inches) traveled. A mower with an FC of 2 clips per inch has a CR of 0.5 (the inverse of 2 is 1/2, or 0.5) and is well suited for mowing turf at 0.5 inch.

The phenomenon of marcelling can be explained by some simple geometry. With clipping, turfgrass shoots are drawn together to form an isosceles triangle (Figure 5.4). After clipping the shoots return to an upright orientation, resulting in a marcelled surface with measurable peaks and valleys. The height difference between peaks and valleys is called the restitution height (R). When CR and MH are known, R can be computed by first determining the hypothenuse formed by shoots rooted farthest from the position where clipping occurs:

$$c^2 = a^2 + b^2, \quad c = (a^2+b^2)^{1/2}$$

R equals the difference between c (hypothenuse) and a (MH):

$$R = c - a = (a^2 + b^2)^{1/2} - a$$

Substituting MH for a and 1/2 CR for b provides the following formulas for determining R:

$$R = (MH^2 + [CR/2]^2)^{1/2} - MH$$

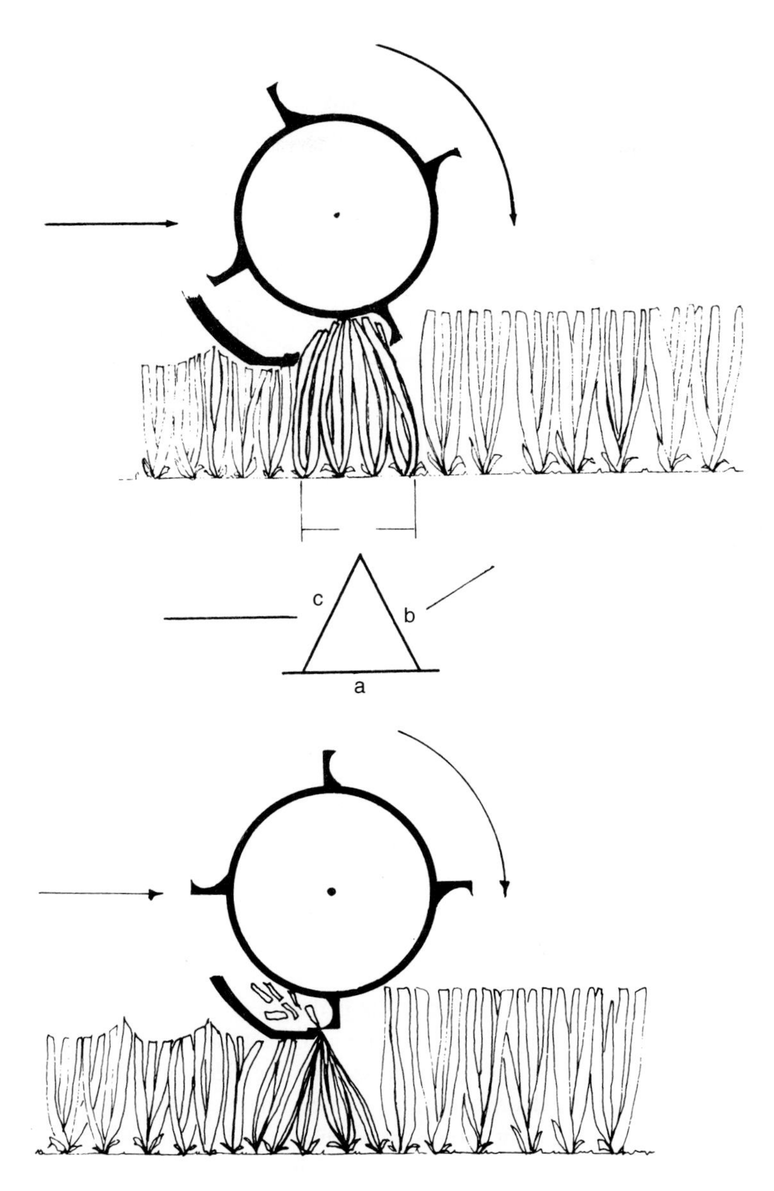

Figure 5.3. Shearing-type cutting action from a reel mower. Cutting action begins as the bedknife pushes grass shoots forward and the reel blade pulls shoots toward the bedknife. Following cutting, the surface of the turf may have a ribbed appearance, depending on the CR/MH relationship.

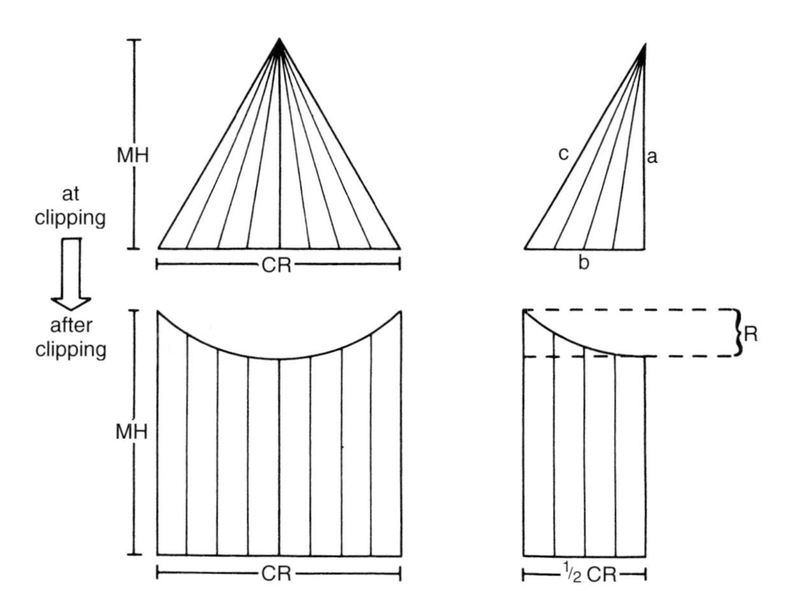

Figure 5.4. The geometrical basis for marcelling where CR > MH.

Finally, to calculate the percent of variation (V) in vertical shoot height:

$$\%V = R/MH$$

Observable marcelling may or may not result, depending on the percent of variation in shoot height from mowing. Where CR = MH the shoot height variation equals approximately 12%—an acceptable level that does not result in a marcelled appearance. Where CR > MH large restitution heights and substantial shoot height variations occur, resulting in observable marcelling.

The energy expended at the bedknife creates a tendency for a reel mower to rotate around the bedknife. Upon striking an object the mower tends to dip and dig into the turf. During turns the mower is depressed on the side opposite the direction of turn. With fast, tight turns, this depression may cause some scalping or severe defoliation of the grass. Where feasible, tight turns should be avoided or, at least, performed slowly. Because of metal-to-metal contact between reel blades and bedknife, lubrication is necessary to reduce wear at the cutting edges. During mowing this is accomplished by "juices" from within the grass leaves. When a reel mower is operated on nonvegetated surfaces (roadways, sidewalks) during transport, heat expansion of the metal may result in severe wear or even seizure. Therefore, reels should be disengaged before transporting equipment between mowing sites.

Reel blades and bedknives should be ground when nicks appear or when the cutting edges become rounded or worn. Periodic lapping with an emery powder slurry (in oil or water plus detergent) applied to the blades while the reel is rotating backward helps to maintain a sharp cut between grindings. The operating manual accompanying the mower at the time of purchase is a good reference for routine maintenance.

The highest possible mowing quality is obtainable with properly maintained and operated reel mowers. For this reason reel mowers are widely used on golf courses, athletic fields, parks, sod farms, and many commercially maintained lawn sites. For utility turfs, however, reel mowers do not provide the versatility often required. They do not work well in high grass, and tough seed stalks may resist mowing with a reel. Utility turfs and home lawns are most often mowed with *rotary* mowers. The rotary cuts by impact as the horizontally mounted blades rotate around a vertical shaft. The cut may not be as sharp as with a reel mower, but it is usually acceptable for lawn and utility turfs cut above 1 inch, providing the cutting edges are sharp and the blade balanced. Under normal conditions, the blade will require sharpening once or twice during the growing season. When hard objects are struck during mowing, resulting nicks should be filed to restore the cutting edge, or mowing quality will be reduced.

The rotary is more versatile than the reel. It can be used to mow tall grass, tough seed stalks, and weeds, to mulch or bag leaves, and to trim along walls. The relatively low cost, ease of operation, and simple maintenance requirements of rotary mowers have made them very popular with homeowners.

A second version of impact-type cutting is found in the *flail* mower. The mowing component consists of numerous small knives hinged to a horizontal shaft. When the shaft rotates, the knives are held out by centrifugal force. Due to the small clearance between the knives and the mower housing, cut debris is recut until it is small enough to clear the housing. The advantage of this type of mower lies in its ability to reduce tall vegetation to a finely ground mulch. The free-swinging knives will fold away upon striking a rock or other hard obstruction. Thus there is less danger from projectiles than with the rotary mower. As with other mowers, sharpness of cutting edges is an important determinant of mowing quality. Flail mowers are used primarily on utility turfs, where mowing is performed infrequently.

Two basic considerations in mowing are area size per mowing height and landscape features that interfere with the efficient operation of mowing equipment. As configurations of mowing equipment are continually changing, these will not be discussed. The reader is encouraged to investigate options that are available locally. Some common interfering landscape features that are likely to be encountered include abrupt changes in topography, ornamental plantings, structures, and buildings. Consideration should be given to changing topographical features that result in repeated scalping during mowing or that necessitate labor-intensive hand mowing or trimming. On small sites abrupt transitions between the lawn and sidewalk that result in regular scalping of the turf or damage to the mower can be corrected with a sod cutter, some soil, and a few hours' work. At the base of trees and other obstacles to mowing, application of a ring of mulch after removing the sod or the use of nonselective herbicides may generate savings in mowing costs. Selective removal of landscape plantings may allow for more flexible mower traffic patterns that, in turn, substantially increase mowing efficiency.

The proper choice of equipment for mowing is based on an analysis of operational requirements and costs. An expensive and complex system of hydraulically operated mowing units may be well suited to an expansive site that is mowed frequently, but totally impractical for home lawns. Relevant factors that should be considered in choosing equipment include size of the area, number and location of obstructions, turfgrass cultural intensity, specific turfgrasses used, frequency and intensity of play or use, equipment maintenance capability, and the cost and technical sophistication of available labor.

Mowing Variables

Several variables in the mowing program influence turfgrass quality. These include the height, frequency, and pattern of mowing. In addition, turfgrass quality and cultural requirements can be affected by the removal or return of clippings during mowing.

Mowing Height

The effective mowing height is the height of aerial shoots immediately following mowing. It is similar, but not identical, to the *bench setting*, or the mechanically set mowing height of the mower. Since the mower rides atop compressed turfgrass shoots, the effective mowing height is expected to be slightly higher than the bench setting. The extent of this difference would depend on the rigidity, elasticity, and size of the turfgrass shoots as well as the weight and configuration of the support elements of the mower. However, when the ground is soft or where thatch is present, the mower may sink in enough to cause the effective mowing height to be the same as or lower than the bench setting. For further discussion it will be assumed that effective mowing height and bench setting are essentially the same, and these will be referred to as the mowing height.

Each turfgrass has its mowing tolerance range within which it can be expected to provide a satisfactory turf. The mowing tolerance range is delimited by the lowest and highest tolerated mowing heights. Below the mowing tolerance range the turfgrass stand would tend to thin during stress periods or be overtaken by weeds. Scalping, or the excessive removal of green shoot tissue resulting in the exposure of unsightly shoot stubble, or even bare earth, occurs well below the mowing tolerance range. Above their respective mowing tolerance ranges turfgrasses may become puffy, limp, or decumbent or otherwise fail to form a satisfactory turf. The exact range of mowing tolerance for a specific turfgrass is difficult to estimate because of the interactions between the turfgrass genotype and climate, culture, and other environmental influences; however, some rough approximations are given in Chapter 3.

Research has shown that, within the mowing tolerance range, reducing the mowing height results in substantial physiological and morphological changes in a turfgrass (Figure 5.5). Some effects of closer mowing include stimulated aerial shoot growth, increased shoot density and smaller shoot size, decreased root and rhizome growth, decreased synthesis and storage of carbohydrates, and increased plant succulence. Closer mowing thus produces a turf that is esthetically more pleasing but that is less tolerant of environmental stresses, more disease prone, and generally more dependent on a

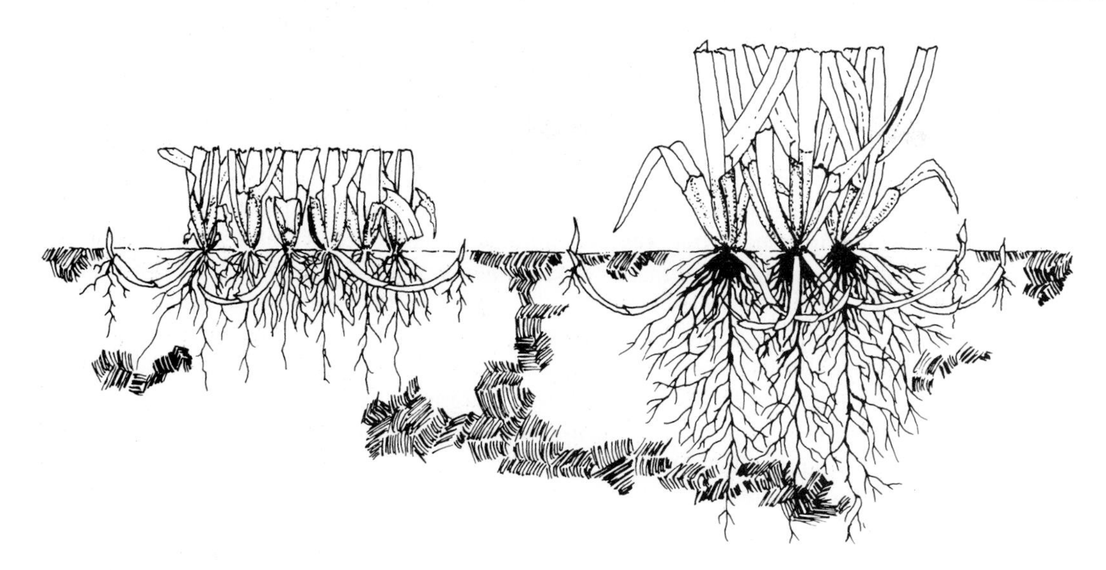

Figure 5.5. Comparison of turfgrass mowed at two heights; closer mowed turfgrass is finer textured and denser, but has less underground growth of roots and rhizomes.

carefully implemented cultural program. The shorter root system requires more frequent irrigation and fertilization to compensate for the plants' reduced ability to secure moisture and nutrients from the soil. The larger number of smaller shoots are under considerably greater competitive stress and thus are less tolerant of other stresses, including heat, cold, drought, pathogens, insects, traffic, and an array of chemicals applied as part of the cultural program. Closely mowed turfs can be successfully sustained, but the technical expertise required of the turfgrass manager is greater than for the same turfgrass maintained at higher mowing heights.

As indicated earlier, the lower limit of mowing tolerance can vary depending on the magnitude of environmental stress. This has been recognized by experienced turfgrass managers for many years, and a common practice with cool-season turfgrasses has been to raise the mowing height during the summer months to compensate for increased heat and drought stresses. Similarly, with warm-season turfgrasses, the mowing height is sometimes raised in the early and late portions of the growing season as a compensating measure for cold stress and reduced photosynthetic activity.

Mowing Frequency

Mowing frequency is the number of mowings per unit of time; it is inversely related to the mowing interval, which is expressed as the number of days between successive mowings. Mowing frequency can vary from daily mowing on greens to several mowings per growing season on some utility turfs. For moderately to intensively cultured turfs, a generally accepted guide has been to remove no more than one-third of vertical shoot growth per mowing; otherwise, an imbalance between aerial shoots and roots may retard growth. Studies indicate that infrequently mowed turf is coarser and more open than turf under a more frequent mowing schedule. It has been claimed that less frequently mowed turfgrasses are able to accumulate sufficient carbohydrate reserves to promote vigorous regrowth following mowing, while more frequently mowed turfgrasses are more likely to deplete their carbohydrate reserves and thus become less vigorous. There is an obvious inconsistency between these two arguments; that is, the "one-third" guide, which dictates fairly frequent mowing to avoid growth stoppages, and the argument for less frequent mowing to allow for accumulation of carbohydrate reserves to support more vigorous growth. The one-third guide would appear to be especially applicable to periods of midsummer stress with cool-season turfgrasses, while violations of the one-third guide are of less consequence during nonstress periods. Presumably, the severity of defoliation would influence the amount of carbohydrates drawn from reserves. Therefore, with less severe defoliation from frequent mowing the magnitude of the carbohydrate draw from reserves would be less than that experienced with infrequent mowing and the accompanying severe defoliation. Until more definitive evidence is obtained to either support or refute these arguments, the one-third guide appears to constitute a sound basis for mowing schedules. Mowing frequencies greater than those established in accordance with the one-third guide should be avoided, however. Very frequent mowing can result in less rooting, reduced rhizome growth, increased shoot density, decreased shoot growth, decreased carbohydrate reserves, and increased plant succulence.

Mowing Pattern

Where the orientation of turfgrass shoots differs in response to mowing direction, light reflection differs as well, producing alternating light and dark green strips such as those seen on some football fields. Golf greens cut with small walk-behind mowers show a similar pattern. Double-cutting at right angles produces a checkerboard pattern that enhances the esthetic appearance of the green. Even home lawns crosscut with small reel mowers can have this pattern for several days following mowing.

Mowing direction should be varied with successive mowings to encourage upright shoot growth. On greens horizontally oriented shoot growth is called *grain;* where severe it can seriously affect the green's putting quality. Mowing in alternate directions is an important measure for controlling grain. Another problem resulting from failure to vary mowing direction is differential compaction, and even ruts, in the soil. On one golf course in the author's experience, parallel lines of annual bluegrass developed in an otherwise predominantly bentgrass fairway where the wheels of the mowing tractor covered the same ground during each mowing. On fairways where mowing is predominantly along the longitudinal axis, turns of the mowing tractor are made repeatedly on the approach to the green. Unless varied, this pattern can result in severe wear and soil compaction. Additional stress imposed by poor drainage and heavy traffic from golfers can make sustaining turf on this portion of the golf course a challenge. Consideration should be given to alternating fairway mowing patterns by lifting outside mowing units periodically to better distribute wheel traffic and by changing mowing direction to side to side and diagonally to minimize wear and compaction.

Clippings

A long-standing concern in lawn culture is whether to catch clippings during mowing. Arguments offered in support of clipping removal include reduced thatching tendency, reduced disease incidence, reduced injury from heavy deposits of clippings, and generally improved turfgrass quality. Certainly on greens, clippings interfere with play and should be removed. On intensively cultured golf course fairways with mixed stands of creeping bentgrass and annual bluegrass, the removal of clippings during mowing (along with conversion to lightweight mowers) has resulted in improved turf quality. On lawn turf, however, clippings need not be a problem as long as they do not remain as objectionable clumps on the surface. Mowing at a sufficient frequency, and avoiding mowing when the grass is wet will usually allow clippings to fall into the turf without reducing turfgrass quality. Where a substantial amount of clipping residue is likely to result from mowing, research in Texas has shown that the use of a mulching mower may be helpful in reducing disease incidence. Reducing clippings to small fragments favors the growth of *Pseudomonas* bacterial populations that, in turn, promote rapid decomposition of residues. Conversely, large clipping fragments may serve as substrates for *Bipolaris cynodontis* fungi that, under favorable conditions, can increase the incidence of leaf blotch disease in bermudagrass.

The contribution of clippings to thatch is believed to be minimal, since leaf blade remnants are readily decomposed. With respect to the association between returning clippings and disease severity, research in Illinois has shown that clipping removal did result in less severe incidence of summer patch disease in Kentucky bluegrass, but only at extraordinarily high rates (about 4 lb/1000 ft^2 in spring) of nitrogen fertilization. Where low (1 lb) to moderate (2 lb) rates of nitrogen were applied, no differences were observed between plots in which clippings were returned or removed.

Clippings are a source of plant nutrients and contain especially large amounts of nitrogen. The nitrogen concentration of dried clippings normally ranges between 3 and 5%. In the course of a growing season this can amount to several pounds of nitrogen per 1000 ft^2 and substantial quantities of other nutrients. Obviously, regular removal of clippings necessitates the application of additional fertilizer to compensate for nutrients that have been removed. The organic matter returned to the soil in clippings is beneficial, as discussed in Chapter 4.

FERTILIZATION

Fertilization is the practice by which essential plant nutrients are supplied as part of a turfgrass cultural program. It ranks with mowing and irrigation as a primary determinant of turfgrass persistence and quality; however, it is one of the least time-consuming and least expensive components of the turfgrass cultural program.

Extensive use of chemical fertilizers is a relatively new addition to the ancient practice of grass culture. In spite of considerable research, plant nutritional requirements for optimum growth are not clearly understood. No simple criterion, such as yield, exists as an appropriate quantitative measure of turfgrass response. Quality is a largely subjective feature, and, depending on fertilization rate and timing, a seemingly high quality level may be followed by severe disease incidence—a condition often attributable to excessive use of nitrogen.

Considering the great complexity of living plant systems, it seems rather amazing at first glance that turfgrass managers were successful at all in sustaining turf at the turn of this century and earlier. However, a plant-soil system is to a large extent self-sustaining. With virtually no knowledge of soil chemistry or plant nutrition the plant culturist relying on nature and some practical experience could be successful as long as environmental conditions were favorable. Today, however, the demands for finer turf, the establishment of turf under distinctly unfavorable conditions, and the severe stresses from traffic and use often require a high level of cultural expertise, including an extensive knowledge of plant nutrition and fertilization.

Nutrient Requirements

Living plants are composed mostly of water with a relatively small amount of dry matter (about 20%) made up primarily of organic compounds. These compounds result from the photosynthetic fixation of carbon (C), hydrogen (H), and oxygen (O) derived from atmospheric carbon dioxide (CO_2) and water (H_2O). The resulting simple carbohydrates are subsequently used to synthesize complex organic compounds containing not only C, H, and O, but several additional elements drawn from the soil.

Currently, thirteen mineral elements within the soil are recognized as being essential for plant growth (Table 5.1). The amounts of these elements found within plants vary considerably; hence, they are grouped into macronutrients, secondary nutrients, and micronutrients, depending on the relative amounts required for growth. The macronutrients—nitrogen (N), phosphorus (P), and potassium (K)—are the ones most often supplied in commercial fertilizers. The annual requirement for fertilizer nutrients varies not only with the specific nutrient, but with other environmental influences as well.

The plant-available pool of each nutrient in the soil is the net result of various input and output processes (Figure 5.6). Inputs include fertilization, natural deposition from the atmosphere, and the nutrients returned to the soil from senescing plant tissues. Some nitrogen is deposited onto turf with rainfall; the amounts are generally very small but significant in the nitrogen balance of the system. As discussed in Chapter 4, some fixation of atmospheric nitrogen occurs due to the activity of certain bacteria. In industrial areas factory emissions contain sulfur and other elements that may eventually be deposited onto turf. Except where clippings are removed, most of the nutrients contained in plant tissues are returned to the soil as plant parts senesce. Outputs include gaseous loss to the

Table 5.1. Essential Mineral Elements for Turfgrass Nutrition

	Element	Chemical Symbol	Available Form
MACRO	Nitrogen	N	NH_4^+, NO_3^-
Largest	Phosphorus	P	HPO_4^{2-}, $H_2PO_4^-$
quantity	Potassium	K	K^+
SECONDARY	Sulfur	S	SO_4^{2-}
Medium	Calcium	Ca	Ca^{2+}
quantity	Magnesium	Mg	Mg^{2+}
MICRO	Iron	Fe	Fe^{2+}, Fe^{3+}
Smallest	Manganese	Mn	Mn^{2+}
quantity	Boron	B	H_2BO_3
	Copper	Cu	Cu^{2+}
	Zinc	Zn	Zn^{2+}
	Molybdenum	Mo	MoO_4^{2-}
	Chlorine	Cl	Cl^-

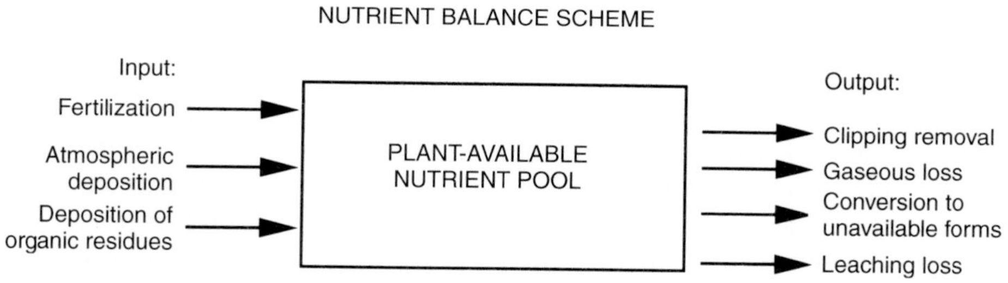

Figure 5.6. Soil nutrient balance listing various inputs and outputs that determine fertilization requirements.

atmosphere, leaching to lower soil depths, and processes by which nutrients are converted to unavailable forms. Where clippings are removed substantial quantities of nutrients are added to the output. Depending on soil conditions, the magnitude of nutrient output can substantially influence fertilizer requirements. Nitrogen and potassium are readily leached in sandy soils. Potassium may be fixed within the clay lattice structure in some soils. Phosphorus deficiencies may limit plant growth even when large quantities of phosphorus exist in the soil, because phosphorus may be converted to insoluble forms. However, where outputs and inputs from naturally occurring conditions are balanced, there may be no fertilization requirement. On many sites turfgrass growth has been favorable with no additions of secondary or micronutrients, and minimal use of phosphorus and potassium has not seriously impaired turfgrass quality. In contrast, newly established turf in sand or infertile subsoils may require careful additions of nearly all essential nutrients in order for the turfgrass community to survive.

The capacity of the turfgrass community to secure available nutrients from the soil also influences fertilization requirements. A shallow-rooted turfgrass can only absorb nutrients from soil regions where roots are growing; hence, results from soil tests can be misleading if the extent of the root system is not considered when estimating the need for fertilization. Important in this analysis is the influence of a thatch layer on nutrient retention and turfgrass rooting. Where the root system is largely confined to the thatch, nutrient input and output processes in the thatch layer may be far more important than the soil nutrient balance until the thatch condition is controlled.

The nutrient required in greatest amounts by turfgrasses is nitrogen. Potassium usually ranks as the second most used nutrient, followed by phosphorus. Determining the exact needs for these specific nutrients is difficult. Generally, it is recommended that nitrogen application be based on turfgrass growth, while phosphorus, potassium, and lime applications should be based on soil tests. While they are valuable as general guides, these recommendations are subject to varying interpretations. Turfgrass nitrogen response is often estimated by coloration, clipping yield, and density. However, aside from a chlorotic condition indicative of nitrogen deficiencies, darker leaf coloration with increasing nitrogen may only be a cosmetic effect and may actually be associated with poor rooting, higher disease incidence, and reduced tolerance to environmental stresses. Increased clipping yield from additional nitrogen often indicates stimulated leaf growth, while rooting and later shoot growth may be adversely affected. Even density is not an entirely reliable index of the nitrogen fertility status of a turf since the growth characteristics of specific turfgrass genotypes can differ substantially even among cultivars of the same species. Nitrogen fertilization, then, is an art as well as a science and is conducted largely by "feel" once sufficient experience on a given turf has been acquired. Density and growth-rate estimates are important, but only qualitative, indices that aid in determining nitrogen fertilization requirements.

Soil tests for determining phosphorus and potassium requirements should, likewise, only be used as guides in developing a fertilization program. No one has yet determined with any precision what levels of these nutrients in the soil are optimum for turfgrass growth. However, soil tests have shown in many instances that where deficiency symptoms occur in plants measured values in the soil are lower than those considered adequate for growth. As with criteria typically used for determining N requirements, soil tests provide important information for determining P and K requirements, but results should be considered along with other information, including past experience with fertilization practices, turfgrass quality, disease incidence, persistence of the turf under stress conditions, thatch depth and condition, and common sense.

In securing soil samples for testing, sampling depth should be limited to 2 to 4 inches. Since nutrient concentration and pH may vary from one small volume of soil to another, the test sample should be a composite of at least twelve individual samples from each test site (lawn, green, fairway, and so on). For routine analyses, the sample should be at least one-half pint (237 ml) in size with subsamples thoroughly mixed.

Numerous studies have been conducted in which the relative concentrations of N, P, and K in plant tissues have been measured to determine fertilization requirements. It is sometimes recommended that, where turfgrasses take up 0.1 unit of P and 0.5 unit of K for each unit of N, a fertilizer ratio of 1:0.1:0.5 (N:P:K) be used. However, this recommendation does not take into account the input-output differences among these nutrients as discussed earlier. If substantial amounts of the applied N are lost due to leaching and volatilization while lesser amounts of P and K are lost from these or other processes, plant uptake ratios alone are inadequate for determining the fertilizer requirements for these nutrients. Differences in short- and long-term availability of nitrogen from fertilizer materials also contribute to the difficulty in establishing ideal ratios. The fate of specific nutrients in turf will be discussed under each nutrient.

Fertilizer Materials

The composition by percent of a fertilizer is called the analysis. Traditionally, this has been expressed as the percent of elemental N, available phosphorus calculated as phosphorus pentoxide (P_2O_5), and potassium calculated as soluble potash (K_2O). The minimum guaranteed analysis of a fertilizer is called the grade. The fertilizer grade is expressed as a three-numbered sequence indicating the percentages of N, P_2O_5, and K_2O. An example is 18-5-9, which is 18% N, 5% available P_2O_5, and 9% water-soluble K_2O. In recent years the fertilizer analysis has also been expressed as the percent of N, P, and K. To convert P_2O_5 to P, multiply by 0.44; for the reverse conversion, multiply P by 2.29. To convert K_2O to K, multiply by 0.83; the reverse conversion is made by multiplying K by 1.20.

In addition to the grade, a packaged fertilizer must be labeled with the weight of the material, manufacturer, and manufacturer's address. Local laws may require additional information, such as the acidifying effect of the fertilizer when added to the soil.

Some primary nutrient carriers used in turf fertilizers are listed in Table 5.2. In addition to the analyses of these materials, information is given that may be important in selecting specific carriers. The salt index measures the effect of fertilizers on the osmotic potential of the soil solution compared to sodium nitrate, which is given a salt index of 100. Dividing the salt index of a fertilizer by the percent of nitrogen or other nutrient (primarily potassium) gives the salt index per unit of nutrient. With high temperatures the potential for foliar burn would be high where high-salt-index fertilizers are used. Similarly, saline soils are likely to become more saline following application of a fertilizer with a high salt index.

The *acidifying effect* of fertilizers is usually expressed as the pounds of calcium carbonate necessary to neutralize 1 ton (2000 lb) of the fertilizer. Generally, fertilizers containing the ammonium form of nitrogen produce acidity, while nitrate-containing fertilizers (potassium nitrate) cause the soil to become more alkaline. Phosphorus and potassium sources have little effect on soil reaction. In regions where basic salts leach from the

Table 5.2. Primary Nutrient Sources in Turf Fertilizers

Source	Formula	APPROX. NUTRIENT PERCENTAGE[a]		
		N	P_2O_5	K_2O
Ammonium nitrate	NH_4NO_3	33	0	0
Ammonium sulfate	$(NH_4)_2SO_4$	21	0	0
Urea	$CO(NH_2)_2$	45	0	0
UF (ureaformaldehyde or methylene urea)	$[CO(NH_2)_2CH_2]_nCO(NH_2)_2$	38	0	0
IBDU (isobutylidene diurea)	$[CO(NH_2)_2]_2C_4H_8$	31	0	0
SCU (sulfur-coated urea)	$CO(NH_2)_2+S$	32	0	0
Milorganite	Complex	6	4	0
Monoammonium phosphate	$(NH_4)H_2PO_4$	11	48	0
Diammonium phosphate	$(NH_4)_2HPO_4$	20	50	0
Superphosphate	$Ca_n(H_nPO_4)_2+CaSO_4$	0	20	0
Treble superphosphate	$Ca_n(H_nPO_4)_2 \cdot H_2O$	0	45	0
Muriate of potash	KCl	0	0	60
Sulfate of potash	K_2SO_4	0	0	50
Potassium nitrate	KNO_3	13	0	44

[a] $P_2O_5 \times 0.44 = P$; $K_2O \times 0.83 = K$.

Table 5.2. Continued

Salt Index[b] Per Unit	Acidifying[c] Effect	Cold-Water[d] Solubility, g/l	Comments
3.2	62	1810	Water-soluble N carrier containing both ammonium ions, which are adsorbed onto colloidal surfaces, and nitrate ions, which are mobile in the soil.
3.3	110	710	Water-soluble N carrier containing 24% sulfur; has the largest soil acidifying effect of any of the listed nutrient sources.
1.7	71	780	Water-soluble, synthetic-organic N carrier with the largest N concentration of any granular N source.
0.3	—	SS	Slowly soluble, synthetic-organic N carrier composed of various-size polymers of urea linked by methylene groups.
0.2	—	SS	Slowly soluble, synthetic-organic N carrier composed of two urea molecules linked by a carbon group through reaction with isobutyraldehyde.
0.7	—	SR	Soluble urea coated with molten sulfur and a sealant to provide slow-release characteristics.
0.7	—	SS	Activated sewage sludge from Milwaukee; N release is through microbial decomposition of the organic complex.
2.7	58	230	Soluble phosphate source containing ammonium ions; frequently used in mixed fertilizers to provide P and supplemental N.
1.7	75	430	Soluble phosphate source with higher N content than in monoammonium phosphate.
0.4	0	20	Source of phosphate for use alone or in mixed fertilizers; contains 12% sulfur in gypsum component.
0.2	0	40	Concentrated source of phosphate; commonly used in mixed fertilizers.
1.9	0	350	Common source of potassium for use alone or in mixed fertilizers.
0.9	0	120	Potassium source containing 18% sulfur; often used in place of KCl to reduce salt index and supply sulfur.
5.3	(−23)	130	Potassium source containing supplemental N.

[b]Relative salinity of salts per unit of nutrient compared to sodium nitrate (6.3): > 2.5 = high, 2.5 to 1.0 = moderate, < 1.0 = low.
[c]Units of $CaCO_3$ required to neutralize 100 units of fertilizer (by weight).
[d]Multiply g/l (grams/liter) by 0.008 to gets pounds/gallon; SS = slowly soluble, SR = slow release.

soil, lime should be applied periodically to maintain a suitable pH. Applications of highly acidifying fertilizers increase the lime requirement even more. However, where soil pH tends to be higher than desired, acidifying fertilizers such as ammonium sulfate can help to reduce the pH to a more favorable level.

The *solubility* of a fertilizer determines, in part, its suitability for liquid application. It also influences the potential for leaching of the nutrients within the soil profile, especially in very sandy soils of low cation exchange capacity.

Fertilizers labeled with only one number in the grade (0-0-60, 38-0-0) are single-nutrient carriers. A complete fertilizer contains each of the three fertilizer elements. However, "complete" does not imply that the fertilizer contains all the nutrients, and in the proper proportion, to sustain the nutritional requirements of turfgrasses. For example, a 5-10-5 fertilizer would not usually be considered satisfactory for use on established turf, while a 24-4-12 might be because of a more favorable balance of N, P_2O_5, and K_2O. Today, high-analysis fertilizers (> 20% of major nutrients) are usually used for fertilizing turfgrasses. To reach such high levels of nutrients per unit of fertilizer, chemical salts of high-fertilizer-nutrient content and high purity must be used. The loss of impurities typically found in older fertilizer formulations also means that many minor nutrients contained in the impurities may not be present in the modern formulations. Therefore, on some turfgrass sites, especially those with very sandy or infertile soils, it may be necessary to apply various secondary and minor nutrients from time to time.

Nitrogen

Nitrogen is the mineral nutrient required in greatest quantities by turfgrasses. It is an essential component of chlorophyll, amino acids, proteins, nucleic acids, enzymes, and other plant substances. The turfgrass plant contains from 3 to 5% nitrogen on a dry-weight basis, except where severe nitrogen deficiencies are encountered. Adequate nitrogen nutrition is necessary for healthy growth. Excessive applications of nitrogen, however, can result in excessive aerial shoot growth; poor root and lateral shoot growth; higher disease incidence; reduced carbohydrate reserves; poor tolerance to heat, cold, drought, traffic, and other environmental stresses; and shifts within the turfgrass community to species or cultivars that are favored by high-nitrogen nutrition. An old rule of thumb has been to "keep the grass a little on the hungry side." Some conditions of turf, such as chlorosis, slow growth, low density, and severe incidence of certain diseases (rust, dollar spot, red thread), may indicate nitrogen deficiency; these can often be corrected with additional nitrogen. Too much nitrogen, however, is not easily remedied, and resulting problems must be weathered until the excess has been naturally depleted.

Nitrogen Carriers

Nitrogen carriers can be subdivided into two groups: quickly available and slowly available. Quickly available or soluble forms include the inorganic salts (ammonium nitrate, ammonium sulfate, ammonium phosphates, and potassium nitrate) and some organic carriers, of which urea is an example. Most are synthesized by reacting ammonia with other compounds as shown in Figure 5.7. The ammonia comes from the conversion of atmospheric nitrogen through reaction with methane (natural gas) under controlled conditions. Some ammonia that is evolved from coke-oven operations is also used, especially for the production of ammonium phosphates. Characteristics associated with quickly available nitrogen carriers include high solubility, rapid but short-term turfgrass response, minimal temperature dependency, high burn potential, and low cost per unit of nitrogen.

Slowly available forms of nitrogen include slowly soluble (UF, IBDU), slow-release (SCU), and natural organic (activated sewage sludge, principally Milorganite) types. The slowly soluble and slow-release types are synthesized from urea as shown in Figure 5.8. UF was first synthesized in 1939 but became popular in the mid-1950s. The nitrogen contained in UF is separated into three percentages based on solubility: CWSN (cold-water-soluble N), CWIN (cold-water-insoluble N), and HWIN (hot-water-insoluble N). At 77°F the CWSN fraction is readily absorbed by turfgrasses. It includes unreacted urea plus some methylene ureas of low molecular weight. Larger methylene ureas must be microbially hydrolyzed to smaller units before they can be used by turfgrasses. Since microbial activity is temperature dependent, solubilization occurs during warmer portions

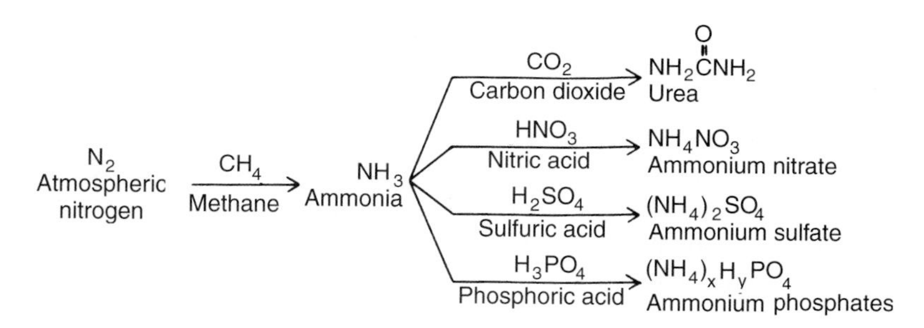

Figure 5.7. Conversion of atmospheric nitrogen to ammonia and synthesis of quickly available nitrogen fertilizers from ammonia.

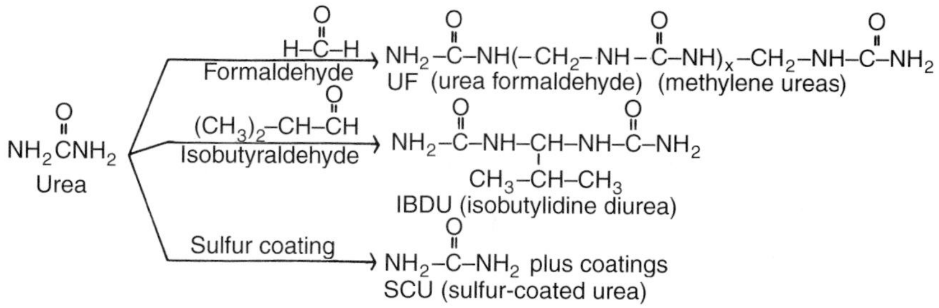

NH_2CNH_2 (C=O)
Urea
H–C(=O)–H
Formaldehyde
$NH_2–C(=O)–NH(–CH_2–NH–C(=O)–NH)_x–CH_2–NH–C(=O)–NH_2$
UF (urea formaldehyde) (methylene ureas)
$(CH_3)_2–CH–C(=O)H$
Isobutyraldehyde
$NH_2–C(=O)–NH–CH(CH_3–CH–CH_3)–NH–C(=O)–NH_2$
IBDU (isobutylidine diurea)
Sulfur coating
$NH_2–C(=O)–NH_2$ plus coatings
SCU (sulfur-coated urea)

Figure 5.8. Synthesis of slowly available nitrogen fertilizers from urea.

of the growing season. Therefore, early spring applications of UF often result in little response until soil temperatures increase sufficiently to favor breakdown of the high-molecular-weight CWIN materials. The percentage of cold-water-insoluble N soluble in hot water is called the *activity index* (AI). The higher the AI, the more rapidly nitrogen is solubilized. A UF carrier should have an AI of at least 40% to have sufficient nitrogen solubilization within the year following application.

Another expression of UF-nitrogen solubilization characteristics is the urea:formaldehyde ratio. Varying the ratio during UF synthesis can substantially influence solubilization. A ratio of 1.3:1.0, typical of many commercially available materials, has approximately 33% CWSN and 67% CWIN. Other commercial fertilizers have ratios up to 1.9:1.0 with CWSN and CWIN percentages of 67 and 33, respectively.

Another slowly soluble nitrogen carrier, IBDU, was introduced in the late 1960s. It differs from UF in that its molecular composition is uniform; each molecule contains two ureas connected by a substituted methyl group from isobutyraldehyde. Solubilization of IBDU-N is only slightly influenced by temperature and depends primarily on particle size, moisture, and possibly pH. Plant-available nitrogen results from chemical, rather than microbial, hydrolysis of the carrier. Small particles with low mass-to-surface area ratios yield available nitrogen faster than do larger particles. Low pH and high moisture also favor solubilization. The CWSN-CWIN-HWIN system for characterizing UF cannot be applied to IBDU because of the uniform composition of this material.

Depending on the CWSN percentage of UF and the application rate, early spring treatment may or may not yield a rapid turfgrass response. Treatment with IBDU at the same time usually results in slow response due to the slow rate at which IBDU becomes soluble. However, when these materials are applied the previous fall, early spring response is usually good with IBDU and slow with UF until soil temperatures warm up sufficiently to favor solubilization.

Sulfur-coated urea (SCU) is a popular slow-release turfgrass fertilizer. It is formed by coating urea granules, or prills, with sulfur and a thin coating of sealant. Some newer formulations of SCU employ thin coatings of polymers atop the sulfur coating to prolong the release of nitrogen from these carriers. Since the nitrogen is already soluble, release depends on its movement out of the sulfur-coated spheres. Water diffuses through micropores in the sulfur coating until sufficient internal pressure builds up to cause breakage. Because of nonuniformity in coating thickness and particle size, some particles break sooner than others to provide the slow-release characteristic. The percentage of total nitrogen considered quickly available from SCU is indicated by the seven-day dissolution rate, which is the percent of nitrogen released through openings in the sulfur coating in water at 95°F (35°C) in one week. Commercially available SCU has a seven-day dissolution rate of approximately 30%. Therefore, initial turfgrass response following an application of SCU is fairly rapid. With this formulation there is a potential for some burn injury from particle breakage on closely mowed turfs due to foot and vehicular traffic. Some preliminary testing should be done, especially on greens, to ensure that burn potential with a particular SCU formulation is minimal.

Natural organic nitrogen carriers include various sewage sludges and plant and animal residues. Of these, Milorganite, an activated sewage sludge produced by the Milwaukee Sewage Commission, is the most widely used for turfgrass fertilization. The raw sewage is inoculated with microorganisms, aerated in tanks to promote flocculation of the organic material, filtered, dried, ground, and screened. In addition to approximately 6% nitrogen, Milorganite contains about 4% P_2O_5 and smaller amounts of potassium and several minor nutrients. Mineralization of nitrogen from natural organic carriers is primarily through microbial decomposition of nitrogen-containing organic compounds.

Characteristics of slowly available nitrogen carriers include slow initial but long-term turfgrass response, low-(IBDU, SCU) to high-(UF, natural organics) temperature dependency, generally low burn potential, and moderate to high cost per unit of nitrogen. An advantage gained with the use of slowly available nitrogen carriers is the reduced loss of nitrogen from leaching and, possibly, gaseous loss due to volatilization and denitrification. In many commercial turfgrass fertilizers, quickly available and slowly available nitrogen carriers are mixed to combine the advantages and reduce the disadvantages associated with each.

Fate of Nitrogen

Once applied to a turf, nitrogen forms can undergo changes in their chemical state and location. Ammoniacal (NH_4^+) and nitrate (NO_3^-) forms add directly to the plant-available nitrogen pool, while other forms are usually transformed to NH_4^+ and NO_3^- before being taken up by turfgrasses (Figure 5.9). Various forces act on the available nitrogen pool to increase or decrease its concentration. Decreases occur with microbial immobilization during organic matter decomposition; leaching, principally as nitrate; clipping removal; gaseous loss due to volatilization of ammonia and denitrification to N_2

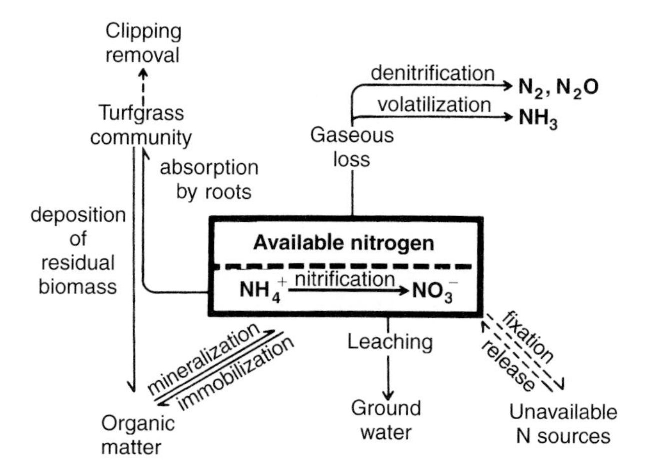

Figure 5.9. The fate of nitrogen in turf.

and N_2O; and possibly some chemical or physical fixation in the soil. Increases occur not only with fertilization and deposition of nitrogen from the atmosphere, but also with the return of organic material to the soil as plant tissues senesce and organic nitrogen is mineralized by microbial activity. The nitrogen balance of the system is thus determined by the effects that atmospheric, edaphic, and cultural conditions have on these processes. Following the application of urea to a turf growing in very sandy soil, substantial loss of nitrogen may occur through leaching and the evolution of gaseous nitrogen to the atmosphere. A clayey soil with a well-distributed turfgrass root system may retain the applied nitrogen so well that much less fertilizer is needed to sustain the turf. Where clippings are removed, additional nitrogen from fertilization is required to compensate for that which is lost. This difference may amount to 2 or more pounds of nitrogen per 1000 ft^2 (100 or more kg/ha) per year. Currently, there is no information on how much of the nitrogen returned with clippings is subsequently available for use by turfgrasses. Presumably, some of this nitrogen is lost to the atmosphere in gaseous forms.

Phosphorus

Phosphorus provides the plant with a means of holding and transferring energy for metabolic processes. Phosphate is an important constituent of ATP (adenosine triphosphate), an organic compound containing high-energy bonds that, when broken, can transfer energy for use in the synthesis and decomposition of various organic compounds. Because of its mobility within the plant, phosphorus is conserved and used repeatedly where needed. Maximum concentrations of phosphorus occur in meristematic tissues where new cell production takes place. It is stored in seeds as a constituent of phytin to provide an internal supply of phosphorus for germination. The phosphorus concentration of dried turfgrass clippings is usually less than 0.5% compared to 3 to 5% for nitrogen.

Phosphorus deficiencies are most evident during turfgrass establishment. Generally reduced growth, dark to reddish leaf coloration, and narrow leaf blades indicate a deficiency in available phosphate in the soil. A soil concentration equivalent to 30 pounds per acre (34 kg/ha) of available phosphate is considered a minimum level for turfgrass growth.

Phosphorus Carriers

Most phosphorus fertilizers used in the United States are derived from rock phosphate ores. Mined rock phosphate is ground, separated from impurities, and treated with acids to form commercial phosphate fertilizers. The two principal types are superphosphate, a combination of calcium phosphates and gypsum formed by treating rock phosphate with sulfuric acid (H_2SO_4), and treble superphosphate, which is calcium phosphates produced from treatment of rock phosphate with phosphoric acid (H_3PO_4). Ammonium phosphates, used in some mixed fertilizers to supply both nitrogen and phosphorus, are products of reactions between ammonia and phosphoric acid. Additional phosphorus fertilizers are produced from various organic sources and by-products of steel manufacturing. (Slag from smelting iron ores of high phosphate content is an important source of fertilizer phosphorus in Europe.)

Fate of Phosphorus

As phosphates are relatively immobile in soil, they do not leach readily from the turfgrass root zone. Many turf soils that have received liberal quantities of phosphorus-containing fertilizers are, thus, high in total phosphorus. However, the plant-available phosphorus concentration in the soil solution is usually a small percentage of the total concentration due to the rapid rate at which phosphorus becomes tied up in insoluble forms. The phosphate ion ($H_2PO_4^-$) combines readily with iron and aluminum cations to form insoluble compounds, especially in acid soils. In many instances the amount of available phosphate in the soil is surprisingly low compared to the amount removed by actively growing turfgrasses. This is apparently due to a dynamic equilibrium between soluble and insoluble forms of phosphorus; as the available pool is reduced by plant uptake some of the insoluble phosphorus becomes soluble at a sufficient rate to sustain turfgrass growth requirements. On many turfgrass sites where clippings are not removed there may be no need for phosphorus fertilizer as long as the residual supply from previous fertilizations is adequate.

Potassium

Potassium is second only to nitrogen in the amounts required to sustain turfgrass growth. Although not a constituent of living cells, it is important in the synthesis of numerous plant components and for regulating or catalyzing many physiological processes. Potassium deficiencies in turfgrasses result in increased respiration and transpiration, reduced environmental stress tolerance, increased disease incidence, and general reductions in growth.

Potassium is absorbed and stored in the ionic (K^+) form. As its concentration within the plant increases, tissue water content decreases and plants become more turgid. Potassium fertilization is often recommended to improve turfgrass wear tolerance as well as survival during periods of cold, heat, and drought stresses. Tissue potassium levels may reach up to 5% of dry weight where liberal quantities have been supplied through fertilization. Normal percentages, however, are about half that much; less then 1% may indicate a deficiency. An N:K ratio of 2:1 in leaf tissue is considered optimum, as excessive potassium concentrations may inhibit the absorption of calcium and magnesium, and possibly other nutrients.

Potassium Carriers

Most of the potassium used for turfgrass fertilization comes from potassium salt deposits that are mined and processed to make muriate of potash (potassium chloride, KCl). Earlier sources of commercial importance were potassium carbonate and other potassium salts from wood ashes. The term potash is derived from pot ashes, obtained by burning wood in pots designed to catch the ashes, leaching the ashes, and drying the leachate down to yield potassium salts.

The potassium chloride generated by mining salt deposits is a widely used source of potassium fertilizer. Reacting potassium chloride with sulfuric acid and nitric acid yields potassium sulfate and potassium nitrate, respectively. These are important potassium fertilizers that also serve as carriers of sulfur and nitrogen. Potassium sulfate is often used in place of potassium chloride to reduce the salt index.

Fate of Potassium

Since potassium salts are readily soluble in water, leaching losses can be substantial in sandy soils of low cation exchange capacity. In most soils with high-clay content, however, potassium is retained in large quantities. Most of the soil potassium is in an unavailable form, occurring as a constituent of soil minerals such as orthoclase feldspar ($KAlSi_3O_8$) and fixed between clay layers of illite and vermiculite in the same way that ammonium ions can be fixed. Plant-available potassium occurs in the soil solution (soluble K) and on exchange sites (exchangeable K). Conversion of unavailable potassium to available forms occurs slowly as minerals solubilize and some fixed potassium is released. Where clippings are removed with mowing, potassium may be depleted from the soil unless replaced by fertilization. Because of "luxury consumption" (plants absorb more potassium than is needed for normal growth as more is supplied), potassium should be applied several times in small amounts rather than in one large application during the growing season, especially to sandy soils or where clippings are removed. On sites where clippings are not removed and nitrogen fertilization levels are low, very little potassium may be required in the fertilization program.

Secondary Nutrients

The secondary nutrients—calcium, magnesium, and sulfur—are absorbed by turfgrasses at levels equal to or just below those of phosphorus. All are constituents of organic compounds within the plant. Calcium is required in large quantities in meristematic regions of the plant for cell production and is an important constituent of cell walls. It also influences the absorption of other nutrients, especially potassium and magnesium, by the roots. In the soil, calcium influences soil structure because of its electrical attraction to negatively charged colloids and the consequent flocculating effect it has on clay particles. As a chemical constituent of lime, it increases soil pH and, thus, influences the availability of other nutrients (Figure 5.10).

Magnesium is a central constituent of the chlorophyll molecule. It aids in the absorption of phosphorus and serves as an important catalyst in enzymatic reactions. Sulfur is a constituent of some amino acids required for protein synthesis. It is also an essential component of several plant vitamins.

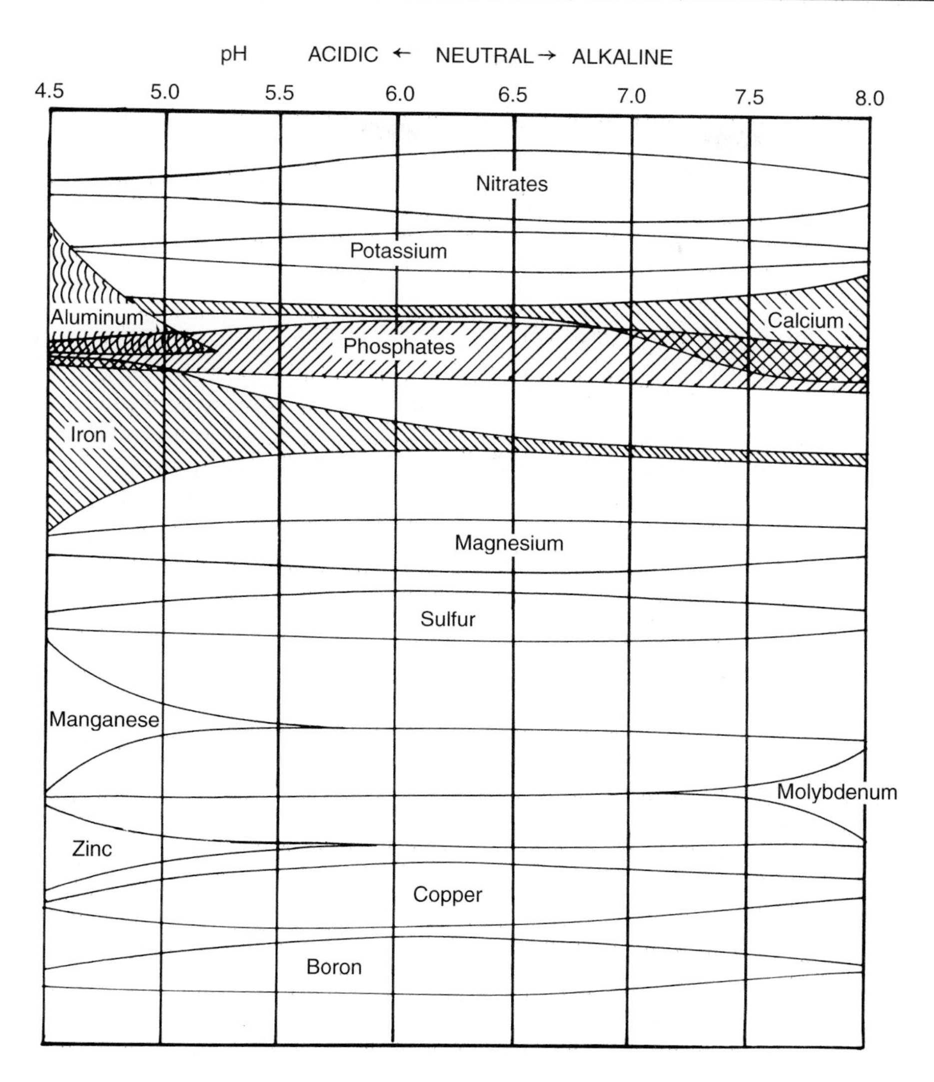

Figure 5.10. The influence of soil pH on nutrient availability.

Ca, Mg, and S Carriers

Calcium is a constituent of some phosphate compounds (Table 5.2), gypsum ($CaSO_4$), and lime [$CaCO_3$, $Ca(OH)_2$, and CaO]. Dolomitic limestone contains both calcium and magnesium carbonates, and thus serves as an important carrier of magnesium where its deficiency is known or suspected. Magnesium is also found as an impurity in several carriers of nitrogen, phosphorus, and potassium.

Sulfur is a component of several major-nutrient fertilizers, including ammonium sulfate, superphosphate, and potassium sulfate. Continued use of high-analysis fertilizers with minimal concentrations of sulfate may necessitate some supplemental sulfur fertilization. Elemental sulfur and some sulfate-containing fertilizers are sometimes used to reduce soil pH when it is excessively high. (Refer to the discussion on pH adjustment in Chapter 4.)

Fates of Ca, Mg, and S

Calcium occurs in the soil as exchangeable ions (Ca^{2+}), carbonates, silicates, and as a constituent of organic matter. Because many calcium minerals are fairly soluble, adequate moisture favors continual solubilization and leaching. The resultant decline in soil pH requires periodic applications of lime as a compensating measure on some sites. Calcium minerals are so prevalent that deficiencies are rare except in very sandy soils or where strong acidity has resulted from intensive leaching. Calcium is often the principal nutrient adsorbed on exchange sites.

Magnesium is second to calcium as the most abundant nutrient adsorbed on exchange sites. As with calcium, it is susceptible to leaching in sandy soils; however, losses are usually not as rapid due to the relatively low solubility of magnesium minerals. High rates of calcium application can, in some instances, induce magnesium deficiencies in plants; however, such deficiencies are uncommon.

In contrast to calcium and magnesium, sulfur occurs in relatively small amounts in the soil. Most soil sulfur is contained within the organic fraction, and its availability to plants is a function of organic-matter decomposition. When used, sulfate-containing fertilizers usually supply adequate amounts of sulfur for turfgrass growth. In some industrialized areas sulfur emissions into the atmosphere are washed in with rainfall or may be directly absorbed by leaves as sulfur dioxide (SO_2). The sulfate ion (SO_4^{2-}), like nitrate, is readily leached from the soil. Therefore, sulfur deficiencies can occur where clippings are removed, soil texture and water percolation favor leaching, and the aforementioned atmospheric and fertilizer sources are not available.

Micronutrients

The term *micronutrients* does not imply that these elements are unimportant; rather, it indicates that the amounts required are relatively low. Fertilizer application of these nutrients may not be necessary where sufficient quantities exist in the soil or are supplied as small impurities in other fertilizers. Micronutrients are sometimes referred to as *trace elements*. With the exception of iron and possibly manganese, direct application of micronutrient fertilizers is rarely necessary or beneficial. Sources of the seven micronutrients from soil and fertilizers are given in Table 5.3.

The importance of micronutrients lies principally in their role as catalysts in enzymatic reactions. Deficiencies are usually associated with conditions that limit the availability of soluble nutrient concentrations. With the exception of molybdenum and

Table 5.3. Sources of Micronutrients

Nutrient	SOURCES	
	Soil	Fertilizers
Iron (Fe)	Slightly soluble minerals Natural chelates[a]	Iron sulfate ($FeSO_2 \cdot 7H_2O$) Chelated Fe^{2+} (FeEDTA)[a]
Manganese (Mn)	Exchangeable Mn^{2+}	Manganese sulfate ($MnSO_4 \cdot H_2O$) Chelated Mn^{2+}(MnEDTA)
Zinc (Zn)	Exchangeable Zn^{2+} Natural chelates	Zinc sulfate ($ZnSO_4 \cdot 7H_2O$) Chelated Zn^{2+}(ZnEDTA)
Copper (Cu)	Exchangeable Cu^{2+} Slightly soluble minerals, Humus	Copper sulfate ($CuSO_4 \cdot 5H_2O$) Chelated Cu^{2+}(CuEDTA) CuO
Boron (B)	Iron and aluminum borates Humus	Borax ($Na_2B_4O_7 \cdot 10H_2O$)
Molybdenum (Mo)	MoO_4^{2+}, humus Soluble salts	Sodium molybdate ($Na_2MoO_4 \cdot 2H_2O$) Ammonium molybdate [$(NH_4)_2MoO_4$]
Chlorine (Cl)	Soluble salts, Cl^-	Potassium chloride (KCl)

[a]Chelates are soluble organic compounds that form bonds with cations to keep them soluble in the soil solution. Some occur naturally and many are artificially synthesized using such chelating agents as ethylenediamine tetraacetate (EDTA) diethylenetriamine pentaacetate (DTPA), and ethylenediamine di-(*o*-hydroxyphenylacetate) (EDDHA).

chloride, micronutrients become highly insoluble in alkaline soils, while, under conditions of excessive soil acidity, solubilities may be so high that they are actually toxic to turfgrasses. Figure 5.10 illustrates the influence of pH on nutrient solubilities. Other conditions that promote micronutrient deficiencies include high soil-phosphate levels, high concentrations of soil organic matter, excessive thatch accumulation, and poor drainage.

Of all the micronutrients, iron is the one most likely to be deficient in turf. Since it is important in chlorophyll synthesis, deficiencies are usually evident as chlorosis, but without an immediate reduction in clipping yield as would occur in nitrogen-deficient turf. Factors conducive to iron chlorosis include frequent removal of clippings, leaching through sandy soils with frequent irrigation, excessive P applications, overliming, and shallow rooting. A common practice is to apply 2 to 3 ounces of ferrous sulfate per 1000 ft^2 as a foliar spray to iron-deficient turfs. Results are often dramatic, but usually not long lasting because of rapid conversion of iron to insoluble forms in the soil. If hard water is used in the spray tank, the iron may precipitate out as ferric hydroxide; treating the water with a small amount of sulfuric acid before adding iron may help prevent this problem. Chelated forms of iron and other micronutrients that are less susceptible to rapidly becoming insoluble in the soil are available. However, chelated forms cost more than sulfate compounds.

Indiscriminate use of micronutrient fertilizers can result in serious turfgrass injury. Except for iron, even moderate applications can cause direct toxicity, and overuse of many micronutrients, including iron, can adversely affect the uptake and activity of other nutrients. Usually, the best ways to avoid micronutrient deficiencies are to ensure that the soil is maintained within a suitable pH range and to limit applications of lime and phosphates to levels necessary for sustaining optimum turfgrass growth.

Fertilizer Application

The frequency and intensity of fertilization and the composition of a turfgrass fertilization program depend on numerous genotypic and environmental factors. The particular turfgrass genotype (species and cultivar) influences the amount of nitrogen and other nutrients required to sustain growth. Some turfgrasses, such as red fescue and centipedegrass, have relatively low fertilization requirements; others grow best where moderate to high rates of fertilizer nutrients are applied. Environmental conditions strongly influence fertilizer requirements because of the interactive relationships between the turfgrass community and its environment and the efficiency with which applied nutrients are used. A shade turf has a lower requirement for nitrogen than does one in full sunlight. Moisture from precipitation and irrigation influences the extent to which nutrients are absorbed or lost through leaching. Temperature determines potential growth rate and, therefore, nutrient requirements. Soil conditions influence such processes as leaching, gaseous nitrogen loss, and insolubilization and, thus, the net requirement for fertilizer nutrients. Mowing height affects the growth of roots and other organs of the turfgrass plant and the consequent capacity of the plant community to absorb available nutrients from the soil. Clipping removal, where practiced, is an important nutrient-output avenue that clearly establishes an increased need for supplemental fertilization. With so many factors influencing the fertility status of turf, it is impossible to construct any one fertilization program for all turfs that will ensure optimum turfgrass growth while efficiently utilizing all fertilizer resources. Each turfgrass site must be studied to determine its particular combination of features before initiating a fertilization program, and an experience bank must be acquired over time to refine and improve the program.

Fertilization Frequency

An ideal fertilization program would be to apply very small amounts of essential nutrients each week or two during the growing season. With such a program, rates could be continually adjusted up or down depending on turfgrass response. Excesses would be avoided and opportunities for testing new materials in small quantities would always be available. Such a program would, however, be extremely expensive and totally impractical on most turfgrass sites. The opposite extreme would be to apply all fertilizer just once each year; this would conserve labor, minimize investment in application equipment, and simplify storage and handling of fertilizer materials. On some turfs sustained at a low intensity of culture, this type of fertilization program has been conducted with reasonable success. Most turfs, however, require at least two fertilizations per year for satisfactory growth and appearance. The number of fertilizations can be minimized with the use of slowly available nitrogen carriers such as UF, IBDU, SCU, and natural organic materials. The suitability of each material depends on the rate at which nitrogen becomes available under different conditions during the growing season. Ultimately, the presence and availability of essential plant nutrients determine the minimum fertilization frequency at which acceptable turfgrass growth and quality occur.

Fertilization Timing

The proper timing of fertilizer applications becomes more critical as the number of applications per growing season is reduced. Where fertilization is performed only once annually, late summer and late spring are usually preferred for cool-season and warm-season turfgrasses, respectively. A second application to warm-season turfgrasses is recommended during early summer to midsummer. For cool-season turfgrasses, a second application has traditionally been recommended for midspring to late spring, following the early flush of growth.

In recent years some researchers have advocated late-season (mid-October to mid-November, depending on location) fertilization of Kentucky bluegrass to minimize disease problems and to promote better color retention in fall and earlier green-up in spring. Heavy applications of quickly available nitrogen in early spring to midspring may encourage more severe incidence of spring and summer diseases in Kentucky bluegrass and some other cool-season turfgrasses. Early summer to midsummer fertilization should be either avoided or minimized to reduce stress levels. Likewise, late summer and early fall fertilization of warm-season turfgrasses may reduce their cold-hardiness and result in some winterkill where cold winters are encountered. Fertilization timing, then, is strongly influenced by two principal constraints: disease and environmental stress tolerance. Within these constraints, a fertilization program must include adequate levels of essential nutrients to sustain growth. Programs for nitrogen fertilization will be provided under Turfgrass Cultural Systems in Chapter 9.

Fertilization Rate

The amount of nitrogen required for optimum turfgrass quality during a growing season depends on many factors already discussed. The maximum amount that may be safely applied at any one time depends on the type of nitrogen carrier, temperature, time of the year, mowing height, and turfgrass. Under favorable growing conditions, a general rule is to apply no more than 1 lb of quickly available nitrogen per 1000 ft^2 (about 50 kg N/ha), as higher rates may cause turfgrass injury or excessive shoot growth. As temperatures increase to stress levels, cool-season turfgrasses should not receive more than about 0.5 lb N/1000 ft^2. Where slowly available nitrogen carriers are used, rates higher than 1 lb may be used; however, exceeding a 3-lb rate is usually considered risky.

Turfgrasses that are mowed at a height near the lower limit of their mowing tolerance range (close mowing) should receive smaller quantities of quickly available nitrogen than where higher mowing heights are used. This is because, with closer mowing, individual shoots are smaller and, therefore, more susceptible to injury from large concentrations of fertilizer nutrients. Also, close-cut grass is often so dense that fertilizer particles cannot fall into the turf; instead, they are held on the foliage. Creeping bentgrass greens usually receive no more than 0.5 lb of quickly available nitrogen per 1000 ft^2. The same turfgrass on tees and fairways maintained at two to three times the mowing height of greens could tolerate two to three times as much nitrogen per application.

Methods of Application

As the root systems of individual turfgrass plants cover a relatively small area surrounding the shoot, it is important to apply fertilizers uniformly to the turf. Spotty or streaked responses to fertilization indicate improper application. Uniformity of application is also a conservation measure; it is wasteful to apply three or four times as much fertilizer to one location to ensure that a sufficient amount has been received by another location. Uniform application requires suitable equipment and good technique.

In earlier times greens and other small turfgrass sites were fertilized by hand. Experienced crew members spread fertilizer with remarkable accuracy and uniformity by careful hand motion and steady walking speed. Today application equipment calibrated to deliver precise quantities of fertilizer has replaced hand application except in occasional small areas. (Hand application must be done with great care to avoid fertilizer burn.)

The two principal types of application equipment are sprayers for liquid application of soluble or suspended fertilizers and spreaders for dry application of granular fertilizers. At low spray volumes (0.5 gal/1000 ft^2 or lower), spraying is called *foliar feeding* because significant quantities of the nutrients are absorbed directly by the turfgrass leaves. Only low rates of fertilizer (about 1/8 lb N or Fe/1000 ft^2) should be applied by foliar feeding to minimize foliar-burn potential. Foliar feeding is usually conducted to supplement a normal fertilization program or in conjunction with routine applications of some pesticides. When higher spray volumes (3 to 5 gal/1000 ft^2) are used the procedure is called *liquid fertilization*. With this method much of the fertilizer is washed off the foliage, and absorption by root uptake increases. Many commercial lawn-care firms use liquid fertilization because of the ease with which fertilizers and pesticides can be applied in a single operation.

Granular fertilizers are applied through either drop-type or rotary spreaders (Figure 5.11). With a drop-type spreader the fertilizer exits through a series of openings at the base. The size of the openings can be adjusted to provide the desired rate of application. In the absence of light wind the fertilizer may be applied in parallel streaks rather than uniformly over the turf. To compensate for this problem, some drop-type spreaders use a baffle board that intercepts and spreads out the flow of fertilizer. Misses between application strips and excessive overlapping are common problems with drop-type spreaders, and operational efficiency is low because of the limited width of distribution. However,

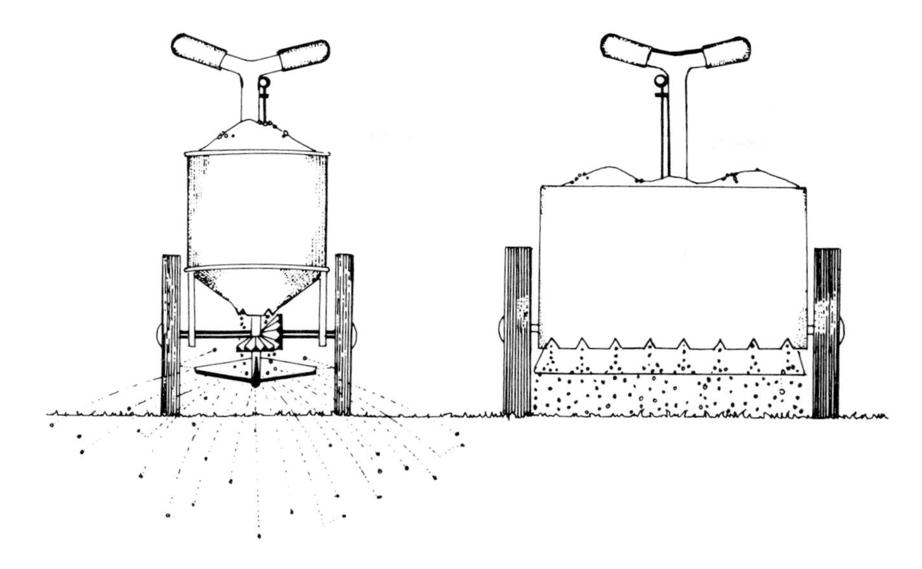

Figure 5.11. Comparison of rotary (left) and drop-type (right) fertilizer spreaders.

with proper operation drop-type spreaders can provide a very accurate and uniform application of granular fertilizers.

Rotary (centrifugal) spreaders are very efficient for fertilizing large turfgrass areas. Fertilizer falls through one or more openings of adjustable size onto a rotating plate at the base of the spreader and is propelled away in a semicircular arc. The distribution pattern is not uniform, but, with a controlled degree of overlapping, adequate uniformity can be achieved. One disadvantage of rotary spreaders is that mixed materials of different sizes tend to be distributed in different patterns; larger, heavier granules are propelled farther than smaller, lighter granules. For this reason materials of substantially different size and weight should not be applied as a mixture with this type of spreader. Separate applications should be made, or a drop-type spreader used, to avoid problems associated with differential distribution patterns.

Fertigation is the application of fertilizers through an irrigation system. Although attractive as a means of combining separate cultural operations, fertigation is often not feasible because of nonuniform coverage by the irrigation system. If two or five times as much water is applied to one area compared to another, the differential will apply to fertilizer application as well. Nevertheless, fertigation is growing in popularity in some locations, especially where dry conditions necessitate frequent irrigations or where rapid leaching of fertilizer nutrients requires that fertilizers be applied frequently. When used, fertigation should be followed by irrigation with water to wash in fertilizer materials to avoid foliar burn and to rinse the fertilizer out of the irrigation system to minimize corrosion.

IRRIGATION

Irrigation is performed primarily to provide an adequate supply of moisture for turfgrass growth. Turf is also irrigated to wash in fertilizers and some pesticides following application, maintain sufficient surface moisture to promote germination of interseeded turfgrasses, and modify turfgrass tissue temperatures on hot days by a practice called syringing.

Irrigation Programming

In many climates irrigation is used to supplement an abundant but inadequately distributed rainfall. Following precipitation or irrigation, water is lost from the turfgrass root zone by evaporation and transpiration processes, collectively called *evapotranspiration* (ET) and by drainage from large (aeration-type) soil pores (Figure 5.12). Some moisture is obtainable from lower soil depths as water moves upward to replace moisture lost through evapotranspiration (that is, in response to a water potential gradient; see Chapter 4). However, upward capillary flow may not proceed fast enough to supply turfgrass requirements, or supplies may be depleted prior to the next rainfall. Thus, without timely irrigation turfgrasses may die or become dormant.

The amount of water consumed by turf is a function of solar radiation; evapotranspiration proceeds at maximum rates during the summer months when solar radiation is most intense. Rooting depth determines the volume of soil that serves as a water reservoir. A deeply rooted turf should, therefore, have a lower irrigation requirement than one with shallow roots; however, the amount of water required per irrigation depends on the volume of the turfgrass root zone. Shallow-rooted turfs require more frequent but less intensive irrigations than deep-rooted turfs.

Numerous environmental and genotypic factors influence rooting depth. Poor soil aeration due to compacted or waterlogged conditions, close mowing, excessive fertilization, and frequent irrigation are sometimes associated with shallow turfgrass rooting. Cool-season turfgrasses tend to lose many of their functional roots during summer stress periods.

Therefore, as the turfgrass community and its environment undergo changes during the growing season, irrigation requirements change as well. A reasonable strategy for sustaining turf at an acceptable level of quality is to encourage deep rooting when conditions favor new root growth and to preserve a well-developed root system as much as possible during unfavorable environmental conditions. Irrigation practices profoundly affect root growth and survival.

Irrigation Frequency

Historically, frequent irrigation was thought to be detrimental to turf. High disease incidence, reduced wear tolerance, low vigor, and high susceptibility to injury from climatic stresses were identified as problems associated with frequent irrigation of turf.

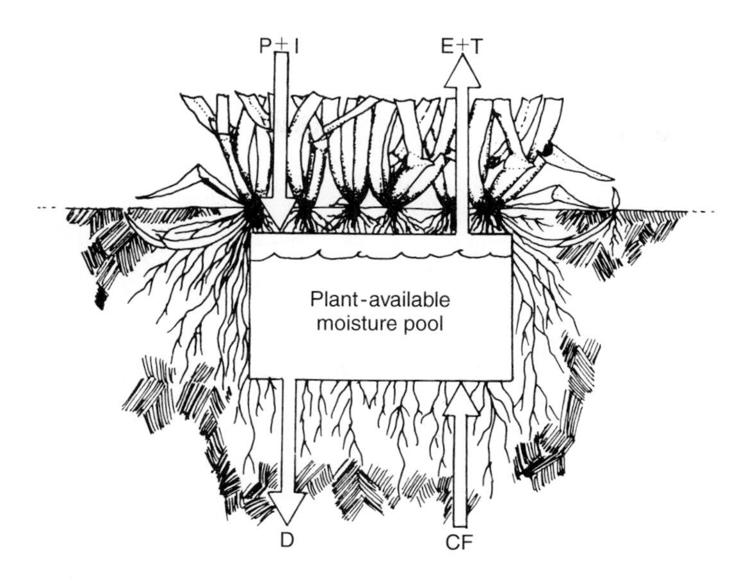

Figure 5.12. Moisture inputs (precipitation [P], irrigation [I], and capillary flow [CF] from lower soil depths) and outputs (evaporation [E], transpiration [T], and drainage [D]) as they influence the plant-available moisture pool in turf.

As irrigation frequency is reduced turfgrasses typically improve until inadequate moisture limits growth. Recent evidence from field research conducted in Michigan and Nebraska suggests that daily irrigation may actually enhance turfgrass growth and quality. This view is certainly more consistent with that of golf course superintendents who practice daily irrigation on greens and other intensively cultured turfs. An explanation of the basis for these contrasting views may be that frequent irrigation at application rates low enough to avoid standing water and in quantities sufficient to replenish soil moisture lost through daily evapotranspiration may actually be beneficial. However, daily irrigation at excessive rates and quantities that result in waterlogging and associated oxygen deficiencies in the surface soil may be detrimental.

Traditionally, the problem has been in determining the optimum irrigation program in which moisture is always adequate but not excessive. There are two approaches for developing a sound irrigation program: one relies on methods for estimating consumptive water use (evapotranspiration), and the other is based on soil moisture measurements. Consumptive use depends on temperature, wind, relative humidity, soil water potential, and other factors; it can be estimated by reference to an evaporation pan placed in an open unshaded area (Figure 5.13). Except where high winds occur, evaporation from the pan approximates evapotranspiration from a well-irrigated turf. As wind velocity increases, pan evaporation exceeds evapotranspiration. Some experimentation is necessary to correlate reductions in the pan's water level with the appearance of turfgrass wilt symptoms following normal irrigation. During cloudy or cool periods water loss from the

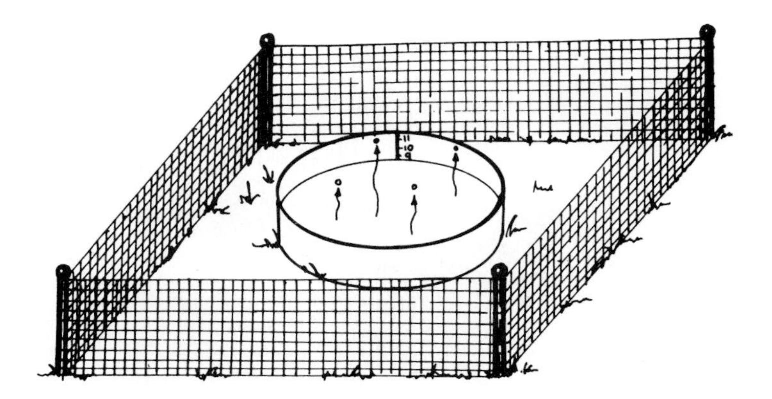

Figure 5.13. An evaporation pan for estimating evapotranspiration from turf.

pan and the turf will be less than during hot, sunny weather, and the interval between irrigations can be lengthened. Changes in rooting depth, thatch thickness, soil bulk density, and the like will result in significant alterations in the evapotranspiration rate, while conditions surrounding the pan do not change; thus, adjustments during the growing season are necessary in order for the system to work. However, several golf course superintendents have been successful in reducing their irrigation frequencies, and thus conserving water and promoting better turfgrass quality, through the use of evaporation pans or reasonable facsimiles.

Soil moisture measurements can be made using several devices; the most useful of these is the tensiometer. A porous ceramic cup at the tip of the tensiometer is coupled to a vacuum gauge at the other end. After the system is filled with water the ceramic cup is inserted into the soil. As the soil dries out, water is pulled from the ceramic cup, causing the gauge to register higher tension (lower water potential) levels, usually in centibars. Depending on the soil, a reading of 10 to 30 centibars (cb) indicates field capacity, and 70 cb indicates reduced plant-available moisture. Proper placement of the tensiometer with respect to topography and soil depth is critical for securing useful readings. Some manufacturers market dual tensiometers that, when correctly installed, provide readings at two soil depths for simultaneous monitoring of soil moisture near the surface (at about 2 inches) and at a lower depth (5 to 6 inches) (Figure 5.14). Using this system, an irrigation program may include several shallow irrigations based on measurements by the 2-inch tensiometer, and occasional deep irrigations based on readings from the second tensiometer. Obviously, considerable experimentation would be necessary to effectively use this system. The opportunity that it presents to develop a sound irrigation program, however, makes it a worthwhile consideration.

The trend toward daily irrigation of turf, especially on golf courses, has been rationalized based on increasing traffic intensity, closer-mowing requirements, and demands for higher turfgrass quality and softer turf. There is also the concern that without daily irrigation the potential for severe wilt or large-scale loss of turf when heavy

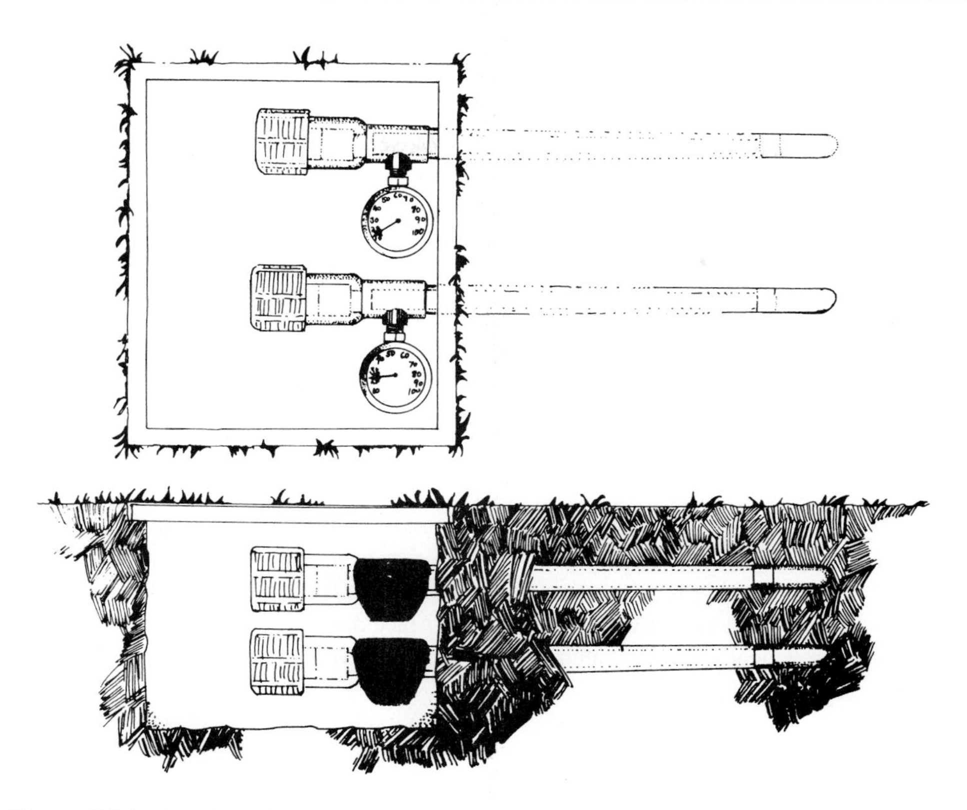

Figure 5.14. Dual tensiometers. Top view (left) shows two units side by side; readings can be made after removing a cover that is flush with the ground. Profile view (bottom) shows the two units at different soil depths.

play precludes irrigating is increased. With this approach, the following cycle of events often develops: daily irrigation at excessive application rates renders the soil more susceptible to the compacting effects of traffic; increased soil compaction limits root growth; reduced rooting results in higher wilting tendency and, therefore, necessitates more frequent irrigation.

Many of these problems can be reduced by altering other cultural practices as well as the irrigation program. Installation of a network of drainage tiles and catch basins can improve soil drainage. Under some conditions an effective program of cultivation improves infiltration and reduces soil compaction. Small, upward adjustments in mowing height can produce a softer, more resilient turf or one with a significantly reduced irrigation requirement. Taken together, the various elements of a sound cultural program, including efficient use of irrigation water, can often provide a turf of adequate playability without excessive irrigations.

In lawn turfs not subjected to intense traffic, irrigation can be programmed based primarily on the appearance of wilt symptoms. Wilting may indicate that soil moisture is so low that turfgrasses are unable to absorb sufficient amounts to offset evapotranspiration losses. With the development of an internal moisture deficit within the plants, cells lose turgidity, stomata close, and leaves exhibit wilt symptoms by turning a shiny, purplish color. Use of a soil probe to extract samples of the turf soil profile provides a means of determining whether wilting is associated with dry soil or with some other problem. Compaction, excessive thatch accumulation, shallow rooting, or poor drainage are conditions that are observable with soil probing and that could account for an unusually high proneness to wilting. Over the short term, irrigations can be scheduled to compensate for these problems; however, corrective measures should be planned so that future irrigation requirements can be reduced.

Irrigation Timing

Irrigation can be performed at any time of the day or night as long as the rate of application ("precip" rate) does not exceed the infiltration capacity of the soil. When the infiltration capacity is exceeded, water will either stand on the surface (*puddling*) or move downslope (*runoff*). On sites where surface drainage is enhanced by shallow sloping and strategically located catch basins, the excess water can be removed (Figure 5.15). These

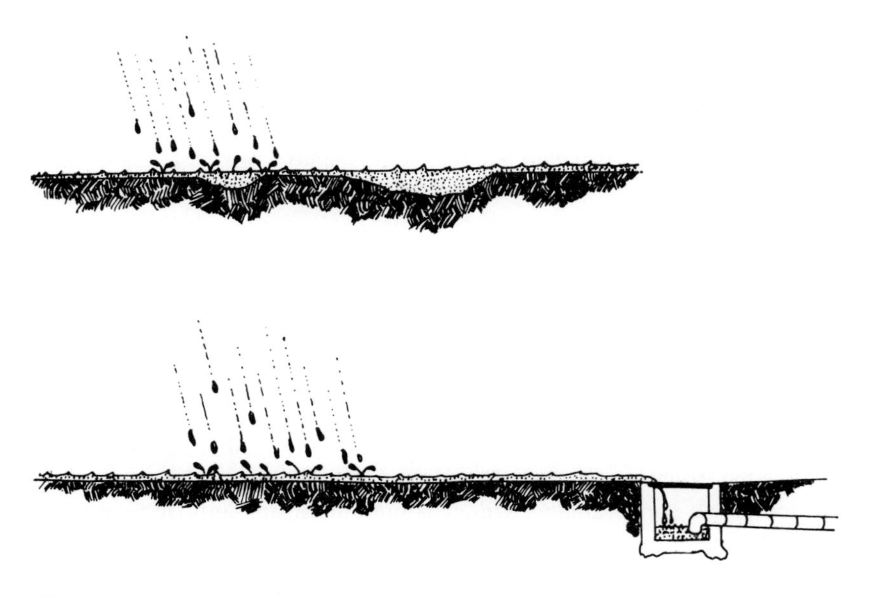

Figure 5.15. Nonuniform surface without slope (above) results in standing water whenever the precipitation rate exceeds infiltration capacity. Sloping surface with catch basin-drain tile system promotes rapid surface drainage at excessive precipitation rates.

are desirable features for handling runoff from heavy rainfall; however, when they are functional during prolonged irrigations, water is being wasted. Without adequate surface drainage, standing water results in a soft surface that will not adequately support traffic. On hot, sunny days, standing water can result in scald injury. This is a loss of turf first evident by a scorched or burned appearance.

A primary reason cited for avoiding midday irrigation during summer months has been the increased incidence of scald injury that sometimes accompanies this practice. Other reasons include interference with use on heavily trafficked turfs, interference with other cultural operations, and reduced efficiency in water use. During midday, substantial quantities of irrigation water may be lost through evaporation. The cooling effect of irrigation on a hot day is due to the conversion of sensible to latent heat (Chapter 4), which requires a change of state from liquid water to water vapor. At night the absence of solar radiation results in much less evaporative loss of irrigation water and, thus, more efficient use of the water for satisfying turfgrass moisture requirements. For the above reasons, nighttime irrigation has become standard on many golf courses and other turfs, especially where automatic irrigation systems have been installed.

An additional benefit from irrigating at night is that there is more time for internal drainage from the irrigated turf. Wet soil is more susceptible than moist soil to traffic-induced compaction. By allowing sufficient time for excess soil moisture to drain prior to heavy traffic, soil compaction can be minimized and a firmer turf provided. On golf courses irrigations are usually initiated during early evening hours and concluded well before golfers start play. On athletic fields irrigation water should be withheld at least one day prior to a game or other intensive use.

A traditional concern over nighttime irrigation has been the presumed higher incidence of diseases caused by prolonged surface moisture on turfgrass leaves. Fungal spores require surface moisture to germinate and cause infection of plant tissue. Therefore, steps taken to reduce periods of prolonged surface moisture, such as irrigating during early morning hours when winds and sunlight can quickly dry the leaves, should reduce the incidence of some diseases. Although there is some field evidence to support this principle, other considerations often outweigh the disease concern and strongly support nighttime irrigation. Where disease on heavily trafficked turf is a concern, fungicides are often used on a regular basis and probably counter most irrigation-related disease influences.

Irrigation Rate

As discussed previously, the "precip rate" (PR), or rate at which irrigation water is applied to a turf, should not exceed the infiltration capacity (IC) of the soil. Where $PR > IC$ standing water or runoff invariably occurs. With surface runoff, the amount of water held by the soil will be less on elevated and sloping sites and proportionately more in depressed areas (Figure 5.16). Turfgrass managers have attempted to compensate for surface runoff by varying the amount of water applied to different sites; sprinklers situated on elevated areas are operated for longer periods, while those situated in depressions are used less. This is sometimes helpful, but results are far less satisfactory than where $PR = IC$.

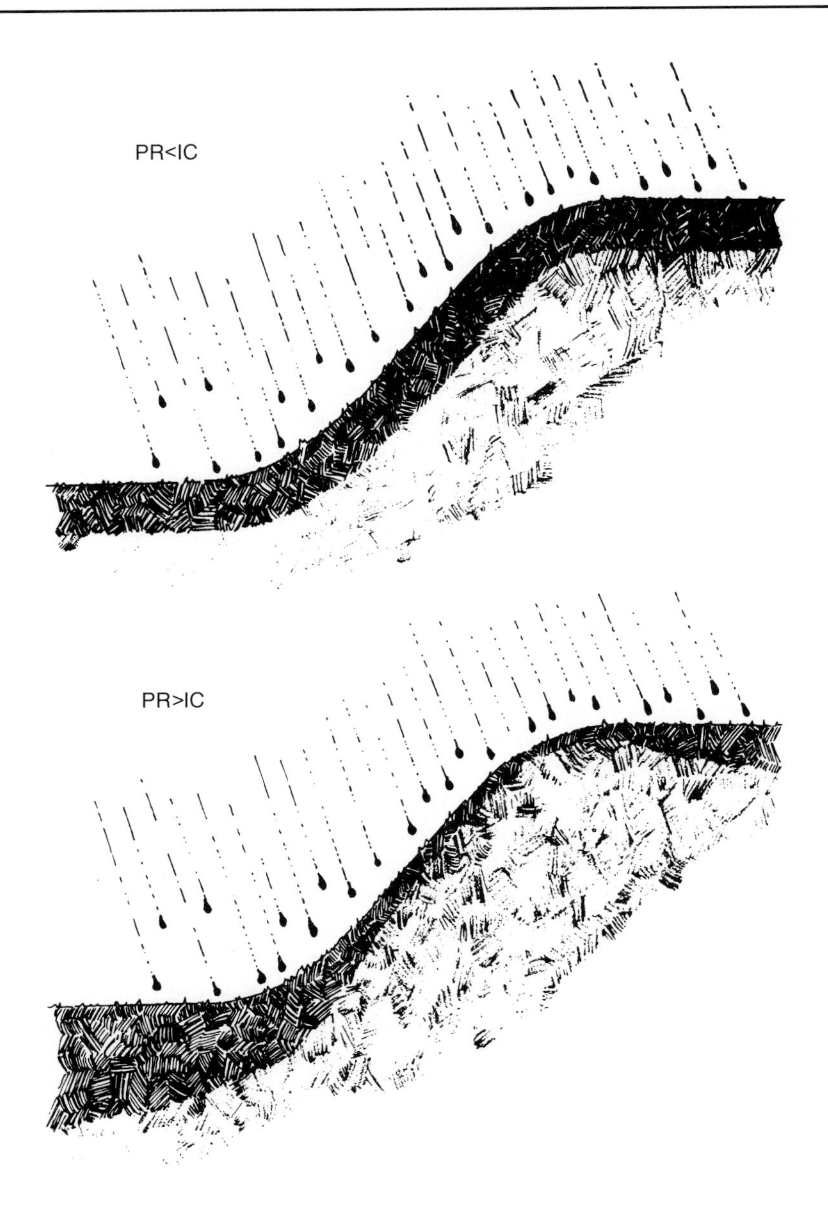

Figure 5.16. Soil moisture condition in sloping turf when precipitation rate (PR) equals or is less than infiltration capacity (IC), compared to where PR exceeds IC.

The soil IC is related to texture, structure, and, perhaps, thatch conditions. Fine-textured, compacted soils may have ICs of 0.1 inch/hour or lower, while most contemporary irrigation systems provide PRs of 0.3 to 0.6 inch/hour. Under these conditions runoff is inevitable unless special precautions are taken. IC can be improved through cultivation practices that expose more surface area or break through compacted surfaces.

PR can be effectively reduced with automatic irrigation systems by employing a procedure called *multiple cycling*. A system is operated for several short periods of time instead of for one continuous period to provide a given amount of water. For example, continuous operation for 40 minutes with PR equivalent to 0.5 inch/hour provides 0.33 inch of water. With multiple cycling, four 10-minute operations spaced 50 minutes apart provide the same amount of water, but with an effective PR of approximately 0.10 inch/hour (4 × 10-minute operations + 3 × 50-minute intervals for a total of 190 minutes, or 3.17 hours to apply 0.33 inch of water). Where high winds occur sporadically, another advantage of multiple cycling is an increased likelihood of achieving uniform distribution of water over the turf. Gusty winds during the early evening hours may seriously disrupt irrigation patterns; with multiple cycling, disruption may be less, as each sprinkler is operated over a longer period within which winds can either change direction or decrease in intensity.

Dry thatch may be difficult to wet under some circumstances and, therefore, may limit the IC of a turf. In some instances the problem may be due to compacted soil underlying the thatch rather than to thatch *per se*. Studies in Illinois showed that where thatch development was induced by pesticides in an otherwise thatch-free Kentucky bluegrass turf, reduced infiltration was due to an altered physical condition of the soil. Restriction of root and rhizome growth to the thatch and the lack of earthworm activity resulted in appreciably higher bulk density and reduced hydraulic conductivity in soil beneath the thatch.

A very porous thatch medium occurring on a sloping site may allow substantial downslope water movement within the thatch layer. This is similar to, but distinguished from, surface runoff in that nutrients, pesticides, and other materials in the thatch are likely to be carried downslope with the water. Consequently, not only is water being wasted, but other materials applied to the turf may be lost.

Irrigation Intensity

The amount of water required in turfgrass irrigation depends to a large degree on the nature of the soil medium supporting the turf. Sandy soils do not retain water as well as finer-textured clayey or silty soils because of rapid drainage from large pore spaces. Also, the soil volume occupied by a given amount of water is relatively small in the finer-textured soils (Figure 5.17). The total porosity of sand is actually less than that of clay; individual sand particles are solid, while clay aggregates have internal porosity. Because of differences in pore-size distribution, however, more water is retained in the small clay pores, while water tends to drain from the large pores in sand and, consequently, moves to lower depths.

Water infiltration is generally faster as soil texture becomes coarser; therefore, water can be applied more rapidly (higher PR), but less water is needed per irrigation to moisten a given soil volume. The total amount of water required during the growing season is greater, however, where a coarser-textured soil supports the turf. Due to differences in pore-size distribution, coarser-textured soils lose more water to drainage and evapotranspiration soon after an irrigation than do finer-textured soils. Thus, even though less

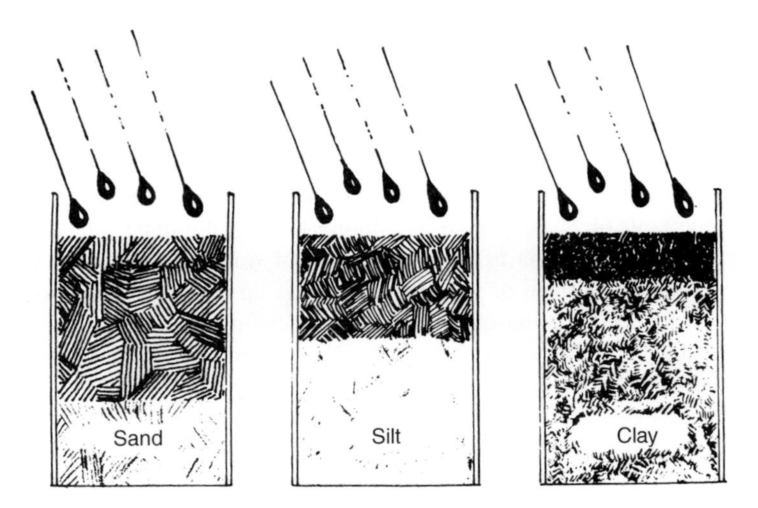

Figure 5.17. Comparison of moistened soil volume from a given amount of water applied to sand, silt, and clay.

water per irrigation is required, more irrigations, and therefore more water, are necessary to sustain turf as soil textures become coarser.

Syringing

Wilting may occur in some turfs even when soil moisture appears adequate. This may be associated with shallow rooting, excess thatch, disease, or poor soil aeration due to compaction or a waterlogged condition. Whenever evapotranspiration exceeds the rate at which water is absorbed by turfgrass roots, wilt can develop because of a moisture deficit within the plants. Midday wilt with seemingly abundant soil moisture is most common with bentgrasses and annual bluegrass, especially where close mowing is practiced.

Syringing is the practice by which water is applied to correct plant water deficits, reduce plant tissue temperatures, and remove substances from the leaves. Since the water is applied principally to the turfgrass shoots, and not to restore soil moisture, only small amounts are necessary. Syringing should be initiated at the first signs of wilt, usually at midday. A single syringing may be adequate to carry the turf through the day; however, under extreme conditions two or more syringings may be required.

Syringing is routinely practiced in the early morning hours on some closely mowed turfs to remove dew and guttation water. Paradoxically, this light application of water actually accelerates drying of the foliage so that subsequent mowing quality is improved and disease proneness is reduced. In recent years early morning syringing has largely replaced the older practice of poling or dragging greens to remove leaf moisture. Higher labor costs and the ease with which syringing can be accomplished with automatic irrigation systems have been responsible for this change in cultural practices.

Syringing is an important practice employed during turfgrass establishment. To avoid desiccation, newly planted turfgrasses, including sod, seed, plugs, and the like, must be kept moist until sufficient rooting has taken place. Several syringings per day may be necessary during the first one to three weeks following planting; afterward, irrigation frequency should be reduced and intensity increased until the turf has become fully established.

Finally, syringing may be practiced to promote survival of turfgrasses following certain diseases and insect-caused injury. Turfgrass desiccation often accompanies summer patch because of deterioration of the root system. Similarly, white grubs feed on roots and thus reduce the turfgrasses' capacity to absorb moisture. Syringing can be important in sustaining damaged turfgrass populations until sufficient new roots have developed.

Water

Water is the most plentiful substance on the earth's surface. Although it occurs in greatest abundance in the oceans, these are not suitable water sources for irrigation because of their high salt content. Irrigation water must come from surface and subsurface sources of adequate supply and suitable quality.

Water Quality

The suitability of a specific water source for irrigating turf depends on the type and concentration of substances dissolved or suspended in the water. Although some sources may provide essentially pure water, others may have such high concentrations of salts, particulate matter, microorganisms, and other materials that the water would eventually cause direct injury to turfgrasses or indirect injury from effects on soil properties. Two important determinants of water quality are total salt concentration and the relative concentration of sodium and other cations. Total salts, or salinity, are measurable as the electrical conductivity (EC) of the water. ECs below 0.25 dS/m indicate low salinity hazard. In the USDA classification of irrigation waters this would be identified as C1 (Figure 5.18). Higher EC classes include C2 (0.25 to 0.75 dS/m), which can be used if a moderate amount of leaching occurs; C3 (0.75 to 2.25 dS/m), which should not be used on soils with restricted drainage and which should be avoided for use with salt-sensitive species on soils with good drainage; and C4 (2.25 dS/m), which is not generally suitable for irrigation. The EC of a saturated soil extract (see Chapter 4 under Salted Soils) is often two to ten times that of the irrigation-water EC because water absorption by plants occurs faster than ion absorption. The soil solution thus develops appreciably higher salt

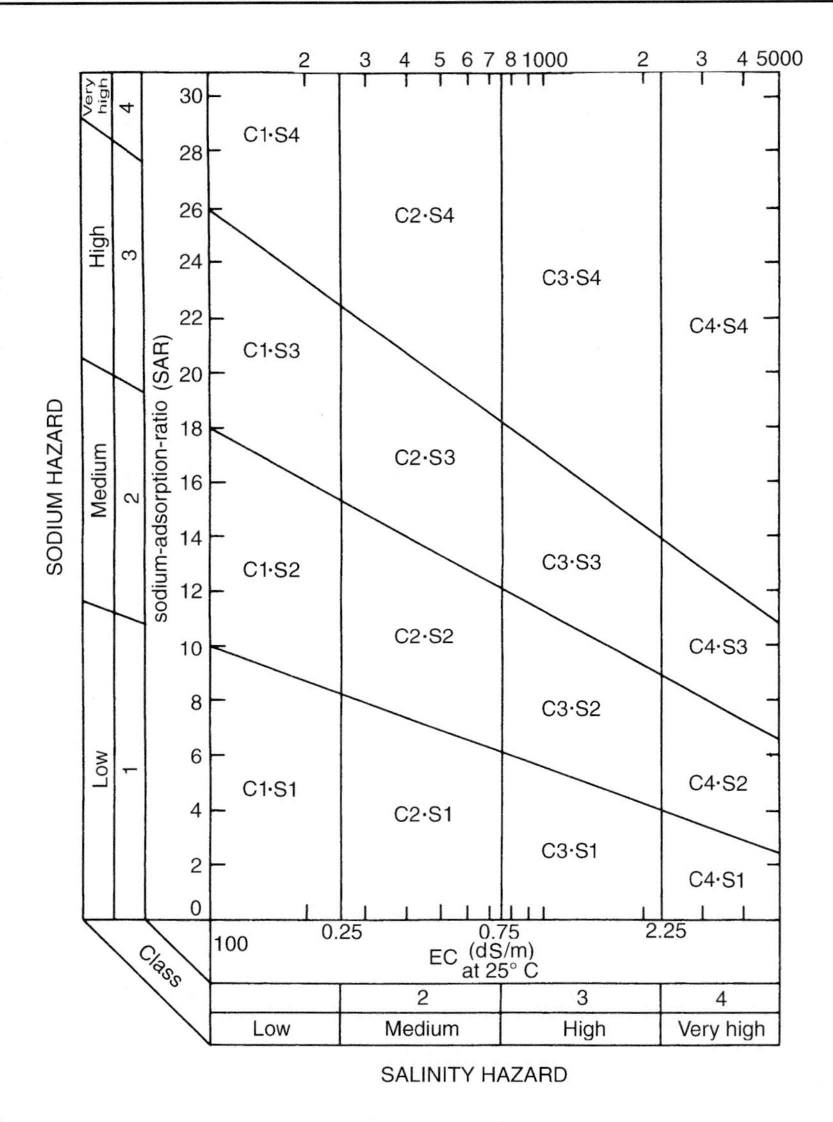

Figure 5.18. Classification of irrigation water based on sodium adsorption ratio (SAR) and electrical conductivity (EC) values, L.A. Richards, ed., "Diagnosis and Improvement of Saline and Alkali Soils," *Agric. Handbook No. 60,* U.S. Salinity Lab. Staff, USDA, Washington, D.C., 1954, p. 80.

concentrations than the irrigation water. Turfgrass species vary in their relative sensitivity to saline soils: Kentucky bluegrass, annual bluegrass, rough bluegrass, colonial bentgrass, and centipedegrass are highly sensitive; Chewings fescue, (strong) creeping red fescue, annual ryegrass, most creeping bentgrasses, and bahiagrass are moderately sensitive; Seaside creeping bentgrass, fairway and western wheatgrasses, perennial ryegrass, (slender) creeping red fescue, tall fescue, blue grama, buffalograss, and zoysiagrasses are moderately tolerant; and alkaligrass, bermudagrass, seashore paspalum, and St. Augustinegrass are highly tolerant.

High concentrations of sodium ions in the soil can adversely affect turfgrasses as well as soil physical conditions. Along with calcium and magnesium, sodium is one of the principal cations found in irrigation water. The relative concentrations of these cations in irrigation water influence its suitability for application to turf. The sodium adsorption ratio (SAR) is calculated as follows:

$$SAR = Na^{+} / ([Ca^{2+} + Mg^{2+}] / 2)^{1/2}$$

where the cation concentrations are expressed in milliequivalents per liter (mEq/l).

For example, if test results show Ca^{2+}, Mg^{2+}, and Na^{+} concentrations of 62, 29, and 56 milligrams/liter (mg/l), respectively, these values must be divided by their milliequivalent weights (Ca = 20 mg/mEq, Mg = 12 mg/mEq, Na = 23 mg/mEq) to calculate SAR:

$$62 \text{ mg/l} \div 20 \text{ mg/mEq} = 3.1 \text{ mEq/l for } Ca^{2+}$$

$$29 \text{ mg/l} \div 12 \text{ mg/mEq} = 2.42 \text{ mEq/l for } Mg^{2+}$$

$$56 \text{ mg/l} \div 23 \text{ mg/mEq} = 2.43 \text{ mEq/l for } Na^{+}$$

$$SAR = 2.43 \div ([3.10 + 2.42] /2)^{1/2} = 1.46$$

Based on Figure 5.18, an SAR value of 1.46 would be low (S1), indicating no harmful effects from sodium. A medium value (S2) would indicate that there was an appreciable sodium hazard in fine-textured soils of high CEC, but that the water could be used on sandy soils with good permeability. High values (S3) indicate that harmful effects could be anticipated in most soils and that amendments such as gypsum would be necessary to exchange sodium ions. A very high value (S4) means that the water is generally unsatisfactory for irrigation.

The sodium hazard increases as the total salt concentration increases. Doubling the concentration of each cation in the above calculations would result in an SAR of 2.06 rather than 1.46. In this way, SAR provides a more reliable indication of sodium hazard than the soluble sodium percentage (SSP) calculated by

$$\text{soluble } Na^{+} \times 100 \div \text{total cations}$$

which is constant regardless of total salt concentration. Generally, as SAR increases, so does the exchangeable sodium percentage (ESP) (see Chapter 4 under Salted Soils).

With high concentrations of the bicarbonate ion (HCO_3^{-}) in irrigation water, there is a tendency for calcium and magnesium to precipitate, with a resultant increase of SAR in the soil solution and ESP on the exchange sites. With a residual $NaHCO_3$ concentration greater than 2.5 mEq/l the water is not considered suitable for irrigation purposes. Below 1.25 mEq/l it is probably safe.

Boron is an important minor nutrient, but where its concentration in irrigation water exceeds one part per million (1 ppm) it may be toxic to turfgrasses. Since significant concentrations of boron frequently occur in various water sources, the water should be checked to ensure that toxic amounts of boron are not present.

Other potentially toxic elements in some water sources include high concentrations of various minor nutrients along with other ions such as chromium, nickel, mercury, and selenium. These are most likely to occur in some effluent waters from municipal and industrial sources.

Various organic and inorganic particles may be suspended in water sources, especially those with moving water such as streams and rivers. These should be filtered where possible to avoid damage to irrigation system components. Also, silt and clay particles applied to some soils can seal the surface and reduce infiltration.

Water Sources

A water source should be adequate to provide sufficient quantities of water during the entire growing season. It should be a dependable source that is preferably not subject to regulation by local governmental agencies. Total reliance on city water may result in less-than-adequate supplies during extended drought periods when water demand is greatest.

The three principal sources of irrigation water are groundwater from wells, stationary surface bodies (lakes, reservoirs, ponds), and flowing surface bodies (rivers, streams). A fourth source of growing importance is effluent water from municipal treatment facilities.

Where it has been determined that sufficient groundwater exists, wells can be developed to provide an independent source of water for irrigation. Well water is usually desirable because of freedom from weed seeds, pathogenic organisms, and various organic constituents. Its quality is fairly uniform over time, in that the concentration of salts and other constituents does not change appreciably during the growing season. Tests for quality should be made to determine if a particular well-water source is suitable for irrigation. Concentrations of sodium, bicarbonate, boron, and chloride may present problems if excessive. With increasing competition for groundwater in some areas, levels may drop and wells may dry up during extended droughts.

Large rivers are reliable sources of water, but the level of pollution may preclude satisfactory use of this resource. Small rivers and streams can be diverted to create local reservoirs as irrigation sources. Annual flow must be sufficient to satisfy needs, and legal restrictions must be thoroughly investigated to ensure that the source is practical. Flowing water often carries particulate matter, which must be screened to prevent clogging of irrigation systems. Where practical, impounding can allow settling of suspended particles prior to actual usage.

Golf courses and some other large turfgrass facilities often have small lakes or ponds that can serve as excellent sources of irrigation water. Where suitably located, these may be fed by springs, surface drainage, and precipitation, and supplemented as needed by city water. Care must be exercised so that stationary water bodies do not become polluted or overrun by algae and aquatic weeds.

Irrigation Systems

The function of an irrigation system is to supply water in sufficient quantity to sustain turf under a specific set of environmental conditions. Ideally, the system should provide uniform application of water at a rate that does not exceed infiltration capacity. An irrigation system's capacity to perform its function is limited by its design, construction, and operation. A poorly designed or improperly installed system can never be satisfactory unless errors are corrected; however, even a "perfect" system is only as good as the person operating it. An irrigation system, regardless of its sophistication or cost, is merely a cultural resource; its use must be managed in a fashion consistent with sound principles of turfgrass culture.

The components of an irrigation system include sprinklers, piping, control systems, valves, and pumps. Components of the system must be mutually compatible and of sufficient capacity to perform adequately.

Sprinkler Heads

A sprinkler head directs water to a specified area of turf. Two types of sprinkler heads are used for irrigating turf: the spray head and the rotating stream head (Figure 5.19). Spray heads discharge water in all directions and are used for narrow areas of turf, as well as for gardens, flower beds, and woody ornamental plantings. Stream heads direct water through one or several nozzles while rotating. The size of the area covered by stream heads is relatively large, while the volume of discharge and consequent precipitation rate are low compared to nonrotating, spray heads. Rotating stream heads are by far the most widely used sprinklers in turfgrass irrigation. The principal drive mechanisms used to effect rotation of the nozzles are of two types: impact and gear. Impact-drive sprinklers employ a spring-loaded drive arm that forces rotation of the nozzle assembly. The water stream deflects the drive arm sideways while the spring pulls the drive arm back to the nozzle assembly, causing the nozzles to rotate slightly. Repeated impact provides continuous rotation of the nozzle assembly. In gear-driven sprinklers pressurized water causes spinning of a high-speed rotor. The rotor is connected to a gear train to reduce rotational velocity and produce a turning torque that slowly and smoothly rotates the sprinkler assembly.

The typical pattern of water distribution from rotating sprinklers is wedge shaped at suitable operating pressures. During rotation the area covered at each increment along the radius increases as the distance from the sprinkler increases. Therefore, the amount of water applied per unit area decreases toward the periphery of the area of coverage (Figure 5.20). Where two adjacent heads are properly spaced their overlapping patterns can provide nearly uniform coverage along a line drawn between the heads. On either side of this

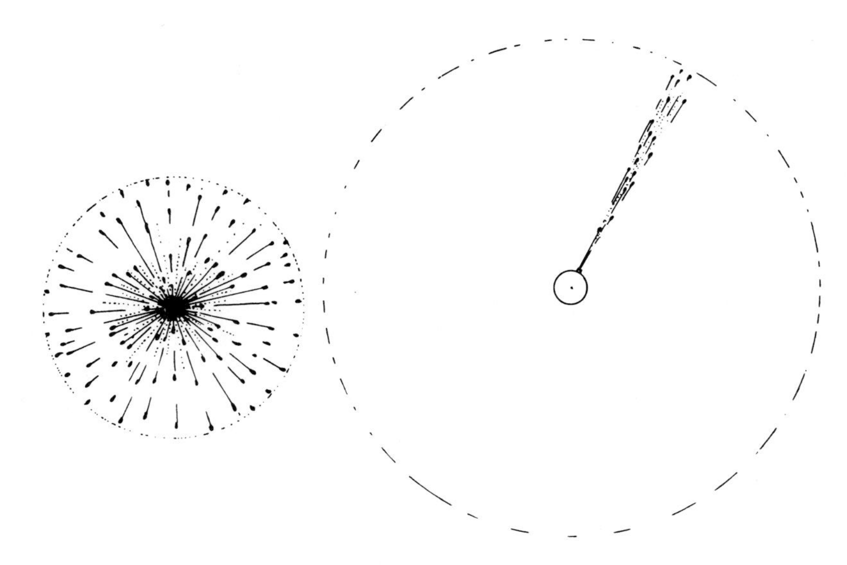

Figure 5.19. Comparison of spray head (left), which shoots water in all directions simultaneously, and rotating stream head (right), which directs water in one or several streams across a longer radius. Spray heads are used more for ornamental plantings, while stream heads are standard for turf.

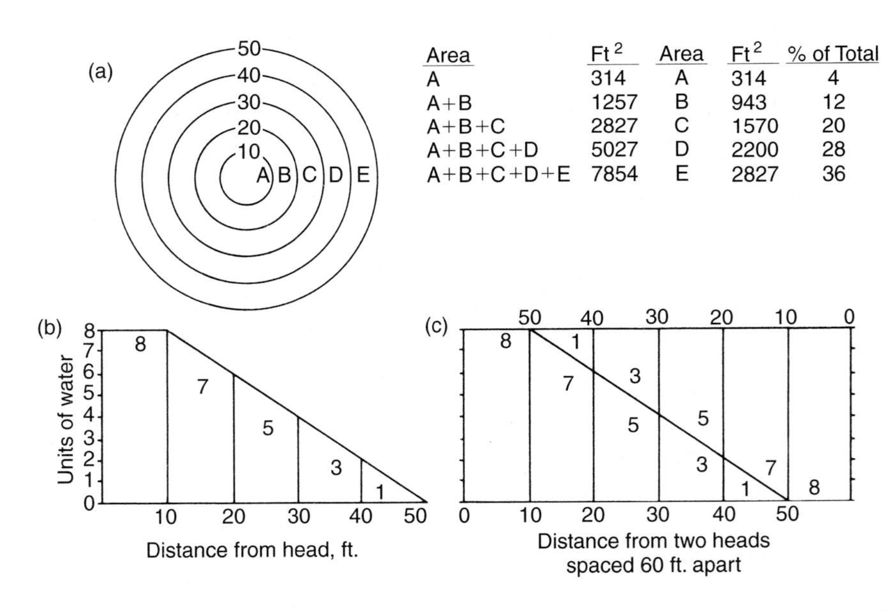

Area	Ft^2	Area	Ft^2	% of Total
A	314	A	314	4
A+B	1257	B	943	12
A+B+C	2827	C	1570	20
A+B+C+D	5027	D	2200	28
A+B+C+D+E	7854	E	2827	36

Figure 5.20. Descending precipitation rates from a rotating sprinkler along the radius of cover are due to the increasing areas per unit of radius from the head to the circumference (A → E), as illustrated in (a). Decreasing units of water applied as the basis for overlapping coverage (c) to achieve uniform distribution between adjacent heads.

line irrigation coverage decreases as distance from the line increases. Thus, a fairway with a single line of sprinkler heads along its center will receive the greatest amount of water in the middle of the fairway and decreasing amounts toward the sides. To provide sufficient moisture to sustain the turf along the sides of the fairway, the middle must be irrigated in excess of its needs.

This type of distribution, common on golf course fairways, can have at least two adverse influences. In sandy soils of rapid water percolation, nutrients can be leached faster where more water is applied, thus influencing turfgrass nutritional responses. In fine-textured soils the potential for compaction from traffic is greater along the more heavily irrigated center of the fairway than along the sides. Nonuniform distribution of water can also induce differentials in disease incidence, persistence and efficacy of pesticides, and physical condition of the turf as it influences play or use. Often, these problems may not develop (or they may not be apparent) because of the modifying effects of natural rainfall, formulation and plant uptake of fertilizer nutrients and other materials applied to the turf, timing and intensity of traffic, and so on. However, the efficiency with which water is used by the turfgrass is certainly reduced where nonuniform distribution of water by an irrigation system necessitates applying excessive amounts to a portion of the area.

Irrigation systems that include multiple rows of properly spaced sprinkler heads offer considerable advantages for achieving uniform application of water. By combining overlapping coverages from heads in line and heads in adjacent lines, an entire fairway, lawn, or other turf can receive maximum uniformity of irrigation.

Among factors that influence the distribution of water from a sprinkler, the most important are nozzle size, pressure, head spacing, and wind. Increasing nozzle size allows increasing amounts of water to be discharged from the sprinkler. Generally, larger nozzles are used to achieve more area coverage per sprinkler and, therefore, reduce the number of sprinklers required per unit area of turf. Proper operating pressure is essential for achieving distribution patterns that can be overlapped for uniform coverage (Figure 5.21). Inadequate pressure results in streams of water that do not break up sufficiently for wedge-shaped distribution. In extreme cases a donut-shaped pattern occurs in which most of the water is confined to a ring surrounding the sprinkler. A series of green rings appears with brown or dormant turfgrass occurring inside and outside each ring. Excessive pressure causes fine atomization of the spray so that the small water droplets are highly susceptible to wind drift. The resultant pattern is irregular and unpredictable.

Given a wedge-shaped pattern from sprinklers operating under correct pressure, proper spacing of sprinkler heads is essential for uniform coverage. Spacing heads too far apart results in less water being applied to areas at midpoint between sprinklers. Similarly, overspacing causes excessive coverage in the immediate vicinity of each sprinkler. Commonly used designs include triangular and square spacing (Figure 5.22).

Wind cannot be controlled by sprinkler construction or system design. The distorting effects of wind are directly proportional to wind velocity. They include a shortened radius of coverage and increased precipitation rate on the windward side of a sprinkler, longer radius of coverage and decreased precipitation rate on the leeward side, and shortened radius of coverage where the stream of water is at right angles to the wind.

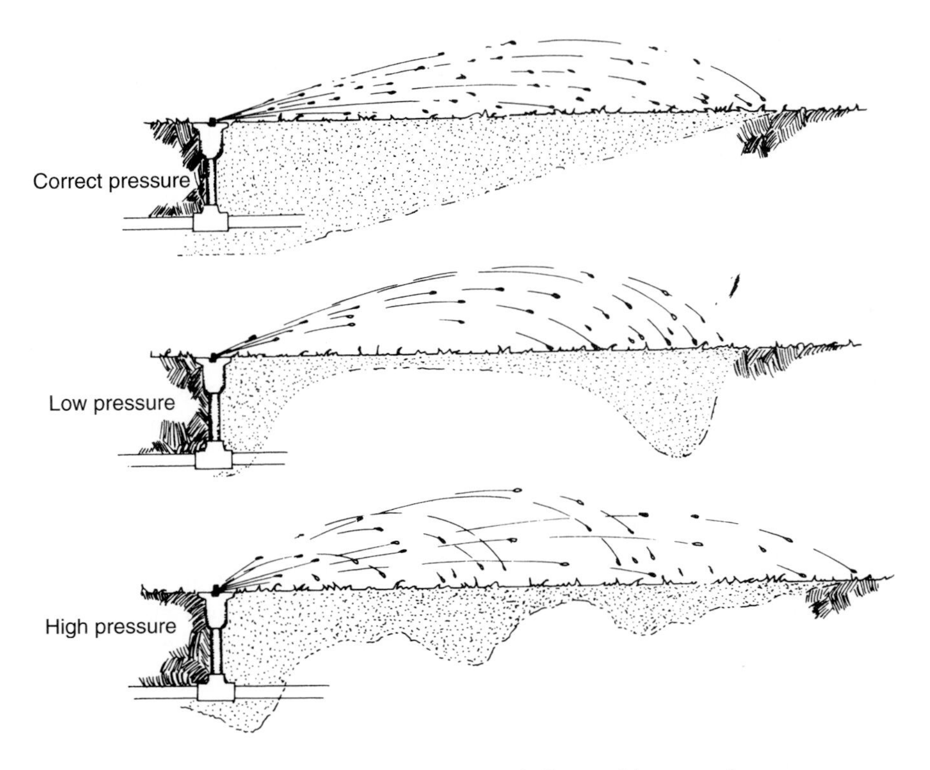

Figure 5.21. Comparison of irrigation patterns as influenced by operating pressure.

One advantage of nighttime irrigation is that wind velocities tend to be lower, thus reducing the distortion of irrigation patterns.

Piping

Piping is the foundation of a turfgrass irrigation system. Pipes transport water from the source to the sprinkler head for distribution to the turf. Proper functioning of an irrigation system, therefore, depends on the type, size, and condition of pipes and fittings used for transporting water. In earlier years metal (cast iron, copper, galvanized steel) pipes were used exclusively. With increased costs and corrosion problems encountered with iron and steel, a transition occurred in the early 1960s to thermoplastic materials; today nearly all pipes used are either PVC (polyvinyl chloride) or PE (polyethylene). In addition, some A-C (asbestos cement) pipe is still used for mains and other large piping in irrigation systems. Where metal pipes were welded, soldered, or threaded at joints, the thermoplastic piping materials are solvent welded (PVC) or clamped to plastic inserts (PE) (Figure 5.23). PVC is a rigid pipe used primarily for lines measuring from 2 to 10 inches in diameter. PE is flexible and is used for 2-inch and smaller lines. Both

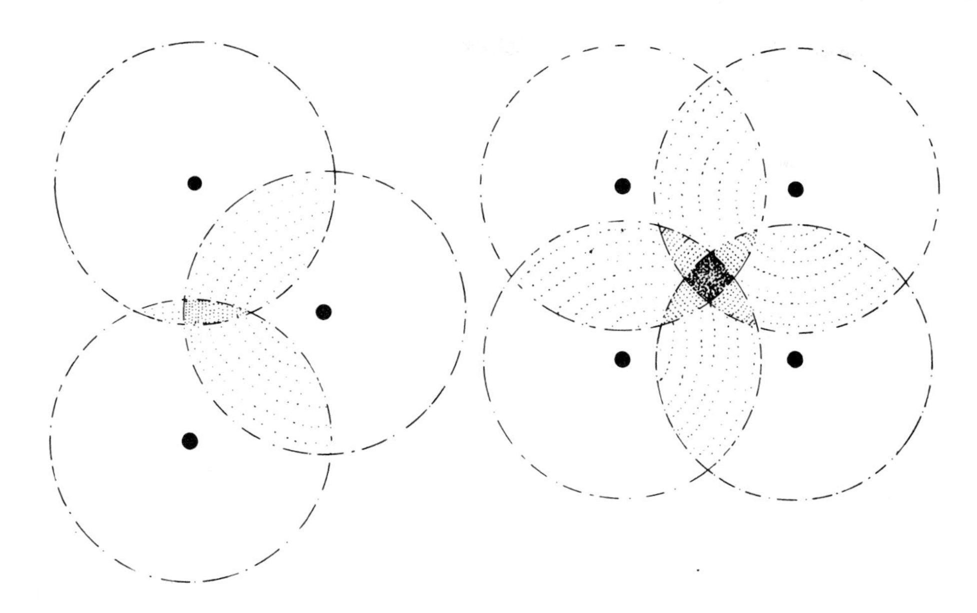

Figure 5.22. Triangular (left) and square (right) spacing of sprinkler heads.

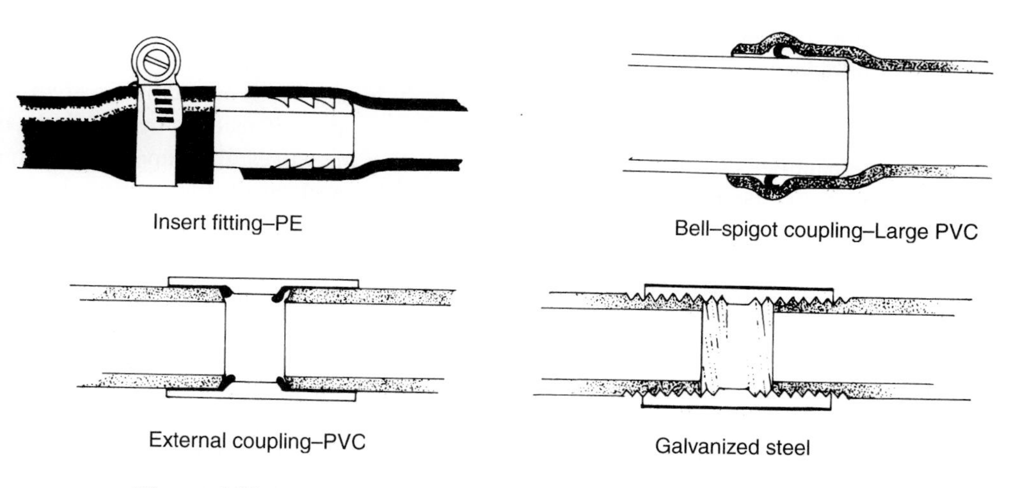

Figure 5.23. Types of fittings used for metal and thermoplastic pipes.

are lightweight, relatively inexpensive, and resistant to corrosion and rust. One disadvantage of thermoplastic pipe is lower strength compared to metal pipe. Where excessive pressure occurs thermoplastic pipes can stretch (creep), causing thinning of the pipe walls. Under extreme conditions the pipes could rupture. However, proper system design virtually eliminates these problems. When water is allowed to remain in pipes during winter in cold climates, freezing could also cause rupture. Therefore, pipes must either be installed below freezing depth or drained prior to cold weather.

Control Systems

Controllers and remote-control valves are included in control systems. Fully automatic irrigation requires control systems that provide sufficient flexibility and control over irrigation programming. Manual irrigation systems do not require controllers; water distribution is actuated by keys or quick-coupler valves.

Controllers are designed for mounting on building walls or free-standing pedestals. They incorporate a clock, a timer, and a series of terminals called *stations*. The clock turns on a timer which, in turn, energizes the terminals in sequence to actuate remote-control valves. The valves, positioned at individual heads or along a pipe connected to several heads, control the flow of water downstream. Thus, when the valve connected to station five is actuated by the controller the downstream head or heads connected to valve five begin sprinkling. The time at which sprinkling is initiated and terminated is set at each station in the controller. The entire progression of energizing stations is called the *cycle*. At the end of each cycle the controller is automatically reset for the next scheduled operation.

Remote-control valves can be actuated by two methods: hydraulic and electric. Hydraulic actuation is of two types: normally open and normally closed. In normally open hydraulic valves pressure from a hydraulic tube forces the valve to close; when pressure is released by action of the controller the valve opens to allow water through to the sprinkler heads. In normally closed hydraulic valves pressure from a hydraulic tube forces the valve to open, allowing water through. In both types of hydraulic valves water in the hydraulic tubes comes from a supply independent of that for irrigation. As only small quantities of water are required for actuating the valves, this supply can be maintained free of debris that might otherwise clog the mechanism. In contrast, electrically operated valves utilize irrigation water for operation. Thus, any debris contained in the water can clog the valve mechanisms. Electric valves are normally closed; current from the controller flows through the solenoid of the valve to open the valve and allow water flow to the sprinklers.

Pumps

Two types of pumps are used in irrigation systems: booster pumps and system-supply pumps. As its name implies, a booster pump boosts in-line water pressure, but without affecting flow rate. For example, a pump rated at 20 gallons per minute (gpm) and 20 pounds per square inch (psi) can be installed in a line carrying water at 20 gpm and 30 psi to boost the pressure to 50 psi. Booster pumps are used in domestic lines with inadequate pressure and in large irrigation systems where pressure losses due to elevation must be compensated for.

System-supply pumps pull water from a water source to supply a specific flow rate at required pressures. Pumps in series provide pressure that is equal to the sum of each pump's capacity, while flow rate is equal to that of a single pump. Pumps in parallel arrangement provide pressure equal to that of a single pump, but flow rates are additive. The suction inlet for the pumps has screens and filters to remove particulate matter from the water. In addition, check valves may be used at the suction inlet to prevent air movement into the pumps and consequent loss of prime when the water level drops below the inlet.

Valves

In addition to the remote-control valves used in automatic systems, numerous other valve assemblies are used in irrigation systems. *Manual valves* are used to isolate sections for repair and drainage of the systems. Some irrigation systems have a *master manual valve* for closing off water flow into the entire system. *Check valves* are used to limit water flow to one direction; suction inlets often require check valves to prevent loss of prime, and sloped pipelines may have check valves to prevent water in the lines from draining out the low head of the section. *Antisyphon valves* are used to protect domestic water supplies against back-flow when irrigation is terminated. *Drain valves* are used in cold climates to allow drainage from lines after irrigation. During operation of the system automatic drain valves are kept closed by water pressure; upon completion of an irrigation cycle, the valve opens to allow drainage. Manual drain valves are used for distribution lines that are continually under pressure; when the system is winterized the valve is opened to drain the pipes. *Pressure-regulating valves* are used to prevent excessive water pressure. In low areas where elevation boosts the pressure too high or at positions close to booster pumps, pressure-regulating valves prevent excessively high pressures from adversely affecting spray patterns from the sprinkler heads.

Manual irrigation systems usually employ *quick-coupler valves* at each sprinkler position. The base of the sprinkler is fitted with a key so that when the sprinkler is inserted into the valve and turned the valve is opened to allow water flow through the sprinkler. Quick-coupler keys can also be screwed into hose endings for irrigating with portable sprinklers.

Irrigation systems must be properly designed and constructed for satisfactory performance. Selecting the correct sprinklers, pipes, valves, pumps, and control systems, along with designing the system, requires a competent irrigation designer. Often a turfgrass manager can assist the designer by clearly establishing the performance criteria desired, but the engineering of the actual system should be left to a professional who understands the capabilities and limitations of system components and who is knowledgeable about the complex and rapidly evolving technology of irrigation design.

QUESTIONS

1. List and compare the types of **mowers** typically used in turfgrass culture.
2. Explain **marcelling** (provide a geometric analysis as part of your answer).

3. List the measurable differences in turfgrass morphology, physiology, and cultural requirements associated with high versus close **mowing** of a turfgrass (but well within its mowing-tolerance range).
4. What effects would increased mowing **frequency** have on turf quality?
5. Provide arguments in favor of returning (versus removing) turfgrass **clippings** during mowing operations.
6. List all the major, secondary, and minor **nutrients** required to support turfgrass growth.
7. Explain the **analysis** of a fertilizer.
8. List the effects of **excessive** nitrogen fertilization on a turfgrass community.
9. Differentiate between quickly available and slowly available nitrogen **carriers** with respect to turfgrass response.
10. Explain what is meant by the following terms in relation to **slowly available** nitrogen carriers: CWSN, CWIN, activity index, and seven-day dissolution rate.
11. Illustrate the possible avenues of nitrogen **fate** in turf.
12. Explain the factors affecting the availability of **phosphorus** and **potassium** in turf soils.
13. What would be the likely results from excessive use of **micronutrients** in a turf fertilization program?
14. Explain the factors influencing the plant-available **moisture** pool in a turf soil.
15. What equipment and procedures can be used to predict turfgrass **irrigation** requirements?
16. Provide arguments for and against **nighttime irrigation** of turf.
17. Differentiate between **precipitation rate** and **infiltration capacity,** and explain their relevance to turf irrigation.
18. Explain the factors influencing the **quality** of turf irrigation water.
19. What are the important considerations in developing and operating **a turf irrigation system?**

CHAPTER 6

Supplementary Cultural Practices

It is sometimes necessary to supplement primary cultural practices with additional operations to sustain turf at a desired level of quality. Supplementary cultural practices become necessary when problems arise, or are anticipated, because of unfavorable developments in the turf.

Excessive thatch development, soil compaction, grain, and convolutions in the turf's surface are conditions that can often be remedied by cultivation, topdressing, rolling, and other practices not included in primary culture. As the effects of these practices are largely confined to the surface 1 to 3 inches of the turf, they may not be completely corrective where highly unfavorable soil conditions exist. There is no easy substitute for proper site preparation prior to planting turfgrasses; however, surface problems can be substantially reduced by timely and properly implemented practices described in this chapter.

CULTIVATION

Cultivation refers to mechanical methods of selective tillage to modify physical and possibly other characteristics of a turf. Cultivation practices are primarily designed to reduce problems associated with excessive thatch and soil compaction; however, some practices are also effective in reducing grain in greens. The principal types of cultivation are coring, drilling, slicing, spiking, vertical mowing, and water-injection cultivation.

Coring

Coring or core cultivation is the practice by which hollow tines or spoons are used to extract cores from the turf (Figure 6.1). Core size varies from 0.25 to 1.0 inch in diameter, depending on the size of the tine or spoon. The vertical length of the cores varies with soil strength and penetration capacity of the coring apparatus. Since soil strength is proportional to bulk density and moisture content, increasing soil moisture facilitates deeper penetration of the tines or spoons. Core lengths of 3 inches are not uncommon where moisture conditions favor deep penetration; longer lengths are possible with high-powered units.

Many core cultivators are commercially available. However, all fall into one of two types: vertical-motion units with hollow tines and circular-motion units with open spoons or hollow tines (Figure 6.2). On greens and other closely mowed turfs the vertical-motion core cultivator provides deep penetration of the turf with minimal surface disruption. Because of the necessity for linking vertical and forward operations, these units are relatively slow, requiring approximately 10 minutes of operation per 1000 ft^2 of turf.

Circular-motion core cultivators employ tines or spoons mounted on a hollow drum or a series of disks or directly mounted on a rotating, horizontal shaft. These units are considerably more efficient in covering the turf, but surface disruption is moderate to severe, and penetration depth is shallower compared to vertical-motion types.

Coring is often called *aerification;* this term implies that soil aeration is directly improved by the procedure. Certainly the coring hole is "well aerated," as all soil has

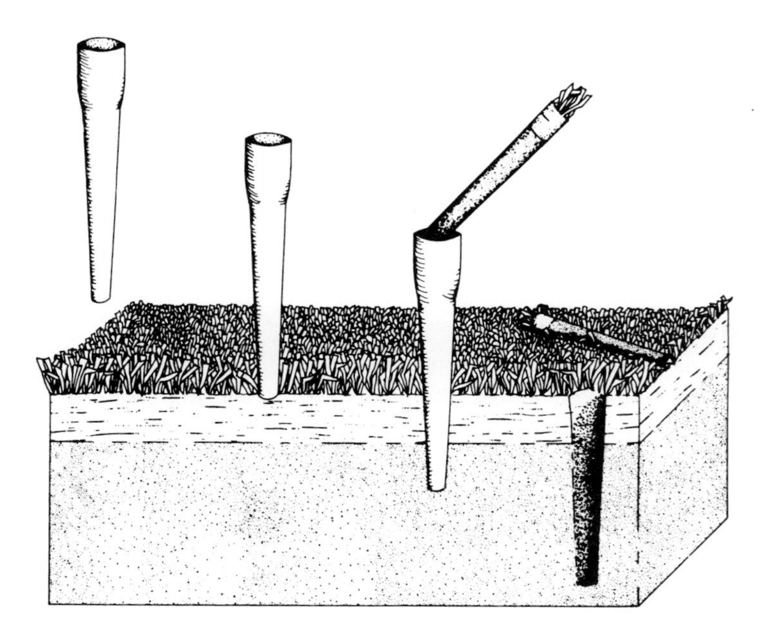

Figure 6.1. Core cultivation.

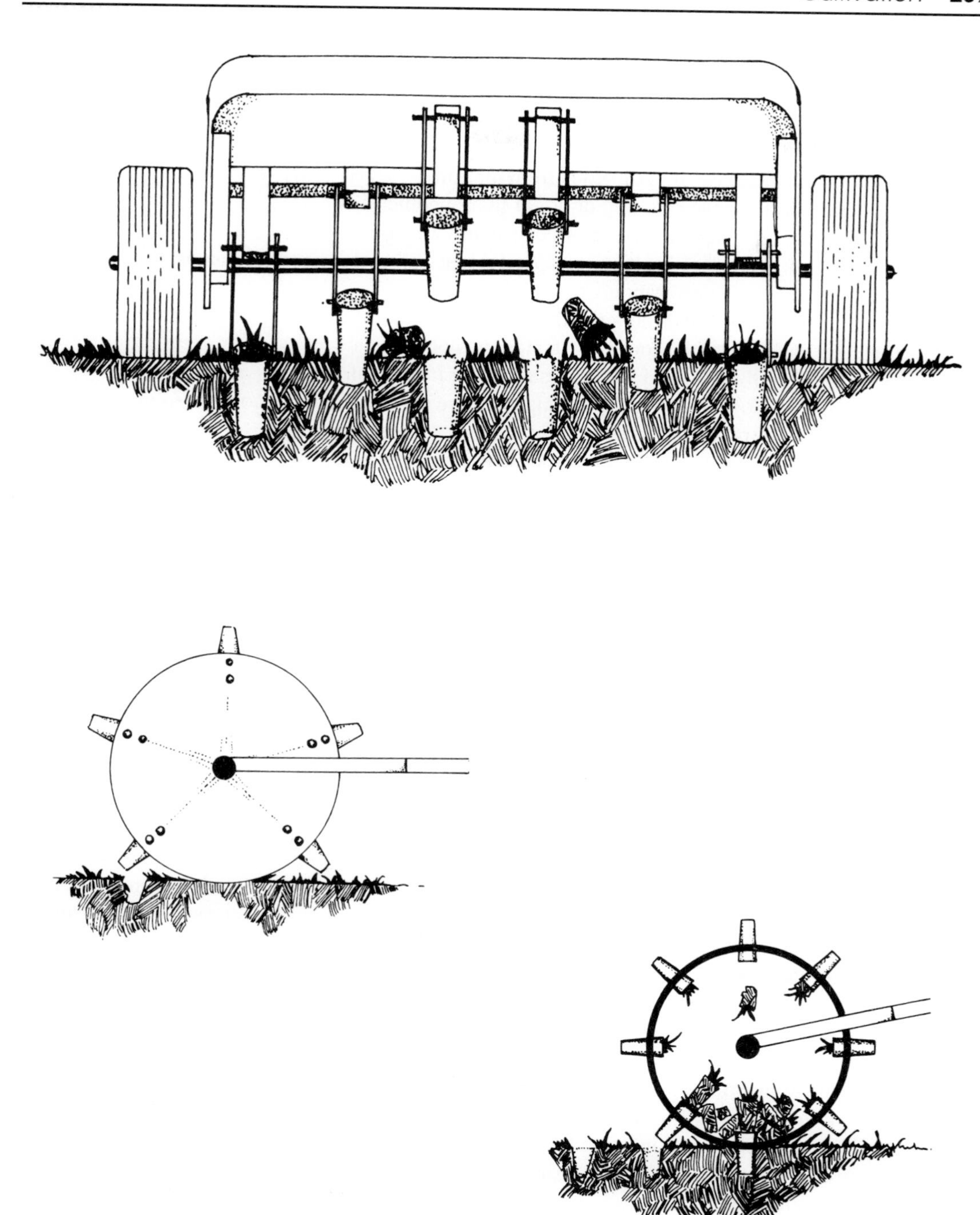

Figure 6.2. Vertical-motion (top), disk-mounted (left), and drum-mounted (right) core cultivators.

been removed. However, aeration of the soil remaining between and below the holes may not be improved. In fact, some additional compaction results from movement of the tines along the sides of the hole and at its base. Benefits of coring include:

1. Release of toxic gases from the soil
2. Improved wetting of dry or hydrophobic soils
3. Accelerated drying of persistently wet soils
4. Increased infiltration capacity, especially where surface compaction or thatch limits infiltration
5. Stimulated root growth within the holes
6. Increased shoot growth atop the holes
7. Disruption of soil layers resulting from topdressing
8. Control of thatch, especially where soil cores are reincorporated or where topdressing follows coring
9. Improved turfgrass response to fertilizers

Observed disadvantages of coring include:

1. Temporary disruption of the turf's surface
2. Increased potential for turfgrass desiccation as subsurface tissues are exposed
3. Increased weed development when conditions favor weed-seed germination
4. Increased damage from cutworms and other insects that reside in the holes

On severely compacted greens a series of circular tufts of bright green grass often develops following coring. Thus, coring can result in an improvement of growing conditions in the immediate vicinity of the holes as long as moisture is not limiting. With repeated operations over several years, localized improvements may become generalized throughout the turf. Under drying conditions, however, severe turfgrass desiccation can result. In the author's experience an entire green of creeping bentgrass was lost following midsummer coring on a hot, dry day. The preferred timing of core cultivation is when the turfgrass is growing vigorously and is not subjected to severe stress from atmospheric conditions. Topdressing and irrigation immediately following coring can reduce desiccation potential, but may not be effective in preventing it entirely under stressful conditions.

The increased infiltration capacity of core-cultivated turf is due primarily to the increased surface area resulting from coring. As shown in Figure 6.3, the increase in surface area can be more than twofold when the combined areas of the walls of the holes are added to that of the turf's surface. If the infiltration capacity of the exposed soil is greater than that of the turf's surface, the net increase in infiltration may be greater than that predicted on the basis of increased surface area. Surface compaction, hydrophobic thatch, and surface layering can severely restrict infiltration; if the coring hole traverses these restrictions, substantial increases in infiltration capacity are possible in direct response to coring.

Fertilizer materials applied to turf vary in their mobility within the soil. Lime and phosphorus are highly immobile and, therefore, may move downward very slowly in the soil profile. Furthermore, the fate of surface-applied nitrogen can be quite different from

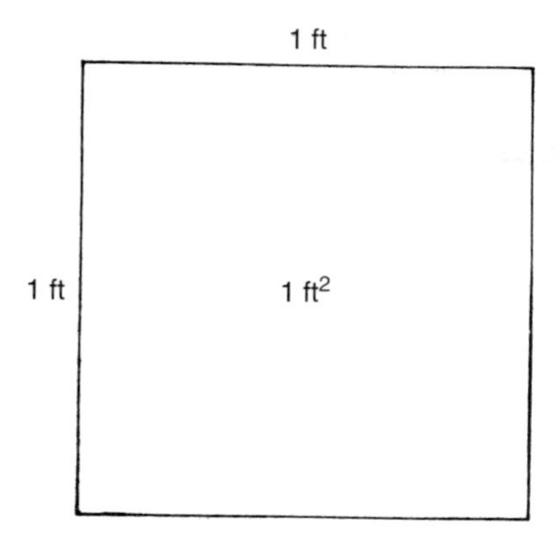

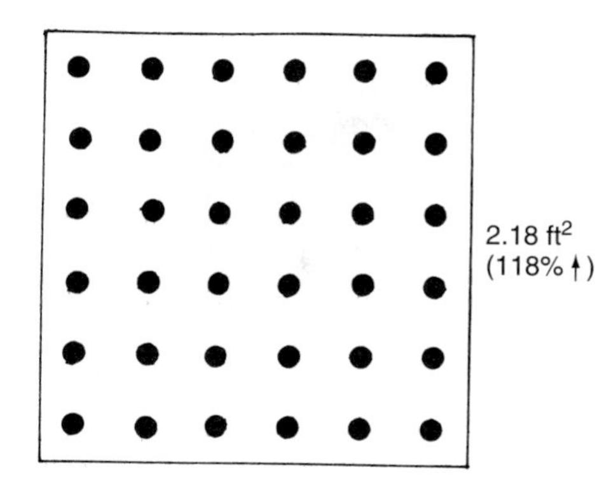

Figure 6.3. The total surface area of a turf following core cultivation to a depth of 2 inches with 0.75-inch-diameter tines on 2-inch centers increases 118 percent.

that of nitrogen applied following coring. At lower soil depths gaseous loss may be reduced, and solubilization of slowly soluble forms may be more uniform. Coring, then, provides a means of inserting fertilizers below the turf's surface for more favorable turfgrass response.

When coring holes are left open they are soon filled by turfgrass roots and adjacent soil. Under traffic and irrigation, sand particles quickly flow into the holes. Wet soil is also susceptible to lateral flow, and redistribution is accelerated by traffic. Thus, many of the beneficial effects of coring may be short-lived unless the holes are filled with suitable materials. Reincorporation of the soil cores is often practiced on large sites because of the logistical problems associated with their removal. Left in place, these soil cores eventually break down and the soil particles flow into the turf. However, most of this soil is not returned to the holes; rather, it adds to the surface soil or becomes integrated with the thatch. Matting or vertical mowing operations can hasten breakdown of the cores and provide a more uniform distribution of the soil over the turf (Figure 6.4). Soil reincorporation following coring has a dramatic influence on thatch because its edaphic properties are altered in such a way that it becomes a more suitable medium for supporting turfgrass growth. The thatchlike derivative thus formed (mat; see Chapter 4) is characterized by:

1. Higher cation exchange capacity (on a volume basis) and improved nutrient retention
2. Improved water retention
3. Accelerated decomposition rate of the organic residue
4. A firmer surface for sports and recreational activities

Topdressing is sometimes performed following core cultivation. Where the topdressing soil differs from that present in the turf the surface medium supporting the turfgrass is changed. With the application of a sufficient volume of material, the coring holes

Figure 6.4. The effects of shallow vertical mowing in breaking up soil cores following core cultivation.

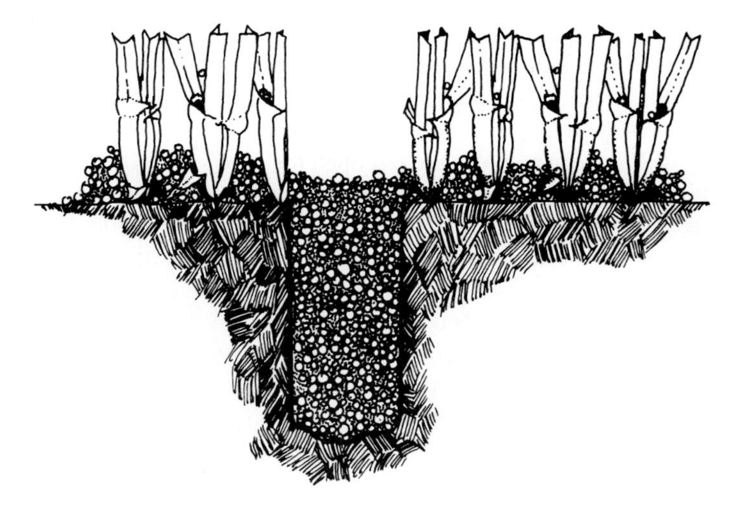

Figure 6.5. Topdressing-soil placement following core cultivation and removal of soil cores.

can be completely filled while the thatch may be thoroughly infused with the topdressing material. In this way, coring holes provide a means of bypassing some surface problems to link the underlying soil with that provided through topdressing (Figure 6.5).

Intensive core cultivation is used to prepare a turf for overseeding. This may or may not be done in conjunction with the application of a nonselective herbicide for total vegetation control (see Chapter 8, Renovation and Reestablishment). Germination and seedling growth of overseeded turfgrasses is enhanced by coring to expose soil and open holes within which seedlings can develop. The development of weeds from seeds present in the turf can also be aided by coring. For this reason, the timing of core cultivations

Figure 6.6. Illustration of drill cultivator with apparatus for backfilling holes with sand.

should be such that weeds of primary concern (annual bluegrass, crabgrass, and others) are not favored. In temperate climates annual bluegrass invasion into closely mowed turfs may be favored by early- and late-season cultivations; similarly, late spring and early summer coring can favor the development of crabgrass and other summer annuals except where the turfgrass is so vigorous that it quickly fills the coring holes.

A variation of coring, or hollow-tine cultivation (HTC), is solid-tine cultivation (STC). While similar to the coring procedure, solid-tine cultivation does not remove soil cores; thus, the potential exists for even more compaction of the soil along the sides and at the bases of the holes. Some solid-tine cultivators can penetrate the soil to depths of up to 16 inches. Results from field trials with these units suggest that as the tines move in and out of the turf the soil may be uplifted and fractured, resulting in a direct increase in soil aeration. This is more likely to occur under relatively low soil moisture levels; in wet soils the tines will tend to compact the soil in the immediate vicinity of the holes.

Drill cultivation was introduced several decades ago but, because of the slow speed of the operation, was not accepted by the turfgrass industry. Drill cultivation has reemerged in recent years with the introduction of machines that can probe deeply into the turf, extract and deposit large quantities of soil onto the surface, and backfill the holes with sand or other coarse-textured media (Figure 6.6). By connecting suitable surface media (developed, perhaps, through intensive topdressing) to gravel blankets or other

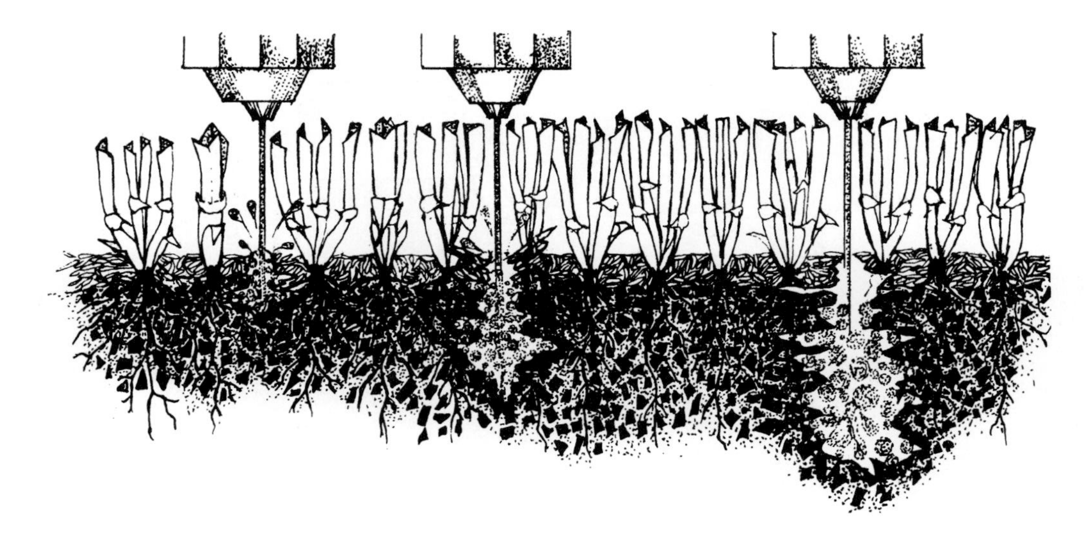

Figure 6.7. Illustration of water-injection cultivation.

porous media located 12 or more inches below the surface, channels are created that can provide "bypass" drainage for conducting large quantities of water through the turf. The number of channels required to adequately improve drainage and turf quality would vary depending on the size of the channels and the severity of the drainage problem. While this could be determined empirically, a 4- to 6-inch spacing would probably be needed for satisfactory results.

Water-Injection Cultivation

Water-injection cultivation (WIC) is a revolutionary concept introduced by the Toro Company in 1990 for improving infiltration and alleviating soil compaction in turf (Figure 6.7). Unlike other cultivation methods, it causes essentially no surface disruption nor any significant damage to turfgrass communities; therefore, it may be employed virtually anytime during the growing season. It works by injecting 10-millisecond pulses of highly pressurized (5000 psi) water into the turf at speeds of up to 600 miles per hour. Results from field trials in Michigan showed improved shoot growth along with reduced bulk density and increased saturated hydraulic conductivity in soils underlying WIC-treated turfs.

Slicing and Spiking

Alternative forms of cultivation include slicing and spiking. Slicing is the process by which a turf is penetrated to a depth of 3 to 4 inches by a series of V-shaped knives mounted on disks (Figure 6.8). Unlike coring, there is no removal of soil cores; therefore,

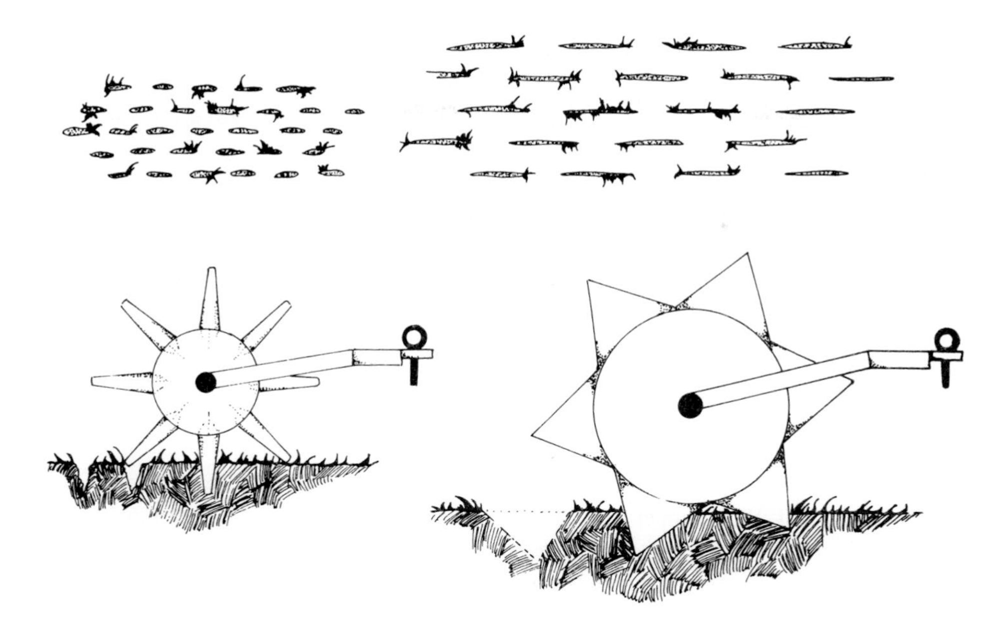

Figure 6.8. Spiking (left) and slicing (right). With spiking, shallow perforations of the turf are made with small knives, whole slicing produces deeper and longer cuts into the turf.

disruption of the turf is minimal. Spiking is a similar process except that penetration is limited to approximately 1 inch and the length of the perforations along the turf's surface is shorter. Slicing is typically practiced on fairways and other heavily trafficked turfs during midsummer stress periods when coring might be too injurious or disruptive. Spiking is primarily practiced on greens. As with coring, slicing and spiking are used to improve infiltration, especially where surface compaction is severe. Because of the severing of stolons and rhizomes, these types of cultivation often result in stimulated root and shoot growth in the immediate vicinity of the perforations. Many turfgrass managers consider slicing and spiking as practices to be employed between core cultivations for achieving similar but less dramatic results. Since they cause only minor disruptions of the turf, they can be practiced as often as weekly during the growing season to mitigate the soil-compacting effects of traffic.

Vertical Mowing

Vertical mowing is a cultivation procedure involving the use of vertically oriented knives mounted on a rapidly rotating, horizontal shaft. Depending on the penetration depth of the knives, different objectives can be met. When the knives are set to just nick the turf, stolons and decumbent leaves are severed to reduce grain on greens (Figure 6.9). Following coring, shallow vertical mowing can break up the cores to facilitate reincorporation of

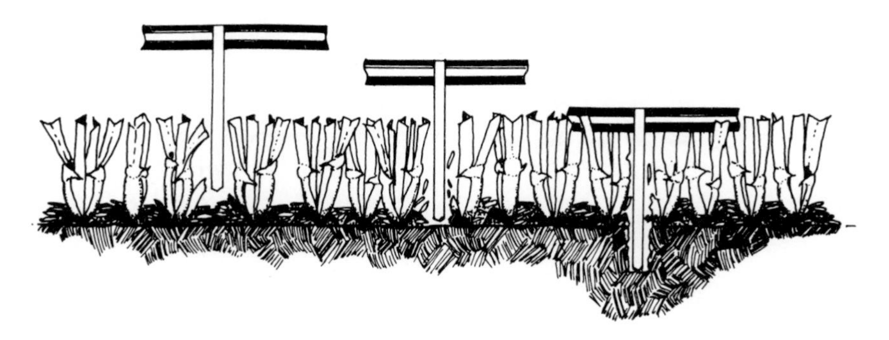

Figure 6.9. Depending on the depth to which vertical-mower blades penetrate the turf, the effects may include grain control (left), thatch removal (middle), and soil cultivation (right).

the soil. With deeper penetration of the knives much accumulated thatch can be removed. Setting the knives to penetrate still deeper results in a cultivation of the underlying soil to alleviate surface compaction. Other names for vertical mowing include power raking and dethatching.

Shallow vertical mowing can be accomplished with a variety of equipment, including triplex greens mowers with interchangeable units that enable the operator to exchange the reels for vertical mowing units. In some small units for home lawns, flexible wire tines are used instead of knives for extracting debris from the turf. Shallow vertical mowing is sometimes employed where earthworm casts disrupt the turf's surface in early spring. When dry the casts are easily distributed over a larger area to smooth out disruptions. In addition to standard vertical mowers, specialized riding units are available with large-capacity catch-boxes that permit simultaneous extraction and removal of organic debris in dethatching operations.

Mechanical removal of thatch can be very destructive to shallow-rooted turfs. Some preliminary testing at the site is recommended prior to thatch removal. Adequate rooting across the thatch-soil interface stabilizes the turfgrass and permits substantial removal of organic debris without uprooting considerable numbers of live plants. The number and distribution of surviving vegetative propagules determine the recuperative capacity of the turfgrass following severe vertical mowing. When the penetration depth of the vertical-mowing knives extends beyond the thatch-soil interface, substantial amounts of soil may be pulled up and into the remaining thatch. The result is similar to soil reincorporation following coring; the edaphic properties of the thatch are altered and decomposition is accelerated. Extracted piles of debris should be removed as soon as possible following vertical mowing to avoid light-exclusion effects, especially during warm weather.

As with coring, the preferred timing of vertical mowing is when turfgrass growth is vigorous, atmospheric stress is minimal, and a sufficient period of favorable growing conditions will follow for recovery of the turf. In temperate climates late summer or early fall is usually preferred for vertical mowing of cool-season turfgrasses. For warm-season turfgrasses late spring or early summer is preferred.

Figure 6.10. A disk seeder with the vertical-mowing unit positioned toward the front and the disk and seeding unit at the rear.

Soil and thatch should be dry when vertical mowing is practiced to minimize the disruption of the turf and facilitate effective cultivation of the soil.

Deep vertical mowing is often practiced as part of a turfgrass renovation procedure. Several vertical mowings can result in a satisfactory seedbed for interseeded turfgrasses. Seeding is sometimes performed prior to removing extracted debris from the surface. Subsequent matting of the area tends to "ribbon" the debris to facilitate its removal while leaving most of the seed in place. The presumed advantage of this sequence is that the seed can be worked into openings in the turf prior to their closure during the matting operation. A popular contemporary practice is the use of a disk seeder, which incorporates a vertical mower and seeder in one unit (Figure 6.10). The vertical mower at the front of the unit cuts grooves into the turf, while the rear-mounted seeder places seed directly into the grooves. Consequently, less seed is used than with broadcasting, while results are often comparable or superior.

ROLLING

Rolling is the practice by which minor disruptions in a turf's surface are corrected. In earlier decades extensive rolling was practiced on sports turfs to create a smooth playing surface. As turfgrass technology progressed, the severe soil-compacting effects of heavy rolling were recognized, and many traditional rolling practices were abandoned. Beginning in the 1990s, however, rolling with relatively lightweight units has enjoyed a resurgence of popularity in the management of greens. Following are examples that illustrate the role that rolling plays in contemporary turfgrass management.

Rolling is important in turfgrass establishment. In the absence of a sufficient period for settling, a tilled soil can be rolled to provide a smooth, firm surface for planting. Following planting, rolling ensures favorable contact between the soil and planting materials (seed, stolons, sprigs, plugs, and sod). Where rolling is not practiced, a fluffy seedbed may show best germination where foot traffic has compressed the surface soil. Desiccation potential of vegetative planting materials is likewise reduced following rolling.

In climates with freezing winters, rolling is employed to press heaved turfgrasses back into the soil. Otherwise, these plants may desiccate or be scalped with mowing. The soil should be moist, but not wet, at the time of rolling. Dry soil resists the pressing of plants into the soil, while excessively wet soils are highly susceptible to compaction from rolling. Although surface disruptions are regarded as undesirable effects of frost heaving in cold climates, they are indicative of the beneficial effects of alternate freezing and thawing on soil structure. Root growth of cool-season turfgrasses in such soils can be extensive during the cool weather of early spring. Heavy rolling destroys soil structure and inhibits root growth and function, thus reducing turf quality throughout much of the growing season.

Where major irregularities in the turf's surface interfere with use or culture, the sod should be removed to allow regrading of the underlying soil. Following this, the sod can be replaced to establish a turf of sufficient smoothness. Moderate surface irregularities are correctable with successive topdressings. Where this measure is employed, care should be taken to secure a topdressing soil that matches as closely as possible the soil underlying the turfgrass.

Athletic field superintendents sometimes employ a light rolling following a game to press uprooted turf back into the soil and, thus, reduce further injury from desiccation. In addition, a light irrigation is also helpful in promoting the survival of injured turf.

Sod growers often roll prior to harvesting to ensure uniform thickness of the sod. With this practice, shallower harvesting depths can be employed to reduce soil removal and produce lighter sod for easier handling. Except for organic soils or sands, which resist compaction, rolling should not be too heavy, as this may impede sod rooting at the transplant site.

Many golf course superintendents and lawn bowling managers employ lightweight rolling to enhance the playability and ball-roll characteristics of greens. Rolling is regarded as an alternative to the ultra-low mowing heights needed to sustain the fast speeds desired for championship play, especially on sand-based greens that resist compaction under traffic.

Three basic types of rollers are used in turfgrass management: drum rollers, triplex-attachment rollers, and dedicated greens rollers. Traditional drum rollers consist of one or more hollow drums attached to a handle for manual operation (Figure 6.11) or mounted to a utility vehicle. The drums may be filled with ballast to achieve the desired rolling weight. Triplex-attachment rollers are mounted onto triplex mowers in place of reels. One commercially available version utilizes vibrating roller bars for increased effectiveness. An advantage of this type of roller is that it is operated in much the same way as a triplex mowing unit and, thus, no additional training is required for personnel selected to operate

Figure 6.11. A water-ballast roller for use on turf.

this equipment. Finally, the greens roller, first introduced in the early 1990s, has among its special features a variety of roller-bar surfaces, (e.g., grooved, smooth, rough-coated, and rubber-coated) that can be selected for specific applications.

TOPDRESSING

Topdressing is the practice by which a thin layer of soil is applied to an established turf or a new turfgrass planting. When used in conjunction with turfgrass establishment its purposes are to partially cover and stabilize the planting material and to retard desiccation. On established turfs topdressing is performed for several purposes, including controlling thatch, smoothing a playing surface, promoting recovery from injury or disease, protecting greens in winter, and changing the characteristics of the turfgrass growth medium.

The topdressing requirement is a function of turfgrass genotype, natural environmental conditions, culture, and the purpose for which the turf was established. A green used for golf, bowling, and other sports must be very uniform to be suitable for play. Moderate surface irregularities that result from traffic, climatic conditions, turfgrass growth, and other causes can be smoothed out by sufficient topdressing. It is believed that topdressing originated specifically for this purpose. Golf superintendents observed that, as a result of topdressing, greens were less prone to develop thatch and associated problems. Where greens were established in unfavorable soils, repeated topdressing with a more favorable soil over many years eventually produced a superior turf with improved drainage, aeration, and resiliency. Thus, contemporary topdressing practices evolved from early attempts to improve playability and from observations of subsequent effects.

Layering

One of the commonly observed problems associated with topdressing is layering within a turf soil profile resulting from the use of different topdressing materials over time. Where a submerged layer of sand or other coarse materials exists within the profile, the soil above the layer is often persistently wet, and root growth is restricted. Even relatively small differences in soil type among layers can have significant adverse effects within the turfgrass root zone. Furthermore, where the topdressing rate and frequency are insufficient to ensure thorough infusion of soil into the accumulated thatch, alternate layers of soil and organic residue may be apparent for long periods. In some instances it may be necessary to "open up" the thatch layer by vertical mowing before topdressing in order to prevent layering.

Coring, followed by reincorporation of the core soil, produces effects similar to topdressing. As long as the soil underlying the turf is suitable, the coring-reincorporation procedure may be just as satisfactory or better than topdressing with a foreign source of soil. However, the amount of soil available for reincorporation following coring depends on coring intensity (size and number of cores) and the amount of disruption allowable in the turf. With topdressing, the amount of soil available is independent of coring intensity, and topdressing may be performed without any cultivation. Where coring and topdressing are performed together care must be exercised to ensure that layering does not result. For example, initiation of a sand topdressing program on greens usually involves intensive coring to allow sand placement in the coring holes. Cores from the holes should be removed, and no reincorporation should be practiced until the sand depth exceeds the maximum penetration depth of the core cultivator. Otherwise, successive layers of sand and soil can result, with predictably unfavorable effects.

Selecting a Topdressing Soil

Selection of a soil for topdressing is one of the most important decisions in turfgrass culture. Where the indigenous soil is favorable, the topdressing soil should be identical to it. For this reason, constructing a new green should involve stockpiling additional soil for future topdressing requirements. The only difference between a construction soil and a topdressing soil is the amount of organic amendment added. Organic amendments may not be required in a topdressing soil, since the turfgrass will usually generate sufficient amounts of organic residues for future needs. In fact, one objective of topdressing is to promote decomposition of this organic material to prevent excessive thatch accumulation.

Where the indigenous soil is unfavorable, the superintendent must decide between rebuilding the green or selecting a more favorable topdressing soil. This decision is based on the severity of problems encountered with the green and the anticipated likelihood of improving the green through topdressing, cultivation, and other practices. (Greens construction and rebuilding will be discussed in Chapter 8.) Selection of a topdressing soil of different composition from the indigenous soil is extremely difficult. Often a somewhat sandier soil is tried, with unpredictable results. For years there has been a trend toward topdressing entirely with sand. Although results have been generally favorable, there are some serious concerns over sand topdressing, including the necessity to continue with sand indefinitely once the program is initiated, the harder playing surface when a sand layer is developed, the increased requirement for water and fertilizer in the cultural program, and the increased likelihood of encountering localized dry spots if the sand becomes hydrophobic.

Topdressing Intensity

The amount of a topdressing soil required per application, or during a growing season, depends on the purpose for which the program is being carried out. Smoothing a playing surface that has large irregularities may require frequent applications of considerable quantities of material. Similarly, initiating a change in the composition of the soil in the turfgrass root zone requires large amounts of topdressing. The rate and number of applications are limited by the capacity of the turf to absorb the material. Excessive amounts of topdressing soil can prevent light from reaching the turfgrass leaves and lead to substantial losses of turf. The playable status of a green also influences topdressing intensity; more topdressing soil can be applied where a green is taken out of play temporarily than where play must continue.

For thatch control the frequency and amount of topdressing required depend on the rate at which thatch is accumulating. Some greens require no topdressing because of low turfgrass vigor or conditions that result in rapid decomposition of organic residues. For greens in which thatch continues to develop, a good rule of thumb has been to topdress when the thatch layer becomes "pencil thick" (approximately 1/4 inch). The amount of topdressing soil required under these circumstances would be about 0.2 yd^3 per 1000 ft^2, which provides a layer of approximately 1/16 inch.

The integration of topdressing soil and thatch results in a medium of positive value in a trafficked turf. Depending on its composition, the topdressing soil can provide firmness, improved water and nutrient retention, and, possibly, some nutrients. The thatch provides resiliency and aeration. Thus, the short-term effect of topdressing is thatch modification. In time, however, thatch decomposition is favored by topdressing, resulting in an increased humus content in the soil.

Where essentially pure sand is used for topdressing, it can be applied at rates as low as 0.1 yd^3 per 1000 ft^2 and as often as every 2 weeks during periods of active growth. A medium-fine (0.25 to 1.0 mm) sand has been recommended by researchers in California to facilitate working it into the turf. One potential problem with sand media is that dust and other particulate matter from the atmosphere or from irrigation water may become incorporated into the sand in sufficient quantities to substantially reduce porosity. For this reason it may be advisable to core-cultivate periodically and apply "clean" sand to the coring holes. In this way layers sealed by particulate matter can be channelled sufficiently to ensure adequate drainage. A possible alternative would be to topdress at a frequency sufficient to dilute the particulate matter and thus avoid sealed layers.

Coring may also be desirable for maintaining the proper depth of media in the green. With the addition of substantial topdressing material the level of the green may rise appreciably after several years. Since the media depth in a USGA green (see

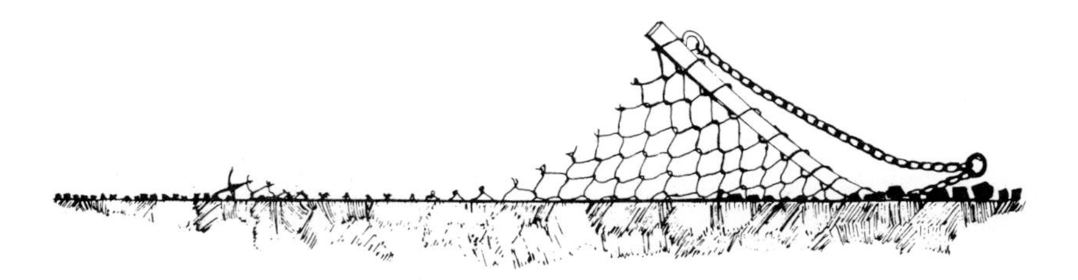

Figure 6.12. A matting operation with a section of chain-link fence being pulled across the turf.

Chapters 4 and 9) is critical to proper performance, increasing this depth may result in a droughty surface. Therefore, any additions of topdressing materials should be compensated for by reducing comparable amounts of media by coring and core removal.

MATTING

Matting is the procedure by which a heavy steel mat or similar device is pulled across a turf (Figure 6.12). Matting is usually necessary after topdressing to remove soil adhering to foliage and to work it into the turf; otherwise it would interfere with mowing, putting, and other activities. Where surface irregularities exist, matting redistributes topdressing soil to fill in low areas. Matting of loose soil can also be accomplished by brushes, brooms, and other devices. Following coring, a heavy steel mat or section of chain-link fence can be used to break up the cores and to work the soil into the turf. Usually, with fine-textured soils, the optimum soil moisture for coring is higher than that for matting. The soil cores should be allowed to dry sufficiently so that they can be easily crumbled between the fingers. When too wet, the soil tends to smear; when too dry, the soil cores are hard and bricklike.

Following seeding onto an existing turf (usually after coring or vertical mowing), the seed is worked into the turf by matting. Where organic debris remains on the turf's surface following vertical mowing, matting can be performed to "ribbon" the debris and facilitate its removal. Prior to the advent of modern herbicides matting was performed to uplift decumbent shoots of crabgrass and other weeds so that they could be mowed more easily.

WETTING AGENTS

Surfactants (surface active agents) include several classes of materials that reduce the interfacial tension between water and solids or other liquids. A wetting agent is a particular class of surfactant that increases the wetting capacity of water in a hydrophobic soil or

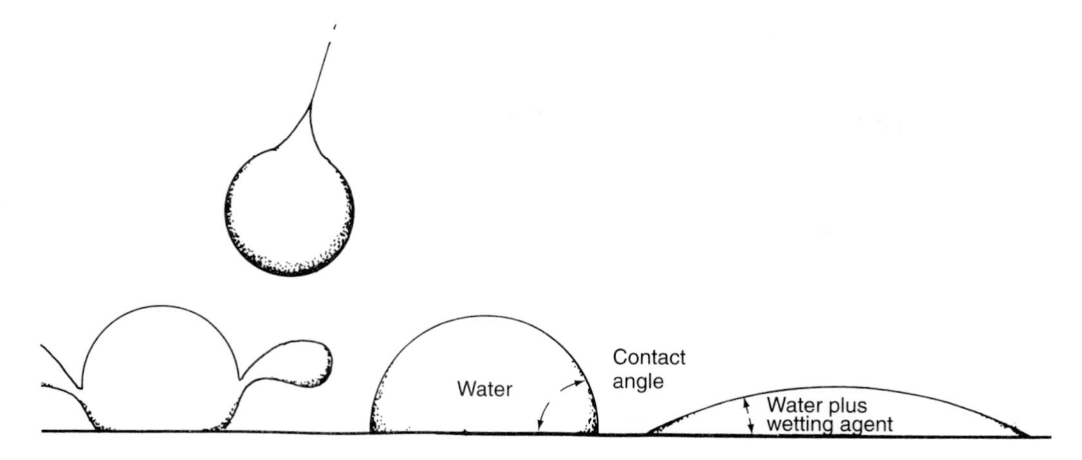

Figure 6.13. The effect of wetting agents in reducing the contact angle between a water droplet and a hydrophobic surface to improve surface wetting.

other growth medium. The surface tension of water is the tendency of surface molecules to be attracted toward the center of the water body. A surfactant-containing water droplet placed onto a hydrophobic surface will flatten out more than a droplet without surfactant; this is measurable as a decrease in the contact angle formed between the edge of the water droplet and the hydrophobic surface (Figure 6.13).

Surfactants differ widely in their chemical composition and effects, yet all have *hydrophilic* (water-loving) and *lipophilic* (soil-loving) components within their molecular structure. At the interface between water and a hydrophobic surface the hydrophilic portions remain within the water while the lipophilic portions adhere to the surface of soil particles to reduce interfacial tension.

Surfactants are classified as anionic, cationic, and nonionic. Anionic surfactants have short-term effects because of their tendency to leach readily within the soil. Conversely, cationic surfactants become bound to negatively charged clay and organic colloids and, once dried, can completely waterproof a soil. Nonionic surfactants have produced the most long-lasting, beneficial effects in turf and, therefore, warrant further discussion. Nonionics include esters, ethers, and alcohols: esters are most effective in wetting sands, ethers are most effective in clays, and alcohols are most effective in organic materials. Nonionic wetting agents made up of blends of different materials provide more effective wetting of a broad range of soil types than do their various components used alone.

An adequate concentration of wetting agent in the soil is necessary to achieve beneficial results. A range of 30 to 400 ppm in hydrophobic soil has been effective in improving wettability. Since wetting agents are susceptible to microbial degradation in soil, two or more applications per growing season may be necessary to maintain sufficient concentrations.

In addition to improved wettability, beneficial effects that have been reported from the use of wetting agents include improved turfgrass growth from increased water and nutrient availability, reduced evaporative water loss, reduced soil erosion during turfgrass establishment and increased seed germination, reduced dew and frost incidence, and correction of localized dry spots. One principal concern over the use of wetting agents in turf has been the potential for phytotoxicity. Turfgrass injury is most likely to occur where a wetting agent is applied at excessive rates or during periods of heat stress. Following application, a wetting agent should be watered in immediately to enhance its effectiveness and reduce the potential for foliar injury. It is entirely possible that turfgrass genotypes could differ in their sensitivity to specific wetting agents. Therefore, as with any new cultural practice, a period of testing on small areas should precede general adoption to ensure that benefits justify the expense associated with the use of wetting agents, and that unacceptable injury is avoided.

TURF COLORANTS

Turf colorants are used for a variety of purposes, including artificial coloration of dormant turfgrasses, cosmetic effects on diseased or discolored turf, and marking areas that have received sprays of fungicides or other materials. In past years the principal use has been for providing green color to dormant, warm-season turfgrasses during the overwintering period. This has usually been in lieu of overseeding with ryegrasses or other cool-season turfgrasses. The color imparted to the turfgrass is a function of the specific material used, its rate of application, the number of applications, and the color of the turf prior to treatment. Colorants vary from blue green to brilliant green; some materials impart a realistic appearance to the turf, while others may produce unsightly results. Without prior experience with a particular material, some testing on small areas is recommended before general use.

Once dry, most colorants will not rub off, and they often provide winter-long color to treated turf. Colorants must be applied uniformly; otherwise, objectionable patterns may be evident. The applicator should walk in front of the spray rather than behind it to avoid tracking. A fine spray applied with sufficient pressure should be used to achieve uniform coverage. The turfgrass should be dry and the temperature above 40°F for best results.

Prior to televised golf matches and other sports events, colorants are sometimes used for cosmetic effect. Unanticipated incidences of disease, insect-induced injury, or damage from other causes may be effectively masked by the correct use of colorants. Of course, colorants are not suitable substitutes for a sound turfgrass cultural program. A deteriorating turf, regardless of its color, will look and play differently from one that is dense and growing at a sufficient rate. However, the appearance of a turf can be improved with colorants when conditions dictate the need for immediate corrective action.

A common practice has been to apply a small amount of colorant with pesticides and other sprayed materials to identify treated areas and provide a check for the spray operator (and supervisor) to estimate accuracy and uniformity of coverage.

PLANT GROWTH REGULATION

Plant growth regulators (PGRs) were introduced more than forty years ago for application to utility turfs to reduce the mowing requirements. As the activity of these compounds was not restricted to meristematic regions within aerial shoots, results were often less than satisfactory. Some undesirable effects included discoloration and thinning of the turf, reduced rooting, reduced turfgrass recuperative capacity, increased weed and disease incidence, inconsistent results in shoot-growth suppression, and differential suppression of turfgrass species in mixed stands.

With the introduction of more effective PGRs, however, their potential utility in turfgrass management has increased. While the potential for phytotoxicity is still significant, some PGRs are being used successfully for species conversion, seedhead suppression, and clippings reduction. Given the importance of uniform application at precise rates and the limited application windows within which PGRs must be used, one should be extremely cautious when using these materials.

There are currently two types of PGRs: Type I and Type II. Type I compounds inhibit or suppress the growth and development of susceptible plant species by causing a cessation of cell division and differentiation in meristematic regions of the plant. They may be fairly effective in reducing seedhead development and emergence. Growth-inhibiting compounds are foliar absorbed and immediate in their effects; examples include maleic hydrazide (Slo-Gro), chlorflurenol (Maintain), and mefluidide (Embark). Growth-suppressing compounds are crown and root absorbed and allow some growth to occur following application; examples include EPTC (Shortstop) and amidochlor (Limit). The activity of Type I compounds is largely restricted to grasses; dicot species are minimally affected.

Type II compounds suppress plant growth by interfering with gibberellin biosynthesis and causing reductions in cell elongation and associated expansive growth of susceptible plants. Examples of Type II compounds include flurprimidol (Cutless), paclobutrazol (Scott's TGR), and trinexapac-ethyl (Primo). Unlike Type I PGRs, these compounds are not generally effective in reducing seedhead development; however, seedhead height may be substantially shortened due to the suppression of culm internode elongation. Compared with Type I PGRs, the Type II compounds tend to suppress vertical shoot growth for relatively long periods. Lateral-shoot (principally tillering) and root growth are less affected and may actually be enhanced. The spectrum of activity of Type II compounds encompasses both grasses and dicot species. One of the most promising uses of these materials is for suppressing annual bluegrass in mixed stands with creeping bentgrass. Trinexapac-ethyl is unique among the Type II materials in that it is not root absorbed; thus, in species conversion programs, treated areas can be overseeded soon after application. While not always providing consistent results, successive applications at low rates have resulted in impressive population shifts in treated greens and fairways.

QUESTIONS

1. Explain what is meant by **cultivation** of turf and characterize the principal cultivation methods in contemporary use.
2. List the advantages and disadvantages of **coring** as a cultivation method.
3. List the important effects of soil incorporation into thatch.
4. Under what conditions would it be appropriate to **roll** a turf?
5. Provide examples of conditions in which it would be desirable to apply **colorants** or **wetting agents.**

CHAPTER 7

Pest Management

A turfgrass pest is any organism causing a measurable deterioration in the esthetic or functional value of a turf. Pests include weeds, disease-causing organisms, some insects, and other destructive organisms. Inevitably, when pests are mentioned pesticides are considered as important means for achieving control. Pesticides are valuable components of a turfgrass cultural program, but pest management includes more than simply selecting and applying the appropriate pesticide to control specific organisms. It also includes selecting turfgrasses that are well adapted to prevailing environmental conditions, following proper establishment procedures, and conducting a cultural program that favors healthy turfgrass growth (Figure 7.1). Weeds and diseases are often indicative of unfavorable growing conditions for specific turfgrasses; their incidence can be prevented or substantially reduced where favorable conditions prevail. Damage from insects and other animals is often greater where the turf is also subjected to other stresses. Many pests are, in effect, controlled without the use of pesticides as long as turfgrass growth is favored. As often as not, pest activity is a reflection of underlying problems rather than the problem itself.

Substantial progress has been made in the development of pest-resistant turfgrasses, but no turfgrass is immune to all diseases, weed invaders, insects, nematodes, and other pests. Even if one could be developed, it is doubtful that it would remain that way indefinitely. Natural forces bring about change resulting in new pests or adaptations of old ones. Some pesticides, therefore, will always be needed to sustain pest-free turf. It is the proper use of pesticides as part of a sound cultural program that ensures high quality turf.

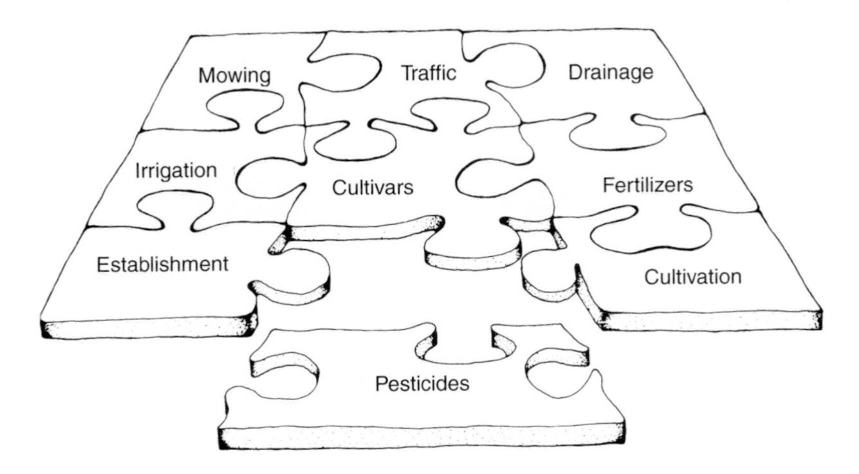

Figure 7.1. An integrated program of turfgrass pest management includes not only pesticides, but all cultural operations for sustaining healthy turf.

PESTICIDES

The term pesticide is derived from the Latin *pesta,* meaning "pest," and *caedere,* meaning "to kill." Pesticides are chemical agents for controlling pests; they include herbicides, fungicides, insecticides, and nematicides. For the most part, pesticides are organic compounds that interfere with some physiological process in the pest organism. To work properly, a pesticide must be present at the site of pest activity at an effective concentration for a sufficient period of time. Desirable attributes of a pesticide include the following:

1. It should effectively control the pest organism.
2. Activity should not extend beyond the target pests.
3. It should not persist beyond the control period.
4. It should present minimal hazards to humans.

Effective pest control does not necessarily mean complete eradication. It simply means reducing the pest population or its activity to a level that does not damage turfgrass or reduce turf quality. For example, symptoms of *Drechslera* leaf spot (formerly *Helminthosporium* melting out) can be observed on nearly all cultivars of Kentucky bluegrasses during the cool, wet weather of midspring, but this disease is not a problem unless substantial thinning and discoloration of the turfgrass occur. A few spots on the leaves do not require fungicide treatment or other control measures. Similarly, the presence of a few insects in the turf does not require the use of an insecticide; it is only when insect populations develop sufficiently to cause damage that a pesticide should be applied. However, some pest problems occur so routinely, and damage is so severe, that a *preventive* approach to control is warranted. Examples include dollar spot and brown

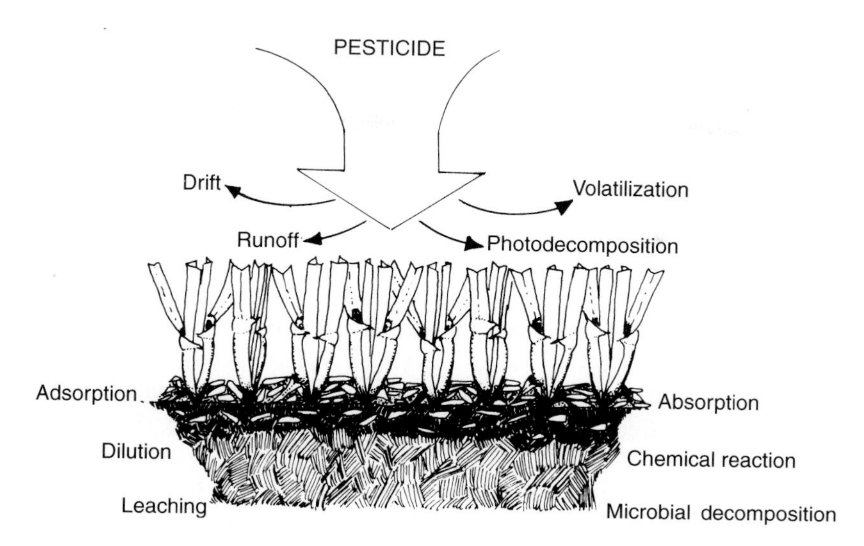

Figure 7.2. Pesticides are subject to many forces operating within the natural environment. The activity of a pesticide toward target organisms is influenced by forces that remove it from the site of application or result in its decomposition.

patch diseases in greens and crabgrass in some lawns. Where a history of these and other pest problems exists, treatment with appropriate pesticides prior to observable incidences of the problems may be advisable. For most turfgrass pest problems, however, pesticide treatment should be withheld until early symptoms indicate that unacceptable damage or reductions in turfgrass quality are likely. This is called the *curative* approach to pest control. Careful monitoring of pest-induced symptoms and timely application of appropriate pesticides can prevent substantial turfgrass deterioration while conserving pesticides and protecting the environment.

Fate of Pesticides

Once applied to turf, pesticides are subject to many forces operating within the natural environment (Figure 7.2). Wind-induced *drift* of spray particles and dusts can result in movement of the pesticide from the application site to a nontarget site. Even where drift is confined to the target site, uniform coverage is reduced. Drift of postemergence herbicides such as 2,4-D, mecoprop, dicamba, and glyphosate can result in significant damage to nearby crops and ornamental plants. To reduce the potential for drift, pesticides should be applied during periods of little or no wind (usually less than four to five miles per hour depending on particle size). Excessive spray tank pressures should be avoided, and spray volumes (amounts of water applied per unit area) should be sufficient to provide spray particles that are large enough to resist drift.

Runoff is the surface movement of a pesticide across the turf. When runoff extends to a nontarget site, serious consequences can result with some pesticides. Runoff usually occurs as a result of surface water movement during irrigation or natural rainfall. The susceptibility of a pesticide to being carried in runoff increases in proportion to its solubility in water. However, even insoluble pesticide particles can move downslope where surface flow of water is sufficient.

Volatilization is the process by which a pesticide changes from a liquid to a gas and is lost to the atmosphere. Volatile pesticides can travel long distances in the gaseous state, causing damage to sensitive plants located downwind. Volatilization can occur during or after application and is therefore a more serious concern than drift. Short-chain esters of 2,4-D are highly volatile; they are not generally used on turf because of their great potential for injuring nontarget plants. Losses from volatilization of soil-applied pesticides can be reduced by watering them into the turf following application.

Photodecomposition is the breakdown of a pesticide under the influence of light. Some pesticides should be washed off of the foliage and into the turf immediately after application to reduce the amount of chemical subject to photodecomposition.

Adsorption is the physical binding of a pesticide onto soil or thatch constituents. With many pesticides, adsorption increases in proportion to the amount of clay or organic colloids in the soil. Rates required for pest control are often higher where adsorption results in partial inactivation of a pesticide.

Dilution and Leaching are processes by which a pesticide is carried downward in the soil. Where the site of pest activity is at or near the soil surface, dilution of the pesticide within a substantial depth of soil may reduce the surface pesticide concentration to below that required for control. Further leaching may carry the pesticide down to the water table, where it can subsequently move in drain tiles to ponds, streams, storm sewers, or other receptacles.

Absorption is the uptake of a pesticide by living organisms. Once absorbed, contact pesticides are usually immobile, while systemic pesticides are translocated within receptor plants. Absorption of a sufficient amount of pesticide by germinating weeds, mature weeds, pathogenic fungi, and destructive insects is essential for pest control. Some pesticide decomposition may occur in plants as their enzyme systems act on the pesticide molecules.

Chemical Reaction results in the transformation of pesticide molecules to usually less active or inactive forms. Hydrolysis, oxidation, and reduction are types of chemical reactions that can occur with some pesticides in some soils.

Microbial Degradation is the breakdown of pesticide molecules by microorganisms in the soil or thatch. Once a pesticide has been introduced into a turf, microbial populations with the capacity to utilize the pesticide as a food source tend to increase until much of the pesticide residue has been degraded; then these populations eventually return to normal levels. Many pesticides are degraded to carbon dioxide or to compounds that naturally occur in biological systems, and therefore pose no long-term threat to turfgrasses or other organisms.

Pesticide Formulations

Pesticides are combined with liquid or dry carriers to facilitate application. Salt concentrates, emulsifiable concentrates, and flowable suspensions are liquid formulations, while dry formulations include wettable powders, soluble powders, and granules. All but baits, dusts, and granular formulations are applied in water through a sprayer.

Baits (B) are pesticides mixed with a pest-attractive substance to control those pests that ingest the formulation. While used mostly for controlling rodents and other mammals, baits are sometimes used to control highly mobile pests such as ants and mole crickets.

Dusts (D) are formulations in which the active ingredient is mixed with fine, dry carriers made from a variety of substances, including talc, chalk, clay, and ash. Some dust formulations are used for seed treatments.

Emulsifiable Concentrates (EC) are formed by mixing oil-soluble pesticides in oil with emulsifying agents. The ester form of 2,4-D, while insoluble in water, is highly soluble in oil. When the pesticide-containing oil is added to water, however, the two liquids do not mix unless a surfactant (emulsifying agent) is also added. As with wetting agents (see Chapter 6), emulsifying agents reduce the interfacial tension between the water and, in this case, oil. The result is a dispersion of tiny oil globules in water, forming a milky substance called an *emulsion.* Emulsions are not as stable as salt solutions; therefore, some agitation in the spray tank may be necessary to maintain uniform dispersion of the pesticide-containing oil throughout the volume of water.

Flowable Suspensions (F) differ from wettable powders in that they are formulated with water. The concentrated suspension is more easily mixed in the spray tank because the powder is already fully wetted. As with wettable powders, agitation is required to maintain a uniform suspension in the spray tank.

GEL is a gelatinous pesticide formulation usually packaged in water-soluble packets or bags.

Granules (G) are a mixture of inert carrier (clay, vermiculite, corn cobs) and a pesticide or combinations of fertilizer nutrients and pesticides. Usually, the pesticide is coated on the surface of the carrier, but some granular formulations may have entrapped or encapsulated pesticide with slow-release characteristics. Since granules are not applied in water, simple fertilizer spreaders can be used for small- or large-area applications.

Microencapsulated (M) formulations are liquid or dry particles of pesticides surrounded by a plastic coating. Mixed with water, these are applied as a spray. The encapsulation process prolongs the active life of the pesticide by delaying its release into the environment.

Salt Concentrates (SC) are formed by combining salt derivatives of a pesticide with water. For example, technical 2,4-D, an insoluble acid, can be made highly water soluble by reacting it with a base to form a salt. When an organic amine or other salt of 2,4-D is then added to a small amount of water, a salt concentrate formulation is produced. Because the salt form of the herbicide (*solute*) and the water (*solvent*) are physically homogeneous substances, the resulting stable mixture is called a *solution.* When a concentrated solution is diluted by its addition to water in a spray tank, the pesticide becomes thoroughly dispersed in the volume of water and requires no agitation to maintain a solution of uniform concentration.

Soluble Powders (SP) are pesticides that are highly soluble in water. When added to the spray tank, soluble powders form a stable solution and require no agitation.

Water-Dispersible Granules (WDG) are, essentially, granular formulations of finely milled wettable powders. The principal advantage of this formulation lies in its dust-free handling properties.

Water-Soluble Packet (WSP) contains a measured amount of the pesticide in packets that dissolve in water when added to a spray tank.

Wettable Powders (WP) are formulated from pesticides that have low solubilities. Most wettable powders contain fillers (clay or talc) of high surface area, the pesticide, and anionic wetting agents to favor wetting and dispersion when added to water. Suspensions of wettable powders in water must be continuously agitated to remain uniformly dispersed.

Pesticide Hazards

Pesticides vary in the hazards associated with their use. Some pesticides, such as nematicides, can be very hazardous and, therefore, should only be used by qualified applicators. Others are so safe that they can be used by almost anyone without any serious risk of personal injury.

The relative hazards posed by exposure to pesticides are indicated on the pesticide label by the precautionary statement. One of three precautionary statements is clearly printed in large letters: CAUTION, WARNING, or DANGER—POISON. The basis for assigning precautionary statements to pesticide labels is the lethal dose required to cause the death of 50% of a population of test animals. This is called the LD_{50} and is expressed in milligrams of pesticide per kilogram of body weight (mg/kg or ppm). If the LD_{50} is 50 mg/kg or less, the pesticide carries the precautionary statement DANGER—POISON with a skull and crossbones illustrated adjacent to the statement. Since a relatively small amount of this pesticide can cause illness or even death in humans and other animals, it should be used with utmost care. Where the precautionary statement is WARNING, the LD_{50} ranges from 50 to 500 mg/kg. Above 500 mg/kg, the word CAUTION is printed on the label.

All pesticides, regardless of which precautionary statement they carry on their labels, should be treated as hazardous materials. They should be stored in a well-ventilated, dry room. The door to the room should be clearly marked PESTICIDES and locked. Access to the room should be limited to qualified personnel properly attired for transporting, measuring, mixing, and applying pesticides. Because some pesticides can be absorbed through the skin and eyes or inhaled, proper attire includes gloves, long trousers, long-sleeved shirts, and, possibly, goggles and face mask, especially during measuring and mixing operations. If the pesticide has been allowed to contact clothing and skin surfaces, clothing should be removed and washed and skin surfaces thoroughly cleaned as soon as practical to minimize exposure time.

Pesticides should always be stored in their original containers with the labels firmly fastened. Only the amount of pesticide required for the specific area of treatment should be mixed in the spray tank. This eliminates the necessity for disposing of unused material. Discarded pesticides should be taken to an approved sanitary landfill and buried, or returned to the supplier.

Pesticide Application

To provide effective control of turfgrass pests and to avoid waste, pesticides must be applied uniformly and at proper rates to the turf. This requires suitable equipment and proper application techniques. Sprayable formulations are applied in water; the concentration of the solution/emulsion/suspension and the amount of water applied per unit area determine the pesticide application rate. Spray volumes used for turf vary from less than 30 to more than 200 gallons per acre. Once a sprayer is calibrated (see Appendix 2) to deliver a known spray volume, the amount of pesticide required for a specific application rate can be determined. For example, a 200-gallon-capacity tank with a spray volume of 50 gallons per acre will cover a total of 4 acres. A pesticide formulated as a 4-pound-per-gallon EC and applied at the rate of 1 pound active ingredient (a.i.) per acre will cover 4 acres for each gallon of pesticide and 200 gallons of water used. A 50% WP formulation requires 2 pounds of product for each pound of active ingredient applied; therefore, at the 1 pound (a.i.) per acre rate, 8 pounds of the formulation in 200 gallons will be required to treat 4 acres. All pesticide application problems are basically this simple, except that some conversions from pounds to ounces, acres to square feet, or gallons to fluid ounces may be required. Conversion tables for both English and metric units are provided in Appendix 1.

Pesticide applicators include both sprayers and spreaders. Sprayers apply water containing diluted pesticides through a system of nozzles, while spreaders apply granular pesticide formulations (and granular fertilizers, see Chapter 5) through openings at their bases. Pesticide applicators should be checked periodically (at least once each growing season) for calibration and uniformity of application.

Spraying units have one or more nozzles, a tank to hold the pesticide-water mixture, and a pumping system to force the mixture through the nozzles. In addition, filters, pressure gauges, pressure regulators, shutoff valves, and connecting hoses are used to complete the unit. Of all components the nozzles are the most important because they influence application rate, uniformity of application, and the potential for spray drift. Other components exist primarily to ensure proper operation of the nozzles. The principal nozzle type used for applications to turf is the flat fan. Where substantial quantities of nitrogen and other fertilizers are applied with pesticides, however, high-volume (flooding) nozzles are used to provide spray volumes of 170 to 220 gallons per acre to minimize foliar burn.

The nozzle converts the pesticide-water mixture into spray droplets. Nozzle size and operating pressure determine the spray volume and droplet size. The correct spray volume depends on the specific pesticide and the purpose for which it is used. Foliar-applied pesticides should thoroughly cover the foliage with a minimum of loss to the underlying soil or thatch; spray volumes of 40 to 60 gallons per acre are usually considered optimum for these applications. Soil-applied pesticides should not be allowed to adhere to the foliage for an extended period; otherwise, some of the pesticide may be lost to photodecomposition, volatilization, runoff, or biological decomposition in or on the foliage. Spray volumes of 100 gallons per acre or higher are preferred for soil-applied pesticides; however, irrigation immediately following application is often recommended to wash the pesticide into the turf, especially where low spray volumes are employed. Where fertilizers and pesticides are applied together, the spray volume used is based on a compromise between simultaneous objectives. The most important consideration is the foliar burn potential from the fertilizers.

Nozzle openings tend to become larger as abrasive suspensions cause wear. This not only increases spray volume, but changes the spray pattern as well. Nozzles should be changed at least yearly, or often enough to maintain uniform application.

WEED CONTROL

A weed is any plant growing where it is not wanted. When it occurs as part of a turfgrass community, its definition can be expanded to an undesirable plant because of its disruptive effect on the esthetic appearance, stabilizing capacity, or overall utility of a turf. A particular plant species may be a weed in some turfgrass communities and a desirable turfgrass in others. Examples of "potential" weeds include tall fescue, creeping bentgrass, and bermudagrass.

Specific weeds are often good indicators of unfavorable environmental conditions for turfgrasses. Large infestations of knotweed frequently occur where severe soil compaction limits turfgrass growth. Ground ivy often invades under trees where insufficient sunlight results in the decline of Kentucky bluegrass and other shade-intolerant turfgrasses. The presence of red sorrel is usually indicative of acid soil conditions. Weed incidence in turf reflects both environmental conditions and growth characteristics of the plants comprising the turfgrass community. These two dimensions of a plant ecosystem are highly interactive. Where a particular turfgrass is marginally adapted, successful culture depends on careful programming of fertilization, pesticide application, and other cultural variables to sustain the turf and avoid large infestations of weeds. In contrast, an especially well adapted turfgrass is less prone to weed invasion.

Although herbicides are important tools for controlling weeds in turf, repeated occurrences of weeds may reflect underlying problems that are not correctable with herbicides. However, the proper use of herbicides can often convert a heavily weed-infested turf into one that is essentially weed free. Thus, turfgrass weed control may be defined as any practice designed either to prevent weed emergence in turf or to effect a shift away from undesirable vegetation and toward desirable turfgrasses.

Weeds

Weed species occur within three botanical groups: annuals, biennials, and perennials. Annuals complete their life cycles from seed within one year; they include both winter annuals and summer annuals. Winter annuals typically germinate in late summer or early fall and die the following summer. Examples include some genotypes of annual bluegrass, common chickweed, and henbit. Summer annuals germinate in spring and usually die with the first hard frost in fall. Some examples of summer annuals are knotweed, spurge, and crabgrass. Biennial weeds, such as bull thistle, live for more than one year, but not more than two years. Perennials live for more than two years and perhaps indefinitely. Simple perennials, such as dandelions and plaintains, are propagated by seed, but severed organs may produce new plants. Creeping perennials can reproduce by stolons, rhizomes, nutlets, and other vegetative propagules as well as by seed. Examples of creeping perennials are ground ivy, quackgrass, and mouse-ear chickweed.

For control purposes, weeds may be divided into three functional categories: annual grasses, broadleaf weeds, and perennial grasses. Some annual grasses can be controlled with preemergence or postemergence herbicides. Although several annual broadleaf weeds can be controlled with preemergence herbicides, the principal herbicides used for their control are the phenoxy (2,4-D, mecoprop) or benzoic acid (dicamba) herbicides, applied postemergence. Herbicides used for controlling annual grasses and broadleaf weeds are, for the most part, selective; that is, they do not seriously affect desired turfgrasses when used properly. In contrast, most perennial grasses cannot be controlled selectively; herbicides do not currently exist for selectively controlling bermudagrass, creeping bentgrass, or most other perennial grasses in Kentucky bluegrass. Consequently, control of most perennial grasses requires spot treatment of isolated clumps or patches with nonselective herbicides. An exception to this is the use of chlorsulfuron as a semiselective herbicide for taking out clumps of tall fescue and perennial ryegrass in Kentucky bluegrass turfs.

Weed Identification

Annual Grasses

Annual Bluegrass (*Poa annua* **L.**) is a winter annual or perennial that may predominate in a turf under moist shaded conditions and on compacted soils. Its growth habit varies from bunch type to stoloniferous. It is frequently observed in dense patches of light green color that grow vigorously during cool weather but often die during summer stress periods. Seed heads are produced throughout most of the growing season but are especially abundant in mid-spring. Closely mowed annual bluegrass can form an attractive turf that persists through the summer months in cool temperate climates as long as adequate disease control and irrigation programs are followed (see Chapter 3).

ANNUAL BLUEGRASS

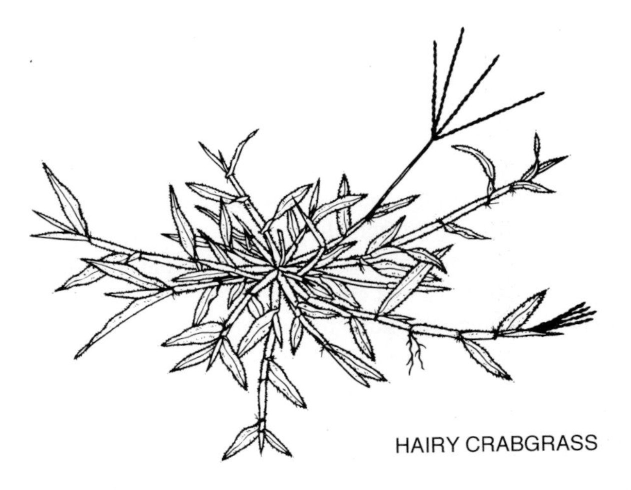

Smooth and Hairy Crabgrasses (*Digitaria ischaemum* [**Schreb.**] **Muhl.** and *Digitaria sanguinalis* [**L.**] **Scop.**) are summer annuals that germinate in late spring and summer on warm, moist sites with moderate to full sunlight. The seed heads appear as several fingerlike projections at the terminals of seed stalks. The spreading growth of crabgrass tends to crowd out desirable turfgrasses in lawns and other turfs. Growth slows or stops during cool weather in late summer and early fall, and plants are killed with the first hard frost, leaving unsightly brown patches in the turf.

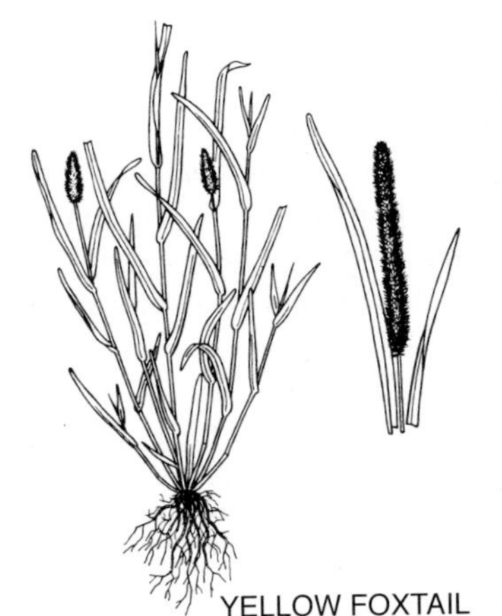

Yellow Foxtail (*Setaria glauca* [**L.**] **P.B.**) is a late-germinating summer annual frequently found in newly seeded lawns. It does occur in some established turfs and is often mistaken for crabgrass, but it is not as widespread. It is identified by the presence of long hairs on the upper surface of the leaf blade near the base and by yellow cylindrical seed heads.

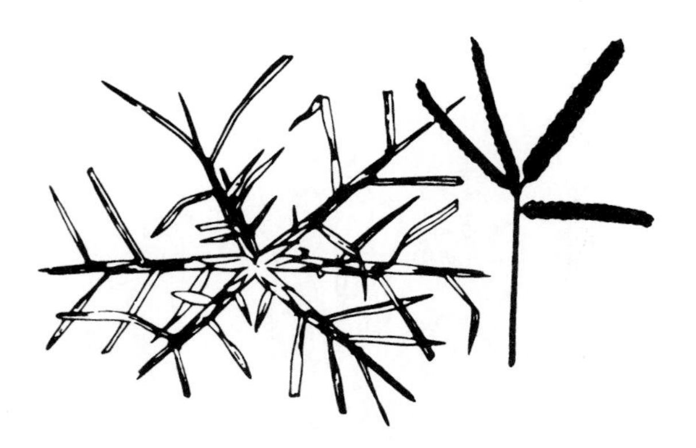

GOOSEGRASS

Goosegrass or Silver Crabgrass (*Eleusine indica* [**L.**] **Gaertn.**) is a summer annual that begins germinating several weeks after crabgrass. It is similar to crabgrass but is usually darker in color, has a center that is a silvery color, and has seed heads that are zipperlike in appearance. It is frequently found in compacted and poorly drained soils in warm temperate and warmer climates.

FALL PANICUM

Fall Panicum (*Panicum dichotomiflo-rum* **Michx.**) is a late-germinating summer annual that can be a problem in new plantings in the fall. It has short purplish sheaths, and the seed head is an open and spreading panicle.

SANDBUR

Sandbur (*Cenchrus pauciflorus* **Benth.**) is a summer annual that occurs in thin turfs, especially in infertile, coarse-textured soils. The seed stalks bear hard spiny burrs that can be irritating in a recreational turf.

Perennial Grasses

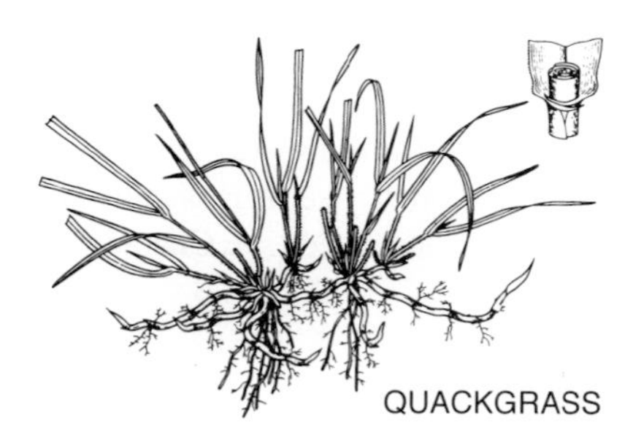
QUACKGRASS

Quackgrass (*Agropyron repens* [**L.**] **Beauv.**) is a perennial grass that spreads by a strong system of rhizomes. It is an especially serious weed in cool temperate climates. It may be identified by its dull green color and long clasping auricles.

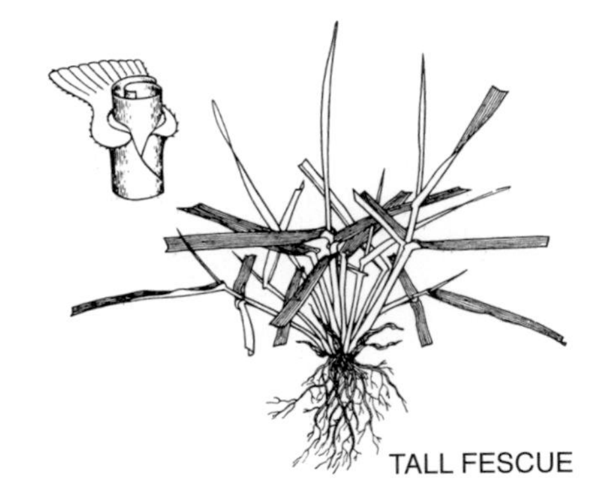

TALL FESCUE

Tall Fescue (*Festuca arundinacea* **Schreb.**) is a coarse-textured, cool-season perennial grass that often forms unsightly clumps in a turf. In pure stands, however, it can form an acceptable turf, especially in transitional climates between temperate and subtropical zones (see Chapter 3).

BERMUDAGRASS

Bermudagrass (*Cynodon dactylon* **[L.] Pers.**) is a warm-season perennial grass grown in subtropical and tropical climates. Its occurrence in warm temperate climates is usually as a weed, but it is used occasionally as a turfgrass in this region (see Chapter 3).

BENTGRASS

Bentgrass (*Agrostis palustris* **Huds.**) is a cool-season perennial grass that spreads by stolons. It forms puffy dense patches that may eventually dominate a lawn turf. Under close mowing and meticulous care, however, it can form an excellent lawn or sports turf; otherwise, it is usually regarded as a serious weed (see Chapter 3).

NIMBLEWILL

Nimblewill (*Muhlenbergia shreberi* **J.F. Gmel.**) is a stoloniferous perennial grass that forms patches resembling bentgrass. It typically occurs on moist shaded sites in warm temperate and warmer climates. The leaf blades are short, generally flat with several broad ridges, and pointed at the tip.

YELLOW NUTSEDGE

Yellow Nutsedge (*Cyperus esculentus* **L.**) is a perennial sedge that reproduces by seed, rhizomes, and small hard tubers called nutlets. It is identified by its triangular stems and yellow green color. The nutlets may persist in the soil for several years, ensuring regeneration of the plants. Aerial shoots first appear in late spring following germination of the nutlets; vigorous growth in the summer results in rapid population increases; then the aerial shoots disappear in the fall followed by overwintering of the nutlets. Yellow nutsedge occurs in warm temperate and warmer climates. A related species, purple nutsedge (*Cyperus rotundus* **L.**) occurs in subtropical and warmer climates.

DALLISGRASS

Dallisgrass or Watergrass (*Paspalum dilatatum* **Poir.**) is a coarse-textured perennial grass that reproduces from seed. It thrives in subtropical and tropical climates and is favored by persistently wet soil conditions. It forms thick clumps that severely reduce the esthetic quality and playability of a turf.

Broadleaf Weeds

DANDELION

Dandelion (*Taraxacum officinale* **Weber**) is a perennial that reproduces by parachutelike seeds. Its long taproot can regenerate the plant when severed. It is easily recognized by its sharply lobed leaves and bright yellow flowers that turn into fluffy white seed heads.

PLANTAIN

Broadleaf and Blackseeded Plantains (*Plantago major* **L.** and *Plantago rugelli* **Dcne.**) are perennials that reproduce by seed. The leaves form a rosette with fingerlike flower stalks protruding upward. They frequently occur in thin stands where inadequate fertilization has limited turfgrass growth.

BUCKHORN PLANTAIN

Buckhorn Plantain (*Plantago lanceolata* **L.**) is a perennial with lancelike leaves and bulletlike seed heads on long slender stalks. It typically occurs in poor turfs of low fertility along with other plantains.

CHICKWEED

Common Chickweed or Starwort (*Stellaria media* **[L.] Vill.**) is a creeping winter annual with small pale green leaves. Its hairy stems branch and take root, enabling the plant to spread over large areas and completely crowd out turfgrasses. White starlike flowers appear during cool seasons. Its presence is indicative of moist compacted soils. Chickweed commonly occurs in greens where disease-causing organisms or insects thin the turf and provide infestation sites.

MOUSE-EAR CHICKWEED

Mouse-ear Chickweed (*Cerastium vulgatum* **L.**) is a perennial that reproduces mainly by seed but also by creeping stems. It is identified by its small, pubescent, dark green leaves and dense growth habit. Like common chickweed, its presence is indicative of moist, compacted soils.

YARROW

Yarrow (*Archillea millefolium* **L.**) is a fernlike perennial weed that spreads by rhizomes. Under close mowing it forms a dense mat and is quite wear resistant and drought tolerant. It typically occurs on dry soils of low fertility.

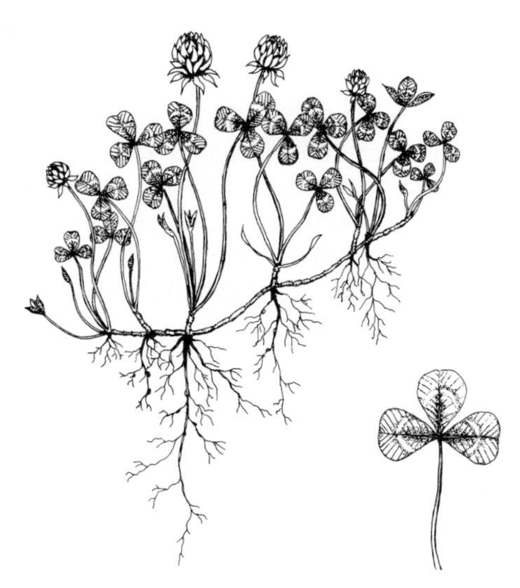

WHITE CLOVER

White Clover (*Trifolium repens* **L.**) is a creeping perennial that competes aggressively with established turfgrasses, especially under moist conditions and low soil fertility. It has a stout taproot and branched rhizomes. It is identified by its three short-stalked leaflets and globular white flowers. At one time white clover was considered an important component of many lawn seed mixtures; today, however, it is generally considered a weed.

BLACK MEDIC

Black Medic or Yellow Trefoil (*Medicago lupulina* **L.**) is an annual that closely resembles white clover. It is distinguished by its yellow flowers and the arrangement of its leaflets on the stem. The middle leaflet is borne on a short petiole, while the lateral leaflets are close to the stem. It typically develops during droughty periods of late spring and summer when turfgrass growth is checked by a lack of moisture.

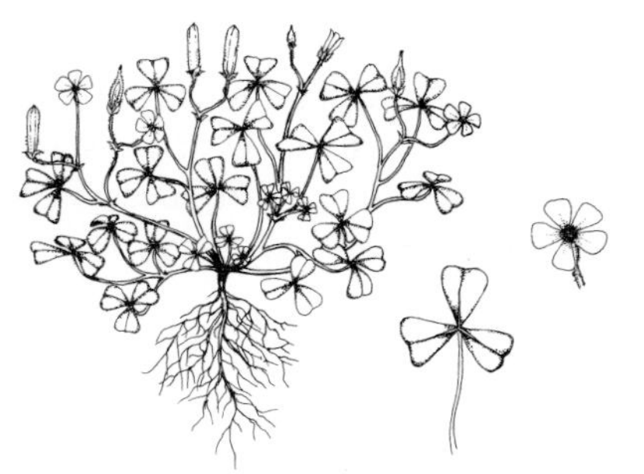
YELLOW WOOD SORREL

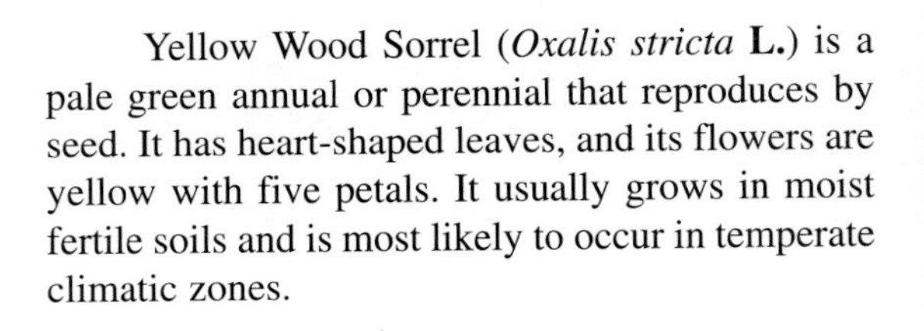
Yellow Wood Sorrel (*Oxalis stricta* **L.**) is a pale green annual or perennial that reproduces by seed. It has heart-shaped leaves, and its flowers are yellow with five petals. It usually grows in moist fertile soils and is most likely to occur in temperate climatic zones.

PURSLANE

Purslane (*Portulaca oleracea* **L.**) is a fleshy summer annual with smooth reddish stems. It grows vigorously in warm moist soils of high fertility, especially where turfgrass competition is minimal. It may be particularly troublesome in new plantings.

KNOTWEED

Knotweed (*Polygonum aviculare* **L.**) is a low-growing annual that first appears in early spring. Its appearance is variable, depending on the stage of maturity. Young plants have long, slender, dark green leaves that occur alternately along the knotty stem. Mature plants have smaller dull green leaves and inconspicuous white flowers. Its long taproot enables it to persist during extended summer droughts. It grows well on heavily trafficked compacted soils in temperate and subtropical climates.

PROSTRATE SPURGE

Prostrate Spurge (*Euphorbia supina* **Raf.**) is a low-growing annual that generally appears in midseason. The small leaves are opposite and frequently have a red blotch in the center. The stem oozes a milky sap when broken.

CURLED DOCK

Curled Dock (*Rumex crispus* **L.**) is a perennial that reproduces by seed. It has a fleshy taproot and large smooth leaves that are crinkled on the edges. It generally grows in moist or wet fine-textured soils of high fertility.

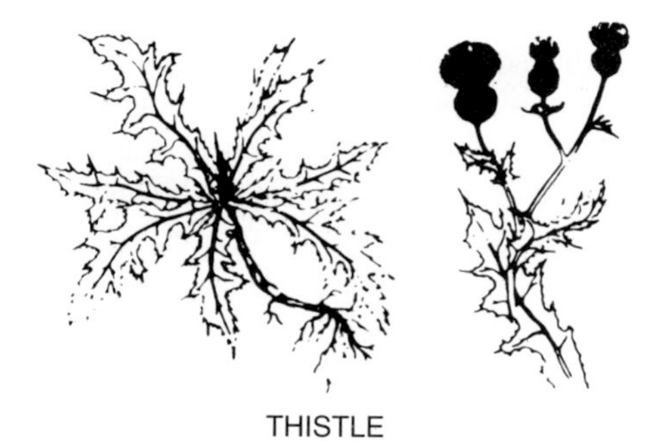

THISTLE

Thistles (*Cirsium arvense* [**L.**] **Scop.** and *Cirsium vulgare* [**Savi**] **Ten.**) are deep-rooted perennials or biennials with spiny and serrated leaves. A rosette-type of growth occurs under mowing. The numerous and sharp spines make these weeds particularly objectionable in turf.

CHICORY

Chicory (*Cichorium intybus* **L.**) is a perennial that reproduces by seed. The taproot is large and fleshy. A rosette of leaves resembling dandelion leaves forms at the base. Bright blue flowers are borne on rigid stalks that resist mowing. It is most likely to be found along roadsides in infrequently mowed turfs of low fertility.

RED SORREL

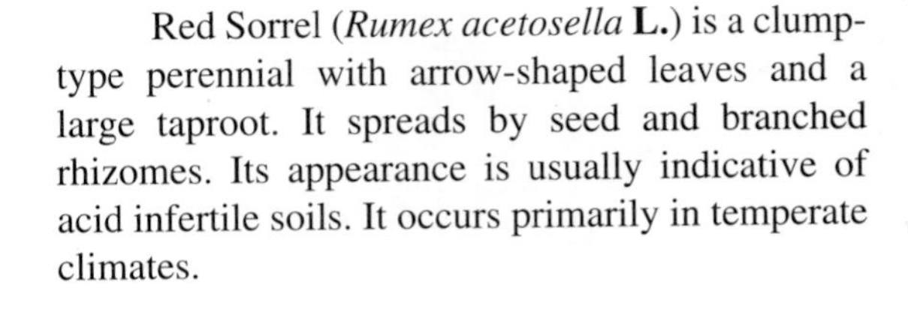

Red Sorrel (*Rumex acetosella* **L.**) is a clump-type perennial with arrow-shaped leaves and a large taproot. It spreads by seed and branched rhizomes. Its appearance is usually indicative of acid infertile soils. It occurs primarily in temperate climates.

CARPETWEED

Carpetweed (*Mollugo verticillata* **L.**) is a summer annual with smooth, light green, tonguelike leaves. Stems branch in all directions forming flat circular mats of growth.

GROUND IVY

Ground Ivy or Creeping Charlie (*Glechoma hederacea* **L.**) is a creeping perennial that forms dense patches in turf. Its bright green leaves are round with scalloped edges. Its flowers are bluish purple and the stems are four sided. It grows well in shady areas where soils are poorly drained. It is largely restricted to cool temperate climates.

HENBIT

Henbit (*Lamium amplexicaule* **L.**) is a winter annual that reproduces by seed. The leaves resemble those of ground ivy but occur opposite along the stem. It is widely distributed, but occurs primarily in moist, fertile soils.

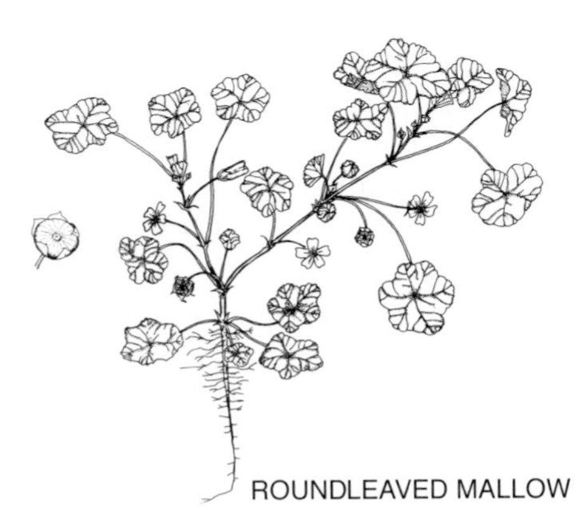

ROUNDLEAVED MALLOW

Roundleaved Mallow (*Malva neglecta* **Wallr**.) is an annual or biennial that reproduces by seed. It has a long taproot and round leaves with five distinct lobes. Its white flowers first appear in late spring and bloom continuously through the season. It is indicative of fertile soils and occurs widely in temperate and subtropical climates.

WILD GARLIC

Wild Onion and Wild Garlic (*Allium canadense* **L.** and *Allium vineal* **L.**) are perennials with slender cylindrical leaves. Wild garlic leaves are hollow; those of wild onion are not. They are propagated chiefly by small bulblets formed in heads with the flowers. They usually grow in fine-textured soils in warm temperate and subtropical climates.

PARSLEY PIERT

Parsley Piert (*Alchemilla arvenis* [**L.**] **Scop.**) is a winter annual with pale green trisect leaves borne on short stalks. It is typically found growing in fertile fine-textured soils in subtropical climates.

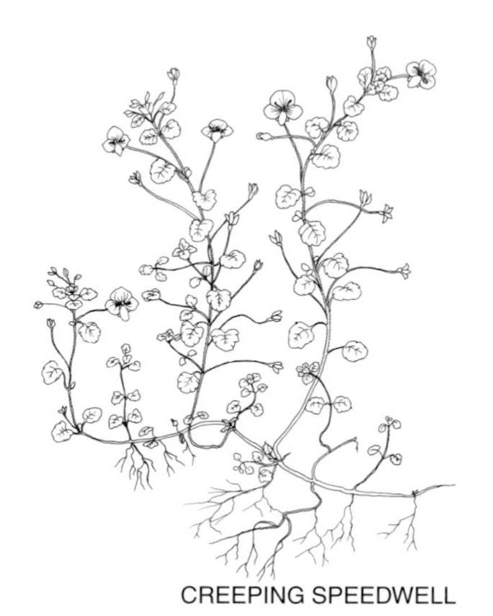

CREEPING SPEEDWELL

Speedwells (*Veronica* **spp.**) include several annual and creeping perennial species that often form dense patches in turf. Because of their attractive blue flowers, some species have been used as ornamental plants in rock gardens. Once they have invaded a turf, some speedwells are difficult to control with conventional herbicides used for broadleaf weed control.

Herbicides

Herbicides vary in chemical structure and method of use; however, nearly all modern herbicides are organic chemicals that break down in the soil or in the plant within several weeks or months. Their breakdown products, called metabolites, are usually smaller organic or inorganic chemicals that in many cases are similar to chemicals that naturally occur in biological systems.

Types of Herbicides

Some herbicides are applied prior to the emergence of target weed species; these are called preemergence herbicides. Once the target weeds have emerged, preemergence herbicides are usually ineffective. The specific date at which a preemergence herbicide should be applied thus depends on the period in which weed germination takes place. For example, in temperate climates crabgrass typically germinates in midspring to late spring; a preemergence herbicide should be applied at least several weeks prior to anticipated germination to ensure control. As one proceeds toward the equator, crabgrass germination usually occurs earlier in the growing season, requiring earlier herbicide application. Postemergence herbicides are applied after weed emergence; applying them before anticipated or actual emergence usually will not result in control. Many annual grasses and broadleaf weeds can be controlled with either preemergence or postemergence herbicides. Perennial grasses are usually treated with postemergence herbicides.

Herbicides are applied to the foliage, where they are absorbed, or to the soil underlying the grass shoots, where they are absorbed by roots or immature organs of germinating seeds. Postemergence herbicides are usually foliar applied, while preemergence herbicides are soil applied. This distinction is of practical importance since a foliar-applied herbicide that is washed off the foliage by irrigation or rainfall shortly after application may not be effective. Conversely, a soil-applied herbicide that is retained on the foliage for an extended period may break down before reaching the soil (or thatch) as required to be effective. Granular formulations of soil-applied herbicides may offer some advantage over sprayable formulations in that they more easily bypass the foliage en route to the soil, especially when the foliage is dry. Where foliar-applied herbicides are formulated as granules, the leaves should be wet at the time of application to increase adherence of the granules to the foliage as an aid to absorption.

Postemergence herbicides are of two types: contact and systemic. Contact herbicides enter and are active in destroying those portions of the weed plant with which they come in contact. For annual weeds, contact herbicides may be quite effective. Perennial weeds often recover following treatment with a contact herbicide because of new growth from belowground regenerative organs. Systemic herbicides are translocated within the plant following absorption and are therefore more effective than contact herbicides for controlling perennial weeds.

Most herbicides used for controlling annual grasses and broadleaf weeds are selective. When applied in accordance with directions on the herbicide container, they control target weeds without seriously injuring desired turfgrasses. Currently, there are no herbicides with adequate selectivity for controlling most perennial grasses; therefore, nonselective herbicides must be used to control these weeds. Since nonselective herbicides will kill or injure all plants, they should be applied only to the target weed as a directed spray to minimize damage to the turf. On small sites nonselective herbicides can be applied with a soft brush or sponge to further limit their contact with plants within the area of application and therefore minimize injury to nontarget plants.

Herbicide Selection

Only about 10% of the current inventory of herbicides is used for weed control in turf. With the exception of nonselective herbicides used for spot treatment or renovation, turfgrass herbicides must provide adequate control of target weeds without causing unacceptable injury to desirable turfgrasses within the plant community. For a particular turfgrass species the list is even more restricted, as some turfgrasses are intolerant of herbicides that can be used on others. A list of currently available herbicides for use in turf is provided in Table 7.1. The information given on turfgrass tolerance is only a guide; a particular turfgrass species may vary in its response to a herbicide depending on the intensity of culture, the physiological condition of the plants, natural environmental conditions, and the specific cultivar or cultivars used. A particular turfgrass under severe stress may be injured by a herbicide to which it is normally tolerant.

Annual Grass Control

Most summer annual grasses can be controlled with preemergence or postemergence herbicides. Preemergence herbicides are often effective with only one properly timed application; however, in areas of severe weed pressure, two applications spaced six to eight weeks apart may be required for satisfactory weed control. Fenoxaprop (Acclaim) often controls summer annual grasses in cool-season turfgrass communities with a single postemergence application, while several applications of the organic arsenicals (two to four applications spaced seven to ten days apart) are usually necessary. On cool-season turfgrasses especially, organic arsenicals can be very phytotoxic; excessive application rates or use under high temperatures (>90°F) can be so injurious that weed invasion is actually encouraged.

A winter annual grass of importance, especially in intensively cultured turfs, is annual bluegrass. It can be controlled in bermudagrass turf with pronamide, but several weeks are required for the herbicide to be completely effective. An application of bensulide in late summer provides some preemergence control of annual bluegrass in bermudagrass turf. By allowing at least ninety days between herbicide applications and overseeding with cool-season turfgrasses, much of the annual bluegrass can be controlled, and temporary winter turf can be developed without interference from herbicide residues in the soil. When bermudagrass is completely dormant, glyphosate can be used to eliminate an existing stand of annual bluegrass. Glyphosate cannot be used, however, where overseeding with cool-season turfgrasses is planned, since overseeding must be initiated while temperatures still permit germination.

In cool-season turfgrass communities annual bluegrass control is more difficult. Preemergence herbicides (except siduron) can provide some control of new plants arising

Table 7.1. Herbicides Used for Weed Control in Turf

APPLICATION[a]				TURFGRASS TOLERANCE[b]										
Name	Name	Site	Timing	KB	CB	FF	TF	PR	BE	ZO	SA	CE	BA	Targets[c]
Asulam	Asulox	F	Po	N	N	N	N	N	I	N	I	N	N	AG
Atrazine	Aatrex	F, S	Pr, Po	N	N	N	N	N	I	I	S	S	N	BL, AG
Benefin[d]	Balan	S	Pr	S	I	I	S	S	S	S	S	S	S	AG
Bensulide	Betasan	S	Pr	S	S	S	S	S	S	S	S	S	S	AG
Bentazon	Basagran	F	Po	S	S	S	S	S	S	S	S	S	S	Nutsedge
Bromoxynil	Buctril	F	Po	S	S	S	S	S	S	S	S	S	S	BL
Chlorsulfuron	TFC	F	Po	S	I	S	N	N	S	N	N	N	S	BL
2,4-D	(Several)	F	Po	S	N	S	S	S	S	S	N	I	S	BL
DCPA	Dacthal	S	Pr	S	I	N	S	S	S	S	S	S	S	AG, Speedwell
Dicamba	Banvel	F	Po	S	I	I	S	S	S	S	I	S	S	BL
Diclofop	Illoxam	F	Po	N	N	N	N	N	S	N	N	N	N	Goosegrass
Dithiopyr	Dimension	F, S	Pr, Po	S	S	I	S	S	S	S	S	S	S	AG, BL
Ethofumisate	Progress	F	Po	I	N	N	N	S	N	N	N	N	N	Annual bluegrass
Fenoxaprop	Acclaim	F	Po	S	N	S	S	S	N	I	N	N	N	AG
Fluazifop-P-butyl	Fusilade	F	Po	N	N	N	S	N	N	S	N	N	N	AG
Glufosinate	Finale	F	Po	—	—	—	—	—	—	—	—	—	—	General, esp. PG
Glyphosate	Roundup	F	Po	—	—	—	—	—	—	—	—	—	—	General, esp. PG
Halosulfuron	Manage	F	Po	S	S	S	S	S	S	S	S	S	S	Nutsedge
Imazaquin	Image	F, S	Po, Pr	N	N	N	N	N	S	S	S	S	N	AG, BL
Isoxaben	Gallery	S	Pr	S	I	S	S	S	S	S	S	S	S	BL
Mecoprop	(Several)	F	Po	S	S	S	S	S	S	S	I	I	S	BL
Metolachlor	Pennant	S	Pr	N	N	N	N	N	S	S	S	S	S	AG
Metribuzin	Sencor	F, S	Pr, Po	N	N	N	N	N	S	I	I	I	I	BL, AG
Metsulfuron	DMC	F	Po	I	N	—	I	N	S	I	I	I	N	AG, BL
Napropamide	Devrinol	S	Pr	N	N	N	N	N	S	S	S	S	S	AG
Organic Arsenicals[e]	(Many)	F	Po	I	I	N	I	I	S	S	N	N	N	AG
Oryzalin	Surflan	S	Pr	N	N	N	I	N	S	S	S	S	S	AG
Oxadiazon	Ronstar	S	Pr	I	N	I	S	S	S	S	S	O	S	AG
Pendimethalin	(Several)	S	Pr	S	S	S	S	S	S	S	S	S	S	AG
Pronamide	Kerb	S	Pr, Po	N	N	N	N	N	S	O	O	O	O	Annual bluegrass
Sethoydim	Poast	F	Po	N	N	N	N	N	N	N	N	S	N	AG, PG
Siduron	Tupersan	S	Pr	S	I	S	S	S	N	S	S	S	S	AG
Simazine	Princep	S	Pr, Po	N	N	N	N	N	I	S	S	S	N	BL, AG
Triclopyr	(Several)	F	Po	S	N	I	S	S	N	N	N	N	I	BL

[a]Application sites include foliage (F) and soil (S); application timing is preemergence (Pr) or postemergence (Po).
[b]Turfgrasses listed are the cool-season species, Kentucky bluegrass (KB), creeping bentgrass (CB), fine fescues (FF), tall fescue (TF), and perennial ryegrass (PR); and warm-season species, bermudagrass(BE), zoysiagrass (ZO), St. Augustinegrass (SA), centipedegrass (CE), and bahiagrass (BA). Tolerances include safe (S), intermediate with possible short-term injury (I), not recommended because of severe injury or death (N), and unknown (O).
[c]Primary targets include annual grasses (AG), broadleaf weeds (BL), and perennial grasses (PG).
[d]Widely used in combination with trifluralin; commercially available as Team.
[e]Organic arsenicals include DSMA, MSMA, and MAMA.

from germinating seed, but they are ineffective against established populations of annual bluegrass. Ethofumisate (Prograss) works exceptionally well for postemergence control of annual bluegrass; however, only perennial ryegrass is truly tolerant, while other cool-season and warm-season turfgrasses may be injured or killed by this herbicide. This problem is further complicated by the fact that annual bluegrass populations often occur as large contiguous patches that are devoid of other turfgrasses. Success in eliminating these populations would result in large bare areas; consequently, desirable turfgrasses would have to be introduced by seeding or other propagation methods. Any herbicide residues in the soil might prevent successful establishment of the newly planted turfgrasses. On the other hand, if the herbicide residues had sufficiently dissipated, germinating annual bluegrass could be very competitive against other turfgrass seedlings, so that large-scale reinfestation of treated sites is likely. In cool-season turfgrass communities, therefore, annual bluegrass control depends more on ensuring that desirable turfgrasses survive and remain competitive than on herbicides.

A recent innovation is the use of Type II plant growth regulators (see the discussion on PGRs in Chapter 6) for differentially suppressing the growth of annual bluegrass in mixed stands with creeping bentgrass. Combined with cultural practices that favor the competitive growth of creeping bentgrass, a PGR-based differential suppression strategy offers several advantages over herbicide-based differential kill strategies because the functional quality of the turf can be maintained. This is especially important where large contiguous populations of annual bluegrass exist within the turfgrass community.

Perennial Grass Control

Most perennial grasses cannot be controlled selectively in cool-season turfgrass communities. Mechanical removal or spot treatment with nonselective herbicides (principally glyphosate or glufosinate) are the only means by which most of these perennial grasses can be removed. (An exception is the use of chlorsulfuron as a semiselective herbicide for taking out clumps of tall fescue and perennial ryegrass in Kentucky bluegrass turfs.) As with annual bluegrass, large voids resulting from the removal of perennial grass populations may require replanting to promote complete coverage by desirable turfgrasses. In most instances, however, perennial grasses do not reinfest the turf from seed and, with glyphosate or glufosinate, no soil residual activity exists to delay replanting. As long as the herbicide has had sufficient time to translocate throughout the target plants (at least three days), replanting operations can be initiated to introduce turfgrasses that are compatible with the rest of the plant community.

In warm-season turfgrass communities some perennial weed grasses can be controlled selectively. Dallisgrass and bahiagrass can be controlled in bermudagrass turf with several applications of an organic arsenical herbicide. Sethoxydim (Poast) can be used for perennial grass control in centipedegrass turf. Glyphosate is used for spot treating some weeds, and it is extensively used for controlling bermudagrass in sandtraps and other sites where it escapes from the turf.

Yellow nutsedge is a problem in both cool- and warm-season turfgrass communities. Although not a member of the grass family, it is a monocot and is usually discussed

Table 7.2. Susceptibilities of Broadleaf Weeds to Selective Control with 2, 4-D, Mecoprop, and Dicamba[a]

Common Name	Latin Name	2, 4-D	Mecoprop	Dicamba
Bindweed, field	*Convolvulus arvensis* **L.**	S-I	I	S
Bittercress, hairy	*Cardamine hirsuta* **L.**	S-I	S-I	S
Burclover, California	*Medicago polymorpha* **L.**	I-R	S	S
Buttercup, creeping	*Ranunculus repens* **L.**	S-I	I	S
Carpetweed	*Mollugo verticillata* **L.**	S	S	S
Carrot, wild	*Daucus carota* **L.**	I	I	S
Chicory	*Cichorium intybus* **L.**	S	S	S
Chickweek, common	*Stellaria media* **(L.) Cyrillo**	R	S-I	S
Chickweed, mouse-ear	*Cerastium vulgatum* **L.**	I-R	S-I	S
Cinquefoil, common	*Potentilla canadensis* **L.**	S-I	S-I	S-I
Clover, crimson	*Trifolium incarnatum* **L.**	S	S	S
Clover, hop	*Trifolium agrarium* **L.**	I	S	S
Clover, white	*Trifolium repens* **L.**	I	S	S
Cranesbill	*Geranium carolinianum* **L.**	S-I	S-I	S
Daisy, English	*Bellis perennis* **L.**	R	I	S
Daisy, oxeye	*Chrysanthemum laucanthmum* **L.**	I	I	I
Dandelion	*Taraxacum officinale* (**Weber**)	S	I	S
Garlic, wild	*Allium vineale* **L.**	I	R	S-I
Hawkweek, mouse-ear	*Hieracium pilosella* **L.**	S-I	R	S-I
Healall	*Prunella vulgaris* **L.**	S-I	S-I	S
Henbit	*Lamium amplexicaule* **L.**	I-R	I	S
Ivy, ground	*Glechoma hederacea* **L.**	I-R	I	S-I
Kapweed, spotted	*Centaurea maculosa* **Lam.**	I	I	S
Knotweed, prostrate	*Polygonum aviculare* **L.**	I-R	I	S
Lambs-quarters, common	*Chenopodium album* **L.**	S	S	S
Lespedeza, common	*Lespedeza striata* **(Thunb.) H. + A.**	I-R	S	S

under the perennial grass category. The traditional control for this species has been the organic arsenicals; however, bentazon has gained acceptance for controlling nutsedge because of its superior efficacy and greater safety to turfgrasses. Purple nutsedge is similar in appearance to yellow nutsedge but has been even more difficult to control; currently, imazaquin is recommended for control.

Broadleaf Weed Control

Most broadleaf weeds are susceptible to at least one of the following herbicides: 2,4-D, mecoprop, or dicamba. All are selective, systemic, foliar-applied, postemergence herbicides. Specific weeds and their relative susceptibilities to these herbicides are listed in Table 7.2. Usually, these herbicides are applied in combinations such as 2,4-D and mecoprop, 2,4-D and dicamba, or all three. The phenoxy herbicides (2,4-D, mecoprop) are relatively

Table 7.2. Continued

Common Name	Latin Name	2, 4-D	Mecocrop	Dicamba
Mallow, roundleaf	*Malva rotundifolia* **L.**	I-R	I	S-I
Medic, black	*Medicago lupulina* **L.**	R	I	S
Mugwort	*Artemisia vulgaris* **L.**	I	I-R	S-I
Mustard, wild	*Brassica kaber* **L.**	S	I	S
Onion, wild	*Allium canadense* **L.**	I	R	S-I
Pearlwort, bird's-eye	*Sagina procumbens* **L.**	I-R	S	S
Pennycress, field	*Thlaspi arvense* **L.**	S	I	S
Pennywort, lawn	*Hydrocotyle sibthorpioides* **Lam.**	S-I	S-I	—
Pepperweed, field	*Lepidium campestri* **(L.) R. Br.**	S	S-I	S
Parsley, Piert	*Alchemilla microcarpa* **Boissier Reuter**	R	S	S
Pigwood, prostrate	*Amaranthus blitoides* **Wats.**	S	S	S
Pineappleweed	*Matricaria matricarioides* **(Less.) Porter**	I-R	I	I
Plantain, broadleaf	*Plantago major* **L.**	S	I	R
Plaintain, buckhorn	*Plantago lanceolata* **L.**	S	I	R
Purslane, common	*Portulaca oleracea* **L.**	I	R	S
Shepherd's-purse	*Capsella bursa-pastoris* **(L.) Medic**	S	S-I	S
Sorrel, red	*Rumex acetosella* **L.**	I-R	R	S
Speedwells	*Veronica* **spp.**	I-R	I-R	I-R
Spurge, prostrate	*Euphorbia supina* **Raf.**	I	I	S-I
Spurweed	*Soliva sessilis* **P.+K.**	I-R	S-I	I-R
Strawberry, wild	*Fragaria vesca* **L.**	R	R	S-I
Thistles	*Cirsium* **spp.,** *Cardulus* **spp.**	S-I	I	S
Violet, wild	*Viola* **spp.**	I-R	I-R	I-R
Wood sorrel, yellow	*Oxalis stricta* **L.**	R	I-R	I
Yarrow, common	*Achillea millefolium* **L.**	I	I-R	S
Yellow, rocket	*Barbarea vulgaris* **(R.) Br.**	S-I	I	S-I

[a] S = Susceptible: I = intermediate, may require several applications for control: R = Resistant.

immobile in the soil and, therefore, pose no serious threat to nearby trees and shrubs from root absorption. However, dicamba, a benzoic acid herbicide, is quite mobile in the soil and can seriously injure woody ornamentals if sufficient amounts are absorbed by their roots. Dicamba applications should be avoided on those portions of a turf that cover the roots of taxus, junipers, and other highly sensitive ornamental plants. Triclopyr, a relatively new addition to this group of herbicides, controls many broadleaf weeds that cannot be controlled with the aforementioned herbicides. Also, dichloroprop, a phenoxy herbicide, has worked well in combination with 2,4-D for controlling various broadleaf weed species.

Usually, the most favorable results from the use of these herbicides applied postemergence for broadleaf weed control occur during periods when the weeds and turfgrass are vigorously growing. At this time, herbicide absorption, translocation, and action in weeds is favored, and turfgrasses are better able to grow into voids remaining after weeds have been controlled.

In warm-season turfgrass communities other herbicides, including the triazines, metribuzin, and pronamide, are used for broadleaf weed control. In St. Augustinegrass and centipedegrass the triazines have been relied on for controlling broadleaf weeds because of the sensitivity of these grasses to the phenoxy herbicides.

Some control of annual broadleaf weeds is obtained from the preemergence herbicides used for controlling annual grasses. This has been a largely unrecognized benefit from these herbicides; however, with the introduction of dithiopyr (Dimension) and isoxaben (Gallery), preemergence control is now an alternative to the traditional reliance on postemergence herbicides for controlling broadleaf weeds.

DISEASE CONTROL

A disease may be defined as an abnormal condition of plants resulting from alterations in their physiological processes and morphological development and caused by some adverse environmental factor. Within this definition, a disease could result from many causes, including such abiotic factors as nutritional deficiencies, poor drainage, and traffic. In a large percentage of diseases, however, infectious microorganisms, called pathogens, enter into the complex and produce characteristic symptoms of disease. Thus, pathogens are typically regarded as causes or agents of disease. More realistically, a disease is the end result of three factors occurring simultaneously: a susceptible host, a pathogen capable of infecting the host, and environmental conditions that favor the development of disease.

Bacteria are causal organisms of some plant diseases, but, until recently, there were no recognized bacterial diseases of turfgrasses. In 1981 a bacterial wilt of Toronto creeping bentgrass was discovered through electron microscopic examination of diseased samples. Further use of this diagnostic technique recently uncovered other bacterial diseases of turfgrasses. Some viral diseases occur—the most important is St. Augustine decline—but these are believed to be relatively minor in other turfgrasses.

Nearly all major infectious diseases of turfgrasses are caused by fungi. Fungi are simple plants that lack chlorophyll and, therefore, the capacity to produce food photosynthetically. Consequently, they must obtain their nutrition from outside sources. Some fungi, called parasites, obtain their nutrition from living hosts; others, called saprophytes, feed on organic residue. Among the parasitic fungi are those that live as saprophytes until environmental conditions become favorable for infecting living plants. Facultative parasites, including many species of such genera as *Fusarium, Pythium,* and *Rhizoctonia,* function primarily as saprophytes, but can become parasitic under certain conditions. Facultative saprophytes function primarily as parasites but can subsist temporarily on organic residue; an example is *Drechslera.* In contrast, obligate parasites, such as *Puccinia* and *Erysiphe,* can complete their life cycles only in intimate association with a living host. When the host plant dies the pathogen normally dies as well. Future survival of obligate parasites therefore requires that hosts survive the disease or that alternate hosts be available.

The process of disease development (pathogenesis) begins with the penetration of the host plant by the pathogen. This occurs following germination of microscopic spores or sclerotia (small compact resting bodies of the fungus) and the growth of germ tubes on the plant surfaces. Penetration takes place through stomates, cut ends of leaves, and other wounds, or directly through cell walls. Within the plant the fungus grows as a multibranched series of hyphae (thin transparent filaments) to form a loosely arranged fungal body called a mycelium. Because of the release of phytotoxins and enzymes by the fungal hyphae, plant cells may lose structural integrity and die. Nourishment and growth of the fungus are accompanied by deterioration of plant tissues as the transfer of food materials takes place. Once the pathogen becomes established within the host plant, the plant is said to be infected. Following an incubation period of several days or weeks, disease symptoms begin to appear. Most fungi reproduce by sexual or asexual spores, which subsequently infect the same or adjacent plants or are carried long distances by wind, water, or other agents to infect other plants. Favorable environmental conditions for disease development promote numerous successive infections, resulting in an epidemic (actually, epiphytotic in plants) with severe losses of turfgrass. Under unfavorable conditions some pathogens can survive in a resting stage as thick-walled spores, or as sclerotia. These persist in the turf until the return of environmental conditions favorable for germination and subsequent infectious activity.

Disease control can be achieved by several independent or interrelated methods. Since to have disease development there must be a susceptible host, a virulent pathogen, and favorable environmental conditions, any method that reduces host susceptibility, controls the pathogen, or modifies the environment so that disease development is not favored may effectively control a disease. Where disease-resistant cultivars are available, these can be substituted for susceptible ones. For example, substituting Warren's A-20 Kentucky bluegrass for Merion can result in a reduction of stripe smut, powdery mildew, stem rust, and, possibly, summer patch diseases. Direct control of the pathogen usually involves the application of fungicides. Susceptible plants can be protected from infection by many pathogens through timely applications of effective fungicides during periods of disease activity. The third method is environmental modification. A specific set of environmental conditions can favor both host susceptibility and activity of the pathogen. For example, *Pythium aphanidermatum,* a water mold, is favored by warm humid weather and high soil moisture. Under these same conditions cool-season turfgrasses may be under considerable stress and, therefore, less resistant to this fungus. Such procedures as improving air movement and soil drainage and avoiding excessive fertilization and irrigation can substantially reduce the likelihood of *Pythium* blight. Since weather conditions cannot be controlled, fungicides may still be required in addition to the aforementioned procedures. Disease control with fungicides is usually most effective where steps have been taken to change environmental conditions to favor turfgrass growth over infection by the pathogen and disease development.

Diseases

The diseases of primary importance in turf are listed in Table 7.3 along with their causal organisms and susceptible turfgrass species. Each turfgrass has its specific complement of diseases; thus, managing different turfgrasses involves managing different disease problems as part of the cultural program. For example, converting from Kentucky bluegrass

Table 7.3. Turfgrass Diseases, Pathogens, and Susceptible Turfgrasses

		SUSCEPTIBLE TURFGRASSES*											
Disease	Pathogen	*KB*	*AB*	*CrB*	*CoB*	*FF*	*TF*	*PR*	*BE*	*ZO*	*SA*	*CE*	*BA*
Anthracnose	*Colletotrichum graminicola*	X	XX	X		XX		XX	X				
Bacterial wilt	*Xanthomonas campestris*	X	X	X				X	X				
Bermudagrass decline	*Gaeumannomyces graminis*								X				
Brown patch	*Rhizoctonia solani; R.zeae*	X	XX	XX	XX	X	XX	XX	X	X	XX	X	X
Copper spot	*Gloeocereospora sorghi*			XX	X								
Dollar spot	*Sclerotinia homeocarpa*	X	XX	XX	XX	X		X	XX	X	X	X	X
Downy mildew	*Sclerophthora macrospora*										X		
Fairy ring	*Agaricus* **spp.,** *Lepiota* **spp.,** *Marasmius* **spp.**, etc.	X	X	X	X	X	X	X	X	X	X	X	X
Gray leaf spot	*Pyricularia grisea*						X	X	X		XX		
Helminthosporium diseases													
melting out/leaf spot	*Drechslera poae* (formerly *H. vagans*)	XX											
leaf spot	*Bipolaris sorokiniana* (formerly *H. sorokinianum*)	XX	XX	XX	X	XX	X	X					
blight (netblotch)	*Drechslera dictyoides* (formerly *H. dictyoides*)					XX	XX						
red leaf spot	*Drechslera erythrospila* (formerly *H. erythrospilum*)			X	X								
brown blight	*Drechslera siccans* (formerly *H. siccans*)							XX					
leaf blotch	*Bipolaris cynodontis* (formerly *H. cynodontis*)								XX				
zonate leaf spot	*Drechslera gigantea* (formerly *H. giganteum*)								XX				
crown and root rot	*Bipolaris tetramera* (formerly *H. tetramera*)								XX	XX			
Necrotic ring spot	*Leptosphaeria korrae*	XX	X			X							
Pink patch	*Limonomyces roseipellis*	X	X	X	X	XX		XX					
Powdery mildew	*Erysiphe graminis*	XX				X							
Pythium blights	*Pythium aphanidermatum: P. ultimum*	X	XX	XX	XX	X	X	XX	X				

Red thread	*Laetisaria fruciformis*	XX	X	X		XX		XX	X				
Rusts													
stem rust	*Puccinia graminis*	XX	X	X		X			X				
stripe rust	*Puccinia striformis*	XX											
crown rust	*Puccinia coronata*			X	X		X	XX					
Bermudagrass rust	*Puccinia cynodonitis*								XX				
St. Augustinegrass rust	*Puccinia stenotaphii*										X		
zoysiagrass rust	*Puccinia zoysiae*										XX		
leaf rust	*Puccinia crandaleii*		X			X							
leaf rust	*Puccinia poae-subeticae*	X	X									X	
leaf rust	*Puccinia recondita*	X	X			X							
Slime molds	*Physarum cinereum: Mucilago spongiosa*	X	X	X	X	X	X	X	X	X	X	X	X
Smuts													
stripe smut	*Ustilago striiformis*	XX		XX	X								
flag smut	*Urocystis agropyri*	XX		X	X								
Snow molds													
Typhula blight	*Typhula incarnata; T. ishikarienis*	X	XX	XX	XX	X	X	XX					
Microdochium patch	*Microdochium nivale*	X	XX	XX	XX	X	X	X					
snow scald	*Myriosclerotinia* (or *Sclerotinia) borealis*	X	XX	XX		X	X	X					
Coprinus snow mold	*coprinus psychromorbidus*	X											
Southern blight	*Sclerotium rolfsii*	X	X	X				X	X				
Spring dead spot	*Leptosphaeria* **spp.** *Gaeumannomyces graminis* *Ophiosphaerella herpotricha*								XX				
St. Augustinegrass decline	SAD virus										XX		
Summer patch	*Magnoporthe poae*	XX	XX			XX							
Take-all patch	*Gaeumannomyces graminis*			XX	XX								
Yellow tuft	*Sclerophthora macrospora*	X	X	X	X	X	X	X					
White blight	*Trechispora* **spp.,** *Melanotus phillipsii*						X						

*Susceptible turfgrasses include: Kentucky bluegrass (KB), annual bluegrass (AB), creeping bentgrass (CrB), colonial bentgrass (CoB), fine fescues (FF), tall fescue (TF), perennial ryegrass (PR), Bermudagrass (BE), zoysiagrass (ZO), St. Augustinegrass (SA), centipedegrass (CE), and bahiagrass (BA). Where a particular disease is of primary importance for a specific turfgrass, the symbol XX is used; X indicates that the turfgrass may be susceptible to the disease under some conditions or is of minor importance.

to perennial ryegrass may result in reduced problems from summer patch, *Drechslera* leaf spot, powdery mildew, stem rust, and smuts, but increased problems are likely from anthracnose, *Drechslera* brown blight, *Pythium* blights, brown patch, crown rust, and *Typhula* blight. Cultivars within these turfgrass species may differ substantially in their relative susceptibilities to various diseases. Also, climate greatly influences the degree to which a particular disease can develop and affect turf quality. Since many new turfgrass cultivars have been introduced in recent years and many more are likely to be developed in the future, local authorities should be consulted for recommendations when selecting new turfgrasses for planting.

The diseases listed below are presented alphabetically, not by their importance or by the time of year when infection occurs. See the inside back cover for illustrations of the foliar and community symptoms of several important diseases.

Anthracnose

Anthracnose, caused by *Colletotrichum graminicola,* a facultative parasite, is typically found in association with various *Drechslera*- or *Bipolaris*-incited diseases. It is characterized by irregular patches of yellow bronze, chlorotic, or blighted turfgrass. On diseased leaves, elongated reddish brown lesions and numerous black spiny fruiting bodies (acervuli) appear and are important diagnostic features. The fungus is a facultative parasite that until recently was not considered a primary cause of disease in turfgrasses. In annual bluegrass the fungus can be active during cool or warm weather, but turfgrass death usually occurs during hot (optimum 80° to 85°F), prolonged moist conditions when the plants are weakened and under stress. Inadequately or excessively fertilized turfs are more likely to be affected by this disease than turf sustained under a moderate level of nitrogen fertilization.

Bacterial Wilt

Bacterial wilt symptoms, caused by *Xanthomonas campestris,* first appear as leaf wilting and shriveling from the tip down. Subsequently, roots and stems turn brown and decompose. Scanning-electron-microscopic examination of infected tissues reveals the presence of numerous rod-shaped bacteria within the xylem vessels. First discovered in Toronto creeping bentgrass in 1980, infectious bacterial diseases have been observed in several other turfgrass species. Treatment of diseased turf with high rates of oxytetracycline, a bactericide, has provided temporary control.

Bermudagrass Decline

Bermudagrass decline, caused by *Gaeumannomyces graminis,* is a newly recognized disease of intensively cultured bermudagrass that occurs during hot wet summer periods. It first appears as chlorosis of the lower leaves, then spreads to the younger leaves. The plants eventually turn dark brown and die. Roots associated with infected plants shorten and turn black. Dark strands of fungal mycelia, called runner hyphae, may be evident during microscopic examination of infected roots.

Brown Patch

Brown patch, or large brown patch, is caused by the common soil-borne fungus *Rhizoctonia solani,* a facultative parasite that attacks many plants, including all turfgrasses. Brown patch is a widespread disease in warm temperate and warmer climates on many turfgrasses. Symptoms appear as roughly circular patches of thinned or blighted turfgrass. In early morning, on close-cut turfgrasses, a dark smoky ring may appear at the periphery of the patch; this disappears as the day progresses so that by late morning the patch is a uniformly light brown or straw color. The pathogen, a facultative parasite, overwinters as small dark brown to black sclerotia and mycelia in living and dead plant tissues and in the top half-inch of soil. Hyphae grow from the sclerotia in all directions. The water-soaked appearance of infected plants is due to the oozing of dead cell contents into intercellular spaces. In this condition, leaves quickly die as wilting and desiccation occur. Saprophytic growth of the fungus can occur on leaf surfaces in the guttation water prior to penetration; thus, removal of guttation water by poling, syringing, or other methods can substantially reduce disease incidence. Although this disease is favored by warm to hot (optimum 75° to 95°F), moist conditions, a "cold-temperature brown patch," also called yellow patch (*Rhizoctonia cerealis*), can occur during late fall, winter, or early spring that resembles summer patch. Poor surface and subsurface drainage and excessive nitrogen fertilization are factors that substantially increase the severity of this disease.

Copper Spot

Copper spot is first evident as small reddish spots (lesions) on the leaves. As the lesions enlarge and coalesce, entire leaves become blighted and small salmon pink to copper-colored patches 1 to 3 inches in diameter appear in the turf. The pathogen (*Gloeocospora sorghi*) overwinters as small black sclerotia that germinate in spring. The disease can spread rapidly during warm (optimum 65° to 80°F), moist weather as large numbers of asexual spores (conidia) are produced and infect other plants. The spores are spread by splashing or flowing water and by equipment.

Damping-Off and Seedling Blight

Damping-off and seedling blight are collective terms applied to several seedling diseases, including those caused by species of *Pythium, Fusarium, Drechslera* or *Bipolaris,* and *Rhizoctonia.* It can occur prior to seedling emergence with seed rotting in the soil (preemergence damping-off) or after the seedlings have emerged (postemergence damping-off). In the latter case the seedlings usually become water soaked, turn yellow to brown, and finally collapse. The end result is thin stands or irregular patches of dying grass. The development of *Pythium* damping-off is usually favored by cool wet conditions, while species of *Rhizoctonia, Drechslera,* and *Fusarium* are usually most serious during warm dry to wet weather. Seed treatment with captan or thiram is helpful in controlling damping-off. Very high seeding rates should be avoided, as exceptionally dense seedling stands are generally more susceptible to these diseases. Proper seedbed preparation, adequate drainage, and avoidance of excessive irrigation are also important.

Dollar Spot

Dollar spot is one of the most destructive diseases of closely mowed turf, especially creeping bentgrass. Symptoms appear as bleached spots about the size of a silver dollar that may be so numerous that individual spots overlap to produce large, irregular areas of sunken, dead turfgrass. On higher-cut turfs the spots may be 4 to 6 inches in diameter. Individual leaves at the periphery of the spots typically have straw-colored bands across the blades with reddish brown borders. The pathogen, *Sclerotinia homeocarpa,* a facultative saprophyte, overwinters as mycelia in previously infected plants and as stromata on foliar surfaces. Under favorable environmental conditions the dormant mycelia or stromata resume growth on leaves and out into the humid air. (This white cottony growth of aerial mycelia observed when dew is present in early morning may be confused with similar growth of *Pythium* and *Nigrospora* fungi.) When the aerial mycelia contact moist foliage they may penetrate the leaf and cause infection. One of the principal reasons for removing the dew on golf greens in the early morning hours is to reduce the severity of dollar spot disease. Prolonged high humidity in the turfgrass canopy is required for fungal growth. Different variants of the dollar spot fungi initiate growth under different temperatures; thus, disease development may occur from late spring to late fall. This disease is most severe on inadequately fertilized turfgrasses; besides regular fungicide use on greens, a major control measure is the implementation of an adequate nitrogen fertilization program.

Downy Mildew (Yellow Tuft)

Downy mildew of St. Augustinegrass appears as white raised linear streaks that develop parallel to the midveins on infected leaves. Streaks appear in the spring and remain throughout the summer, giving the leaves a chlorotic appearance with some necrosis toward the tips. Early symptoms may be confused with the viral disease St. Augustinegrass decline; however, virus symptoms are more yellow in color and more mottled than striped. The name yellow tuft is used for this disease when it occurs on cool-season turfgrasses. Early symptoms include slightly stunted growth and slightly thickened or broadened leaf blades, without any discoloration. When the disease is severe small yellow patches appear in the turf. Each patch contains a dense cluster of excessively tillered yellowed shoots with few roots.

The pathogen, *Sclerophthora macrospora,* is an obligate parasite that persists as systemic mycelia in infected plants. In moist weather, the fungus produces fruiting bodies called sporangiophores that protrude through the leaf stomates. Sporangia germinate to release zoospores that swim in moisture films, then encyst and germinate to produce hyphal strands that infect plant tissues. As the fungus depends on free moisture for movement of infectious spores, the disease is primarily associated with poorly drained or intensively irrigated turfs.

Fairy Rings

Fairy rings appear as circles or arcs of dark green grass, often with thin or dead grass just inside or outside the rings. Fairy rings can be caused by any one of about sixty species of soil-inhabiting fungi that feed on decaying organic matter. Sometimes mushrooms appear in the rings, especially following rainfall or intensive irrigation. The dark green color is due to the release of nitrogen from fungi-induced decomposition of soil organic matter. As the fungal mass grows the soil can become hydrophobic, resulting in desiccation of the turfgrass. There is no easy or inexpensive way to control this disease. The most effective measure involves stripping the sod, fumigating the soil, and replanting with clean sod.

Fusarium Blight

Previously, the name *Fusarium* blight referred to a *Fusarium*-incited patch or "frog-eye" disease of Kentucky bluegrass occurring primarily during the summer months. Recent investigations of this disease revealed that the patch diseases are caused by *Magnoporthe poae* (now summer patch), *Leptosphaeria korrae* (now necrotic ring spot), and possibly other fungi. This is in spite of the ubiquitous presence of *Fusarium* species associated with the final stages of disease development. With respect to summer patch, symptom development may require subsequent infection by *Fusarium* species in some instances.

Fusarium blight is still a valid name for the crown and root rot disease incited by this fungus; however, this disease is distinct from the patch diseases cited previously. In addition, *Fusaria* are recognized pathogens for *Fusarium* leaf spot and seedling diseases (see damping-off).

Gray Leaf Spot

Gray or *Piricularia* leaf spot of St. Augustinegrass, caused by the fungus *Piricularia grisea,* first appears as brown to gray dots on the leaves and stems that enlarge to form round to elongate spots with gray centers, reddish brown to purple margins, and outer rings of chlorotic tissue. Numerous lesions can develop on a single leaf, resulting in a scorched appearance of the grass. The pathogen overwinters as dormant mycelia and free spores (conidia) in older infected leaves and debris; new infections occur during warm moist weather in spring and continue into the summer and fall months. Disease activity is favored by excessive nitrogen fertilization, prolonged periods of high moisture, and temperatures between 70° and 85°F.

Helminthosporium Diseases

The fungi that cause *Helminthosporium* diseases are now recognized as species of *Drechslera* and *Bipolaris;* these are facultative saprophytes that cause several diseases of primary importance in cool- and warm-season turfgrasses. With one major exception, *Drechslera* fungi cause diseases of cool-season turfgrasses only; these include leaf spot and melting-out of Kentucky bluegrass, blight (netblotch) of ryegrasses and fescues, brown blight of ryegrass, and red leaf spot of bentgrasses. The exception is zonate leaf spot of bermudagrass.

Leaf lesions from *D. poae* on Kentucky bluegrass first appear as small water-soaked areas that soon become reddish brown to purplish black. These lesions are often surrounded by a yellow zone that fades into the adjacent green tissue. As the lesions enlarge, their centers change to brown and then whitish in color. Older leaves are more susceptible to attack than are younger leaves. When lesions extend across an entire leaf, the leaf or tiller dies and drops from the plant. In severe cases numerous leaves and tillers are killed, causing severe thinning of the turf; thus the name melting-out. Where roots and stems are invaded directly, wilting and chlorosis of the shoots occur followed by death of the plants. Leaf spots occur during cool wet weather in spring and fall, while melting-out occurs mainly during warm dry weather or during wet periods immediately following dry periods.

On fine-leaf fescues leaf lesions caused by *D. dictyoides* appear as small reddish brown spots. The lesions quickly girdle the fine-textured leaves, causing yellowing and dieback from the tip. Small brown patches of dead shoots commonly occur over large areas when the disease is severe. On tall fescue and perennial ryegrass, *D. dictyoides* causes a fine network of short brown streaks on infected leaves; thus, the name netblotch on these coarse-textured grasses. Eventually, the netlike patterns coalesce to form dark brown solid spots. Heavily infected turfs turn yellow, tip dieback occurs, and ultimately the infected areas turn brown. Direct infection of subterranean plant parts results in general wilting, chlorosis, and finally death of the plants.

On ryegrasses *D. siccans* produces numerous small chocolate brown lesions; as these enlarge, the centers become tan to white. Girdled leaves turn yellow and die back from the tip, causing severe thinning (brown blight) of the turf. As with other *Drechslera*-incited diseases, direct infection of subterranean plant parts can cause wilting, chlorosis, and death of shoots.

On bentgrass *D. erythrospilum* produces small brown to reddish brown leaf lesions. As these coalesce the turfgrass develops an overall reddish cast; thus, the name red leaf spot. Girdling of infected leaves may cause the appearance of drought-stressed turf regardless of soil moisture. As with other *Drechslera*-incited diseases, direct infection of roots and stems may also occur.

On bermudagrass *D. gigantea* produces tiny brown spots that, as they enlarge, may give rise to concentric bands of bleached and brown tissue; thus the name zonate leaf spot. Severely diseased turf becomes thinned as the leaves are killed. This disease typically occurs during the fall, when bermudagrass growth is slowed by cool temperatures. *D. gigantea* may also infect several cool-season (festucoid) turfgrasses during the summer, when growth is slowed by warm to hot temperatures.

With the exception of zonate leaf spot and red leaf spot, which occur during warm weather on cool-season turfgrasses, *Drechslera*-incited diseases are most active during cool wet weather of spring, fall, and mild winters.

With one major exception, *Bipolaris* fungi cause diseases of warm-season (panicoid and eragrostoid) turfgrasses only; these include leaf blotch of bermudagrass and crown and root rot of bermudagrass and zoysiagrass. The exception is leaf spot of numerous cool-season turfgrasses. Leaf lesions from *B. cynodontis,* the causal organism for leaf blotch, are brownish green to black. Severely infected plants turn straw colored, and may occur in large irregular patches. *B. tetramera,* the causal organism for crown and root rot disease, produces purple to black lesions on stems and a general thinning of the turf. *B.*

sorokiniana causes leaf, crown, and root diseases that are essentially identical in appearance to *Drechslera*-incited diseases on bluegrasses, fescues, ryegrasses, and bentgrasses. Leaf infections in *Bipolaris*-incited diseases of warm-season grasses occur during cool wet weather from fall through spring, while stem and root rots occur during warm summer weather. *B. sorokiniana* causes leaf spot on cool-season grasses during warm wet weather in midsummer.

Mushrooms and Puffballs

Mushrooms (commonly referred to as "toadstools") and puffballs are the fruiting structures of saprophytic, basidiomycetous fungi. Their appearance is indicative of decaying organic matter in the soil, especially buried tree stumps, dead roots, logs, boards, or thick thatch. They are most likely to develop following heavy rainfall or intensive irrigation. Although it may be possible to strip the sod and remove the organic debris serving as the fungal food source, the most practical control for this problem is usually to mow or cut off the fruiting structures as they appear, while allowing the fungus to decompose the organic material. The mushrooms and puffballs will disappear when their food base has been exhausted.

Necrotic Ring Spot

Previously called *Fusarium* blight, necrotic ring spot occurs over a broad temperature range, but appears to be more prevalent during cool weather in spring and fall (in contrast to summer patch, which is a warm-weather disease). Like summer patch, symptoms appear as typical "frog-eyes" or crescent-shaped areas of dead turfgrass; however, the pathogen, *Leptosphaeria korrae,* also known to cause spring dead spot of bermudagrass, is highly pathogenic on all Kentucky bluegrasses tested to date.

Pink Patch

Previously, pink patch was synonymous with red thread disease, with *Corticium fruciformis* as the common pathogen. These are now recognized as separate diseases, with pink patch incited by *Liminomyces roseipellis,* a basidiomycete with distinctive clamped mycelial connections. The pink patch pathogen forms a pink to reddish film on infected leaves, but it lacks threadlike mycelial growth from leaf tips and the pink cottony flocks of arthroconidia (sporelike mycelial segments with the capacity to cause infection) that occur with red thread. Otherwise, it is similar to red thread in disease cycle and epidemiology.

Powdery Mildew

Powdery mildew is a serious disease that affects many Kentucky bluegrass cultivars growing in shaded environments. It first appears as small superficial patches of white mycelia on leaf surfaces. As the mycelial cover increases the leaves become chlorotic and then die. An infected turfgrass appears dull white as if dusted with flour. The causal fungus (*Erysiphe graminis*), an obligate parasite, overwinters as mycelial mats on live plants or as cleistothecia on dead plant tissue. The disease is active during cool (optimum 65°F) humid cloudy weather in the spring and fall, when days are mild and nights are cool, and is especially severe on moderately to heavily shaded sites. Fungicides are effective for controlling this disease. The use of shade-adapted turfgrass species and cultivars, however, is preferable for sustaining disease-free turf.

Pythium Diseases

Pythium blight (also known as grease spot, spot blight, cottony blight) is of primary importance on cool-season turfgrasses and bermudagrass, especially those sustained under intensive culture. Among the most destructive of turfgrass diseases, *Pythium* blight can completely destroy a turf in a single twenty-four-hour period under conditions favorable for disease development. In subtropical climates overseeded grasses are highly susceptible to *Pythium* blight during the first few weeks following planting in late fall. In temperate climates the disease usually occurs during summer periods of hot (optimum 80° to 95°F) wet or very humid weather. The disease first appears as small circular spots (up to 6 inches in diameter) or elongated streaks in the turf. In early morning infected plants are water soaked and dark colored, often supporting a cottony growth of gray to white mycelia. As the grass dries the mycelia disappear and the grass leaves collapse, turning reddish brown and finally straw colored. On higher cut turfgrasses, spots tend to be larger and individual leaves at the periphery often show a white or straw-colored banding across the blades, which usually collapse. Species of *Pythium* are facultative parasites that overwinter as dormant mycelia and thick-walled oospores in previously infected plants. The disease spreads primarily by rapid mycelial growth from plant to plant or by transport of mycelia by mowers and other equipment, traffic, and surface-water movement. Although the primary *Pythium* species, *P. ultimum* and *P. aphanidermatum,* are most active in hot humid weather, *P. graminicola* and other species of *Pythium* can be highly destructive during cooler weather in midspring. The symptoms of cool-weather *Pythium* are similar to those of red leaf spot; the turf is severely thinned and infected shoots are reddish in color. As noted earlier, species of *Pythium* can be destructive to germinating seeds and young seedlings. *Pythium* blight is especially severe in poorly drained turfs where air circulation is reduced by adjacent densely growing trees and shrubs. Excessive nitrogen fertilization in summer can predispose Kentucky bluegrass, bentgrasses, and other turfgrasses to severe incidences of *Pythium* blight.

Red Thread

Red thread symptoms, caused by *Laetisaria fruciformis* (formerly *Corticium fruciformis*), typically appear on turfgrass plants as water-soaked, then tan to bleached leaves, yellowing and shriveling from the tips, and matted together by pink gelatinous mycelia. On colonized leaves, the fungus frequently grows beyond the tips of the blades to form red threadlike or antlerlike mycelial growths. Pink cottony flocks of mycelia, up to 1/2 inch in diameter may also be produced on infected leaves.

The pathogen overwinters as mycelial fragments on living and dead plants. The disease is spread by mycelia from infection centers during cool moist weather. It is more severe on slow-growing turfgrasses than on turfgrasses where growth is vigorous.

Rusts

The rusts include several important diseases caused by obligate parasites of the genus *Puccinia.* They were formerly regarded as minor diseases of turfgrasses. With the introduction of susceptible cultivars of some cool- and warm-season turfgrasses, rusts became important diseases of turf. Leaf rust, caused by *P. poae-subeticae* and *P. recondita,* and stem rust, caused by *P. graminis,* can be serious diseases of some Kentucky bluegrasses. Stripe rust, caused by *P. striiformis,* also affects Kentucky bluegrass, especially in cool temperate oceanic climates. *P. cynodontis* is the primary causal organism for bermudagrass rust, while rust on zoysiagrasses is caused by *P. zoysiae.* Tall fescue and especially the ryegrasses are susceptible to crown rust, caused by *P. coronata.*

Early symptoms of rust disease are observed as light yellow flecks on leaves. With further disease development, the cuticle and epidermis of the leaves rupture and the lesions develop into reddish brown or orange pustules. Eventually, the leaves turn yellow to brown and the entire stand of turfgrass may be thinned and weakened.

The cycle for development of a rust disease is quite complex due to the number of spore stages and alternate hosts (woody and herbaceous plants) involved. Generally, rust is more serious on turfgrasses with reduced rates of growth due to inadequate nitrogen fertilization, insufficient water, or other growth-limiting factors. However, during severe epidemics, nitrogen fertilization plus adequate irrigation may not be sufficient to control the disease.

Slime Molds

Technically, slime mold is not a disease; the causal organisms are surface saprophytes (epiphytes) that use leaf surfaces merely to support their reproductive structures. However, some turfgrass injury may result from extensive prolonged coverage by the fruiting bodies. The slime mold fungi, chiefly *Physerum cinerem* and *Mucilago spongiosa,* appear initially as a creamy white to greasy black slimy growth. Eventually, this overgrowth changes to a powdery ash gray, bluish gray, tan to orange, black, or creamy white growth that is easily rubbed off. Mechanical removal by irrigation, mowing, brushing, raking, or other practice is usually all that is necessary for control.

Smuts

Leaf smuts, including stripe smut (*Ustilago striiformis*) and flag smut (*Urocystis agropyri*), are important diseases of several cool-season turfgrasses. The symptoms of both diseases are identical in mowed turf and both fungi may infect the same plant or leaf simultaneously. Infected leaves tend to be stiff, upright, and stunted in growth. Symptoms appear initially in cool weather (optimum 50° to 60°F) as long yellow green streaks along the leaves. As the disease progresses, the streaks become dull gray in color. The cuticle and epidermis covering the streaks soon rupture and expose the blackish brown spore masses of the pathogen. The leaf subsequently splits and curls from the tips downward, turns brown, and withers. Under summer conditions infected plants often lack stress tolerance and die in irregular patches. The pathogens overwinter as dormant mycelia in the crowns and nodes of infected plants and as resting spores (chlamydospores) in the thatch and soil. The spores are disseminated by wind, rain, mowing, irrigation, and other cultural practices. Spores are also transported on seed and in infected sod that does not show symptoms. The spores germinate in a film of water and invade actively growing belowground grass tissues directly. After penetrating the host the mycelia grow systemically throughout the plant. Within Kentucky bluegrass and creeping bentgrass cultivars vary widely in their susceptibility to the leaf smuts. Futhermore, a previously nonsusceptible cultivar may in time become susceptible to these diseases as new pathogenic races of the fungus develop. Specific races of the pathogen are highly host specific. Where possible, therefore, blends of Kentucky bluegrass cultivars and mixtures with other compatible turfgrasses should be used as a measure to prevent the devastation of a turf that may occur where a particular cultivar becomes susceptible. As with summer patch, *Rhizoctonia* brown patch, and several *Helminthosporium* diseases, excessive nitrogen fertilization applied in spring can result in greater disease severity.

Snow Molds

The snow molds include several diseases that occur from late fall to midspring. Some require snow cover, while others develop during cool wet periods in the absence of snow.

Microdocium patch (pink snow mold) is caused by *Microdocium nivale,* a facultative saprophyte that is active during prolonged cool wet weather from fall to midspring. The disease is common in temperate oceanic climates in the absence of snow. Infection and disease development occur most rapidly at temperatures of 32° to 45°F (maximum at about 65°F). On closely mowed turf symptoms appear as small roughly circular spots commonly 1 to 6 inches in diameter. Diseased areas are first tan, reddish brown to blackish brown in appearance, and then turn tan or white as the shoots die. In cool wet weather the leaves are matted together and are covered with a whitish pink mycelial growth that is slimy when wet. When exposed to sunlight the spots may exhibit a pink coloration; hence the name pink snow mold. The pathogen survives unfavorable conditions as dormant mycelia or chlamydospores in living or dead grass plants and debris. Infection results from mycelial growth and also from the germination of asexual spores (conidia) that invade through the stomates. Cultural factors that favor disease development include nitrogen fertilization that stimulates lush growth in fall and winter, mulching turfs for winter protection, and allowing the development of excessive thatch.

Typhula blight (gray snow mold) is caused by *Typhula incarnata* and related species. This disease occurs in cool temperate and subarctic climates. It normally develops under a deep and prolonged snow cover at temperatures of 30° to 45°F. Symptoms appear initially as patches of bleached brown to straw-colored foliage up to 2 feet or more in diameter and matted together with a fluffy white to bluish gray growth of mycelia. At other times a silvery membranous crust develops over the damaged turf. Numerous minute yellow to tan or dark brown sclerotia form on or in the shoot tissue. In warmer regions the damage caused by the pathogen is often superficial, and recovery of the turf is rapid once warm and dry conditions exist. The pathogen, a facultative saprophyte, survives unfavorable conditions as minute sclerotia in the thatch. Germination occurs in late fall or winter during cold wet weather to form fruiting bodies that fuse and form mycelial colonies in the thatch. Under snow cover, and at temperatures slightly above freezing, the mycelia overgrow the grass and produce infections. The disease is quickly checked by increasing temperature and sunlight and decreasing moisture.

Coprinus snow mold, formerly known as winter crown rot or low-temperature basidiomycete (LTB) snow mold, is now known to be caused by strains of *Coprinus psychromorbidus*. *Coprinus* snow mold occurs primarily in subarctic climates where prolonged periods of deep snow cover favor disease development. Receding snow reveals circular or irregularly shaped patches with white to gray mycelial growth. Slight variations in symptoms are produced by sclerotial and nonsclerotial strains of the pathogen. The nonsclerotial strain is characterized by rapid cottony mycelial growth with clamp connections and relatively large (6- to 12-inch-diameter) patches. The sclerotial strain is similar, but also produces sclerotia that are white initially, then become brownish black, irregularly shaped, and flattened. Infected plants have rotted water-soaked leaves or leaf lesions that become pale brown when dry and have dark reddish brown margins.

Snow scald is another subarctic snow mold disease that is favored by prolonged deep snow cover atop frozen soil. The causal organism, *Myriosclerotinia borealis,* produces gray mycelia and dull black sclerotia. Infected leaves are initially water soaked and covered with sparse gray mycelia. They become bleached to white and then die. Black sclerotia form in the tissue near the soil line.

Southern Blight

Southern blight symptoms first appear as small circular dead areas during hot humid weather in midsummer. Usually, some green grass remains in the center of infected areas, resulting in a "frog-eye" appearance. Continued infections may result in dead patches measuring up to 3 feet in diameter. Abundant white mycelia occur near the soil surface on infected plants at the edge of the patch. The pathogen, *Sclerotium rolfsii,* produces abundant sclerotia that serve as survival structures in soil and thatch. Sclerotia germinate during hot weather when moisture is abundant. Infections proceed rapidly as long as environmental conditions are favorable. Recolonization of diseased patches by adjacent turfgrasses occurs during cool fall weather.

Spring Dead Spot

Spring dead spot is a serious disease of bermudagrass turfs, especially in cool subtropical climates where fall and winter temperatures favor a prolonged period of winter dormancy. Symptoms typically appear as bleached circular patches in spring when the dormant grass resumes growth. Patches vary in size from a few inches to several feet in diameter. Small patches may develop into rings with green centers (frog-eyes), perhaps coalescing with other patches to form serpentine arcs of dead grass. Roots and stolons of infected plants are severely rotted. Regrowth in the patches is generally slow and, where regrowth does occur, plants may remain stunted, suggesting the presence of persistent toxins in the soil.

Gaeumannomyces graminis, Ophiosphaerella herpotricha, and *Leptosphaeria* species have been identified as pathogens for spring dead spot. These fungi grow most actively in spring and fall when bermudagrass root growth is extremely slow; thus, the pathogen has a competitive advantage over the turfgrass during cool weather. Spring dead spot is a disease of mature turfs that are intensively managed; it is less severe or absent in turfs sustained at low fertility levels.

Summer Patch

Previously called *Fusarium* blight, summer patch is a warm-weather disease of Kentucky bluegrass that appears as crescent-shaped areas of dead turfgrass or circular patches with green centers (frog-eyes). The subterranean portions of infected plants are seriously rotted and appear brown to black in color. The pathogen, *Magnoporthe poae,* forms dark ectotrophic (surface-inhabiting) mycelia on infected roots and stems. Symptom development for this disease appears to require stressful conditions (high temperatures and light) and possibly subsequent infection by facultative parasites, including *Fusarium* species.

St. Augustine Decline

St. Augustine decline (SAD) is a relatively new and important disease caused by a virus. The initial symptoms are pale green or chlorotic spots, blotches, and stippling, followed by a chlorotic mottling of the leaf blades. In subsequent growing seasons, plants are bright yellow and stunted; the turfgrass becomes thin and eventually dies. Extensive losses of St. Augustinegrass can occur by the end of the third year, especially from winter kill of diseased plants. The virus is spread by mowing infected grass and then healthy grass.

Take-All Patch

Take-all patch, caused by the fungus *Gaeumannomyces* (formerly *Ophiobolus*) *graminis,* is a serious disease in cool temperate climates, especially in the cool moist coastal areas of northwestern United States and northwestern Europe. It is first observed as depressed more or less circular patches of blighted turfgrass that can extend to 2 feet or more in diameter. The centers of the patches are frequently invaded by other resistant grasses or weeds, thus creating a frog-eye appearance. The pathogen overwinters as dormant mycelia in living and dead plant tissues. The fungus forms dark brown to black ectotrophic runner hyphae on roots and stems of host plants. Infections result where the fungus penetrates subterranean plant tissues and spreads to adjacent plants by growing along roots and lateral shoots. The disease is most active during cool wet periods, but symptoms become most noticeable during drier conditions in early summer to midsummer. Applications of ammonium sulfate to lower the soil pH have helped reduce the severity of this disease.

White Blight

White blight is a disease of tall fescue caused by *Melanotus phillipsii* that occurs primarily during hot humid weather. It first appears as blighted patches measuring from a few inches to more than a foot in diameter. The patches are white with pink to salmon-colored borders. Leaf blades within the patch are covered by a network of grayish white mycelia. Necrosis begins at the leaf tips and proceeds toward the sheath. Frequently, fungal fruiting structures called basidiocarps form on the leaf blades.

Yellow Ring

Yellow ring is a disease of Kentucky bluegrass caused by the fungus *Trechispora alnicola,* a pathogen belonging to the class of mushroom fungi responsible for fairy rings. As the name implies, the disease appears as distinct yellow rings in infected turfs. Where the symptoms occur, large amounts of white fungal mycelia are usually evident in the thatch. Infected turfs usually recover quickly as soon as conditions are favorable for growth.

Miscellaneous Diseases

In addition to the diseases covered in this chapter, dozens have been observed and reported and numerous fungal species have been identified as disease-causing organisms. While some "miscellaneous diseases" can have devastating effects, many of them reflect the turf's response to environmental stress. Inadequate moisture, phytotoxic pesticide residues, and poor rooting are factors that enable an otherwise weak pathogen to cause significant damage to a turf. Also, given the dynamic nature of the environment in which turfgrasses grow, new diseases are likely to develop as new fungal strains evolve or as new turfgrass genotypes are introduced.

Fungicides

Fungicides include both organic and inorganic compounds with the capacity to control pathogenic fungi. Some are absorbed by plants, while others remain as surface residues on plant foliage. Applied prior to anticipated disease incidences, many fungicides prevent infection. Following the development of initial symptoms of a disease, application of fungicides can often prevent further disease development and promote recovery of the turf. A list of turfgrass fungicides and the diseases controlled by them is provided in Table 7.4.

Table 7.4. Fungicides and Turfgrass Diseases Controlled by Them

FUNGICIDES																					
Common names	*Trade names*	*Anthracnose*	Brown patch	Copper spot	Damping-off	Dollar spot	Downy mildew	Gray leaf spot	*Helminthosporium* disease	Necrotic ring spot	Pink patch	Powdery mildew	*Pythium* blight	Red thread	Rusts	Smuts	Snow mold (*Typhula* blight)	Snow mold (*Microdocium* patch)	Summer patch	Take-all patch	Yellow tuft
Anilazine	Dymec 50, Dyrene, Pro Turf F-111		X	X		X			X					X	X			X			
Azoxystrobin	Heritage	X	X				X		X	X			X	X			X	X	X		
Benomyl	Tersan 1991, Pro Turf DSB	X	X			X		X			X		X				X		X		
Cloroneb	Pro Turf F-11, Teremec SP, Tereneb SP,														X				X		
Chlorothalonil	Daconil 2787, Pro Turf 10-IV	X	X	X		X			X	X					X	X		X			
Cyproconazole	Sentinel		X			X		X			X						X			X	
Ethazol (Etridiazole)	Koban, Terrazole													X							
Fenarimol	Rubigan		X			X		X			X						X			X	
Flutolanil	Prostar		X									X			X						

Fosetyl-Al	Aliette						X							X						X
Iprodione	Chipco 26019, ProTurf F-6		X			X		X	X		X			X				X		
Mancozeb	Fore, Formec 80	X	X	X					X					X	X			X		
Maneb (and zinc)	Tersan LSR	X	X						X						X					
Metalaxyl	Subdue						X						X							X
Myclobutanil	Eagle		X			X		X		X							X			
Oxadixyl	Anchor					X		X	X	X		X		X	X	X				
PCNB (quintozene)	Terraclor 75		X			X			X								X			
Propamocarb	Banol												X							
Propiconazole	Banner	X				X		X		X	X	X		X	X	X			X	
Thiophenate-methyl	Fungo 50, Cleary's 3336 ProTurf Systemic	X	X	X		X		X		X		X		X		X		X		
Thiram	Chipco Thiram		X		X	X														
	75, Spot-Trete		X		X	X														
Triadimefon	Bayleton ProTurf F-7	X	X	X		X		X			X	X		X	X	X	X	X		
Vinclozolin	Curalan, Touche, Vorlan		X			X			X					X				X		

Types of Fungicides

The two basic types of fungicides are contact and systemic. Contact fungicides are sprayed onto leaves and stems to prevent fungi from infecting the plants. As unprotected leaves are continually emerging and older portions of the shoots are removed with mowing, it is necessary to apply contact fungicides frequently (at seven- to fourteen-day intervals) during periods of disease activity. Many contact fungicides control a broad array of infectious fungi that attack foliage; however, they do not protect the belowground organs of turfgrasses. Chemical classes of contact fungicides include triazines (anilazine), substituted aromatic compounds (chlorothalonil, PCNB, and chloroneb), thiazoles (terrazole), and dithiocarbamates (maneb and mancozeb).

Systemic fungicides are absorbed by turfgrass roots or other organs and are translocated primarily upward. Thus, even newly emerging leaves contain some fungicide and are protected from infection. Once inside the plant a fungicide cannot be washed off the leaves or completely removed with a few mowings. Thus, systemics provide much longer control of diseases (approximately four to six weeks) than do contacts. Systemic fungicides are more specific acting than contacts. Since systemics tend to block only one enzyme in the metabolism of a fungus, the likelihood of encountering resistant strains of the fungus is far greater than with contacts, which inhibit several enzyme systems. Chemical classes of systemic fungicides include benzimidazoles (benomyl and thiophanate methyl), demethylation inhibitors or DMIs (fenarimol, propiconazole, and triadimefon), dicarboximides (iprodione and vinclozoline), acylalanines (metalaxyl), carbamates (propamocarb), and organic phosphates (fosetyl-Al).

Fungicide Application

Fungicides can be applied either prior to infection (preventive) or after the development of disease symptoms (curative). A preventive fungicide program is preferred for controlling diseases that are almost certain to occur under a specific set of environmental conditions. On golf courses, dollar spot, brown patch, snow molds, and *Pythium* blight occur with such regularity that preventive fungicide applications are often employed for control.

Since curative treatment with fungicides depends on the development of disease symptoms, some savings may be realized from this type of application program. During some seasons disease pressure may be so light that only a few applications are required. Higher rates of fungicide are usually required for curative than for preventive treatment, while temporary, and perhaps serious, reductions in turf quality from disease activity are inherent in a curative program.

Contact fungicides must be applied with sufficient water to provide thorough coverage of leaves and stems where infections occur. This usually requires between 40 and 120 gallons per acre (approximately 1 to 3 gallons per 1000 ft^2). Systemic fungicides can also be applied in this way, but their persistence and activity tend to follow those of the contact materials, and they will be ineffective against such diseases as stripe smut and summer patch. For a systemic fungicide to be absorbed by turfgrass roots and translocated upward in the plant, irrigation of the turf (1/2 to 1 inch of water) is required immediately after application to wash the fungicide into the root zone. Furthermore, the application should be preceded by irrigation to reduce adherence of the fungicide to plant and soil surfaces. Granular formulations of systemic fungicides can be quite effective, as the granules tend to bypass the foliage enroute to the soil. However, they also should be irrigated in to promote absorption of the fungicide by turfgrass roots.

Fungicide Resistance

In the early 1960s resistance to the cadmium fungicides by the causal organism for dollar spot was reported. Cadmium-resistant strains of the fungus were also resistant to the mercury chloride fungicides. In later years resistance of the fungus to benomyl occurred. Since benomyl is a benzimidazole-type systemic fungicide, the related thiophenate fungicides were also ineffective against this strain of the fungus. This reflects the occurrence of a common metabolite, methyl benzimidazole carbamate (MBC), an intermediate compound in the decomposition of both fungicides. This "cross resistance" within chemical class has also been reported for dicarboxymides and DMI fungicides. Given the natural variation that exists within a fungal population and the opportunities for mutation and genetic recombination, the potential for the development of fungicide-resistant strains, especially to systemic fungicides, is high. For this reason several fungicides are often used in carefully prescribed sequences to prevent or delay a specific strain of the fungus from becoming dominant. Some proposed sequences for delaying dollar spot resistance to the DMI fungicides rely on contact fungicides—or contact-benzimidazole or benzimidazole-dicarboximide fungicide combinations—for midsummer treatments when disease pressure is high, while restricting the use of DMI fungicides to the spring-early summer and late summer-fall periods when disease pressure is low. Simply alternating or combining contact and systemic fungicides during a season of high disease pressure, however, is likely to accelerate the development of resistance. Increasing application rates and shortening the interval between fungicide applications also tend to effect shifts in fungal populations that favor the eventual dominance of resistant strains. Eventually, new fungicides may be needed to control fungal strains resistant to the current inventory of materials.

Biological Control

An issue of emerging importance is the potential role of microbial inoculants and organic amendments in combating turfgrass diseases. Many products are now available claiming to suppress turfgrass diseases by either supplying antagonistic microorganisms or promoting their sustained growth within the turf. While many of the claims are without foundation, there are some promising signs that reliable biological disease control may be around the corner. Selected strains of bacterial (e.g., *Enterobacter cloacae*) and fungal (e.g., *Trichoderma harzianum*) species have significantly reduced the incidence of dollar spot, brown patch, and other diseases in field trials. Clearly, biological disease-control agents, if reliable, would be welcome alternatives to chemical pesticides wherever they might be incorporated into turfgrass cultural programs.

NEMATODE CONTROL

Nematodes are probably the most abundant form of animal life in the soil. Most species that occur in soil feed on fungi, bacteria, or small invertebrate animals, but many are parasites of higher plants, including turfgrasses. Turfgrass species and cultivars vary in their relative susceptibility to specific nematodes; however, a particular turf can be so severely infested with parasitic nematodes that applications of pesticides (nematicides) may be required to reduce population levels. Although this condition commonly occurs in sandy soils in subtropical and tropical climates, the role of parasitic nematodes in the decline of turfs under other conditions has not been fully ascertained.

Nematodes

Nematodes are minute largely translucent wormlike animals. Adult forms average about 1 mm in length and less than 0.03 mm in diameter, although some types become quite swollen at maturity. Plant-parasitic nematodes feed on turfgrass roots and other organs by puncturing the cells with a hollow needlelike structure called a stylet (Figure 7.3). Digestive enzymes are injected into the cells, and the liquified contents are withdrawn through the stylet. Nematodes seldom kill plants directly, but they are capable of greatly curtailing the growth of plants. Affected plants may be stunted or chlorotic, or show symptoms of wilt under stress. The roots may die or be altered in terms of stunted growth, excessive branching, or the development of lesions, galls, or other deformities. Damaged plants

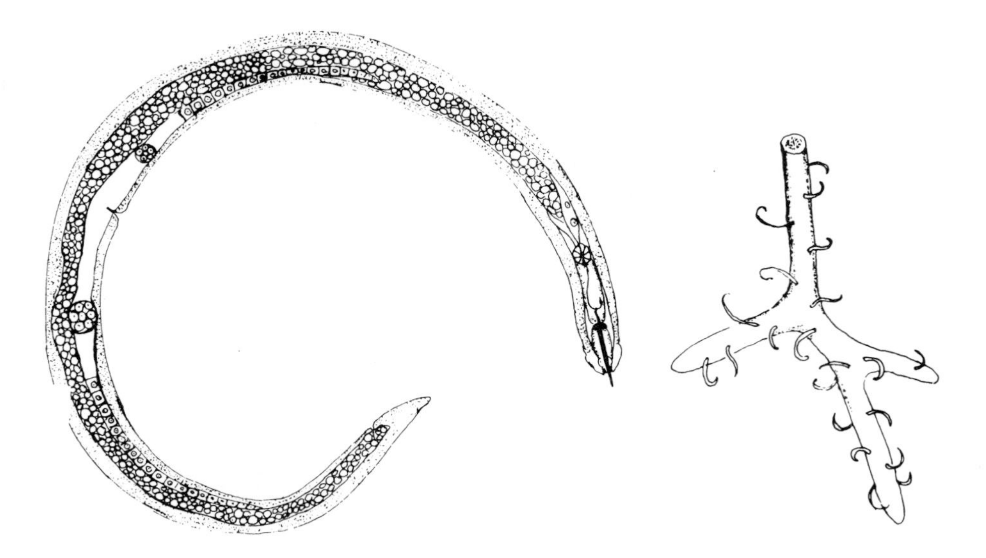

Figure 7.3. An ectoparasitic nematode (left) and a nematode-infested turfgrass root (right).

may also be highly disease prone, as they are more easily penetrated and infected by pathogenic fungi. Even when nematodes are found in association with turfgrass roots, fairly high populations must be present for damage to occur. Heavily infested turf does not respond normally to water and fertilizer.

Nematodes begin their life cycles as eggs. Upon hatching into larvae they undergo four molts to reach adulthood, usually in twenty to sixty days. With each molt a cuticle is usually shed. Mating of adult nematodes (where necessary for reproduction) and subsequent egg laying commonly initiate the next life cycle; however, males are unknown for some species.

Based on their feeding position, nematodes are placed into two major classes: endoparasitic and ectoparasitic.

Endoparasitic Nematodes

Endoparasites, which include the cyst, cystoid, root-knot, lesion, and burrowing nematodes, invade roots and feed from within the host.

Cyst nematodes include various species of *Heterodera*. These nematodes cause a reduction of root growth and may cause stunting and chlorosis of the shoots. Small pearly white female bodies of the nematode are attached to the roots of affected plants. After the lemon-shaped females die, their bodies turn tan, then brown in color, forming a cyst. Overwintering occurs as eggs develop inside the cyst. As soil temperatures increase in spring, the eggs hatch and the wormlike larvae emerge and enter the plant root to feed and later to mate. Eventually, they break through the root's epidermis as their bodies swell. Control of cyst nematodes is difficult without destroying the turf. Certain soil fumigants such as methyl bromide, dazomet, and metham can be used prior to planting in severely infested soils.

Cystoid nematodes include only two species of the genus *Meloidodera*. They are similar to cyst nematodes except that the swollen females do not change color after death. Control is the same as for cyst nematodes.

The root-knot nematodes (*Meloidogyne* **spp.**) penetrate the tips of roots and produce characteristic swellings or galls that vary in shape and size. Eggs of the root-knot nematode overwinter in the soil and in roots of infested plants. Second-stage larvae escape into the soil and move to another section of the same root or to different roots. Penetration is usually at or near the root tips and may result in excessive root branching and consequent stunting of plants. Feeding occurs in the stele region within "giant" cells. With successive molts the females increase in size with a corresponding increase in the size of cells surrounding the head region, forming a "knot." Eggs are extruded in a gelatinous mass at or near the root surface. The life cycle from egg through four molts to adult takes about one month. Control is the same as for cyst nematodes.

Lesion nematodes (*Pratylenchus* **spp.**), which are always vermiform (worm-shaped) and invade the cortex, cause brown to black lesions in turfgrass roots. Colonization of affected tissue by fungi and bacteria can eventually result in a girdling and death of the roots. With large populations of this nematode, entire root systems may be destroyed.

Lesion nematodes enter roots behind the tips. They migrate and feed freely in the stele and cortex regions of the roots. Eggs are laid both inside and outside the roots. The life cycle usually requires thirty to sixty days to complete. Control is the same as for cyst nematodes.

Burrowing nematodes (*Randopholus similus*), which occur only in tropical and subtropical climates, cause injury similar to that produced by lesion nematodes. Complete destruction of turfgrass roots is possible with large populations. Control is the same as for cyst nematodes.

Ectoparasitic Nematodes

Ectoparasites feed mostly on the surface of young roots while their bodies remain in the surrounding soil. The life cycles of ectoparasitic nematodes are relatively simple and closely parallel the description given in the introductory discussion under Nematodes. In general, they cause a decline in turfgrass vigor; symptoms are not very specific.

Spiral nematodes of the genera *Helicotylenchus, Rotylenchus, Peltanigratus,* and *Scutellonema* cause a reduction in turfgrass vigor, chlorosis, and a loss of roots. They are called spirals because of their tight C shape at rest. Some species greatly enhance summer dormancy of Kentucky bluegrass in cool temperate climates.

Sting nematodes (*Belonolaimus* **spp.**) are the largest nematodes that feed on turfgrasses. They are considered the most important pest of turfgrasses in Florida. Their effects include severe growth restrictions and malformations of roots and chlorosis.

Stunt nematodes (*Tylenchorhynchus* **spp.**) cause shrivelling and shortening of roots, but lesions are absent. They are highly persistent in warm moist soils, even in the absence of host plants.

Ring nematodes (*Criconemoides* **spp.**) sometimes cause lesions at the tips and along the sides of roots. Where extremely large populations are active extensive root rotting can occur.

Pin nematodes (*Paratylenchus* **spp.**), the smallest of the plant-parasitic nematodes, occasionally cause stunting and excessive tillering in affected plants. Distinct lesions are present in the excessively branched shortened roots.

Stubby-root nematodes (*Trichodorus* **spp.**) are relatively small animals that cause chlorosis and greatly reduced vigor in affected plants. Deep, dark, irregular lesions are observed in the roots, especially near the tips. Root injury is caused by reduced cell division at the tips rather than by destruction of existing cells.

Dagger nematodes (*Xiphinema* **spp.**) cause stunting and chlorisis in affected plants. Severe root rot and reddish brown to black root lesions can develop where high nematode populations are present.

Lance nematodes (*Hoplolaimus* **spp.**), named for their strongly developed stylet, cause dark root lesions and, eventually, sloughing of cortical tissue. Shoots may be stunted and chlorotic. They are among the most commonly found nematodes in turf.

Needle nematodes (*Longidorus* **spp.**) cause stubby thickened roots and devitalized root tips.

Awl nematodes (*Dolichodorus* **spp.**), like many ectoparasites, cause devitalized root tips, cortical lesions, and stunting.

Damage from ectoparasitic nematodes can be reduced by supplying sufficient water and fertilizer to the turf to sustain vigorous growth. Although some differences in the susceptibility of turfgrass species and cultivars to nematodes exist, these have not been explored adequately to develop specific recommendations in most instances. Chemical control with nematicides offers the most dramatic solution where ectoparasitic nematode populations are high and damage is severe.

Nematicides

Control of nematodes in turf is difficult because of their distribution throughout the root zone and their position on or in roots. Therefore, complete chemical eradication is impossible, but reductions of large populations of the highly destructive ectoparasitic nematodes usually are achievable with proper use of nematicides. The two types of nematicides are soil fumigants and contacts. Fumigants are generally most effective because of the rapid distribution of the gas in soil; however, they are highly toxic to turfgrasses and can only be used prior to planting. These include methyl bromide, dazomet, and metham. Contact nematicides are nonvolatile and must be washed into the root zone of established turfs. Control depends on the nematicide coming into contact with the nematodes. These include ethoprop (Mocap, Pro Turf Nematicide) and fenamiphos (Nemacur). Liquid formulations should be irrigated in immediately after application to reduce foliar-burn potential. Granular formulations are generally safer to the turfgrass as well as for the applicator. A minimum of 300 gallons of water per 1000 ft^2 should be applied following application of a nematicide. The preferred timing of chemical application is at the initiation of spring growth or in late fall when the soil temperature is 55° to 60°F. Core cultivation of turf prior to nematicide application may improve results. The manufacturer's directions printed on the container label should be carefully followed. Nematicide-insecticide combinations are highly toxic and should be used only by personnel who are qualified to handle highly toxic materials.

To correctly diagnose nematode damage it is necessary to see and identify the nematodes associated with injured turfgrasses. Soil samples collected from turf where nematode infestation is suspected must be analyzed in a specially equipped nematology laboratory. Nematodes extracted from samples must be identified by trained personnel and the numbers of each form estimated.

Soil samples can be collected at any time of the year. Sampling should be done in the early morning or late afternoon to avoid overheating the samples in bright sunlight, and efforts should be made to keep the samples cool and moist until processing. Approximately ten to twenty cores should be taken to a depth of 6 inches from the margins of affected areas, using a soil probe. The cores should be placed in a strong airtight package; freezer bags work well for this purpose. These, in turn, should be placed in a strong container (metal can, rigid plastic container, mailing tube, or the like). This should be sent along with detailed information on symptoms observed, size of area sampled, cultural practices, and the turfgrass species and cultivars affected. If the sample must be stored temporarily, it should be in a refrigerator at above-freezing temperatures.

INSECT CONTROL

Insects are among the most abundant animals inhabiting the earth. Many are beneficial because they decompose organic matter or function as predators of other, sometimes harmful, organisms. Some, however, are serious pests of turfgrasses and other plants and animals.

Insects and other related arthropods have their skeletons positioned on the outside of their bodies; this results in a characteristically segmented (head, thorax, and abdomen) body and appendages with jointed sections. Because of the exoskeleton, insect growth occurs in stages in which the skeleton is shed and a new larger one is produced. Each shedding of the exoskeleton is called a molt, and the interim period between successive molts is called an instar.

The series of changes that insects undergo during their life cycles is called metamorphosis. The two principal types of metamorphosis are complex and simple. Although nearly all insects begin life as eggs, those that undergo complex metamorphosis hatch into larvae or wormlike organisms and then transition to the adult form through a quiescent form called a pupa (Figure 7.4). The larval stages of beetles, moths, and flies are called grubs, caterpillars, and maggots, respectively. Larvae feed voraciously, except during overwintering, until they form pupae. The adults emerging from the pupae may or may not feed, depending on species. Often, the principal activity of adult insects is reproduction. Most insect pests of turfgrasses undergo complex metamorphosis.

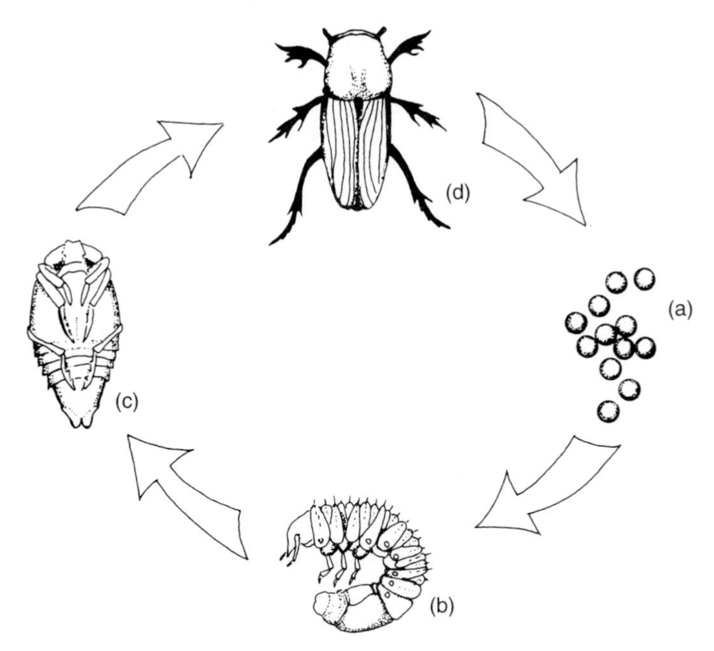

Figure 7.4. Complex metamorphosis with illustrations of (a) eggs, (b) larva, (c) pupa, and (d) adult.

Insects that undergo simple metamorphosis do not go through a larval stage; rather, the eggs hatch into small wingless insects, called nymphs, which resemble adults (Figure 7.5). Through successive molts the nymphs gradually change into the adult form and then reproduce. Examples of insects with simple metamorphosis are aphids, chinchbugs, leaf hoppers, mole crickets, and scale insects.

Insects feed by one of two principal mechanisms: chewing or sucking. Chewing insects have mouthparts that enable them to devour plant tissues or organic residues directly. Sucking insects possess piercing mouthparts that, when inserted into plant tissue, enable them to suck plant juices. They may also release toxic substances that can adversely affect the plants. Both chewing and sucking insects can cause severe damage to turfgrasses by feeding on leaf, stem, and root tissues.

Insect-induced injury tends to be more severe in turfs sustained under moderate to high intensities of culture. Close mowing, frequent irrigation, and abundant supplies of fertilizer result in more succulent plants that attract insects.

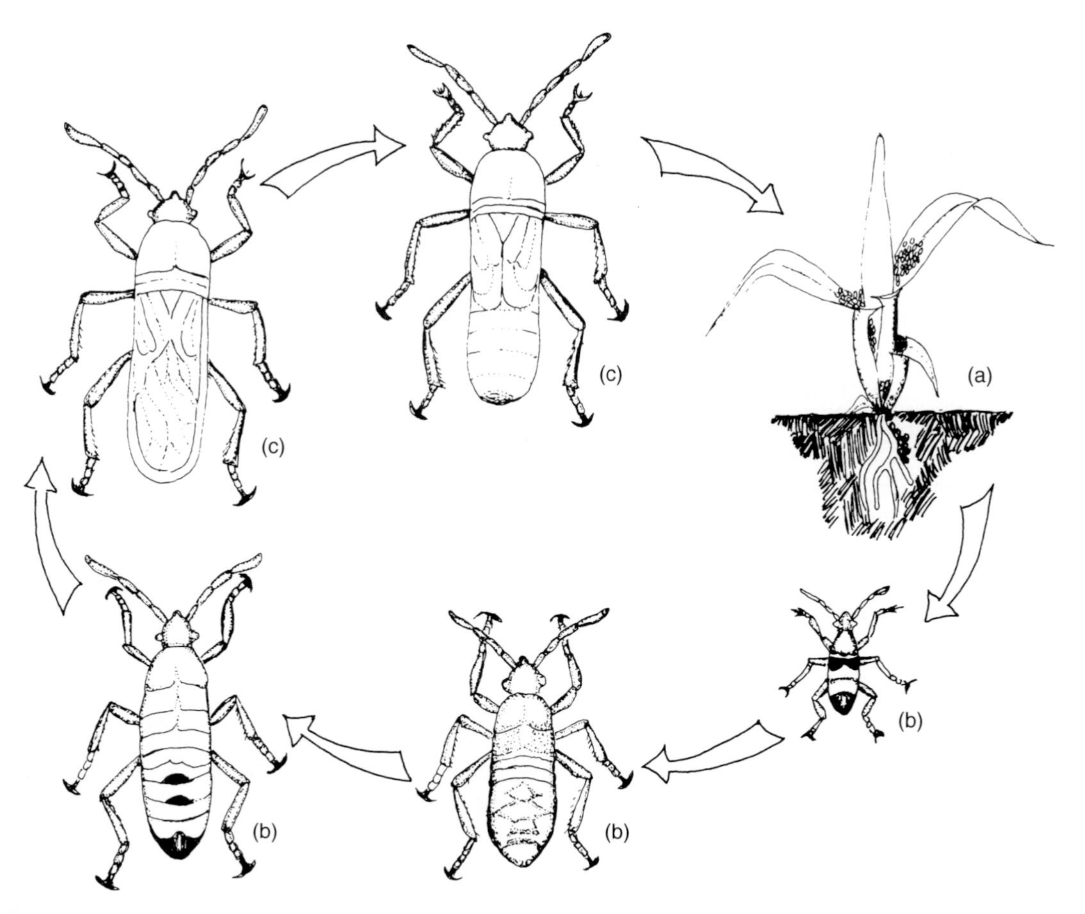

Figure 7.5. Simple metamorphosis with illustrations of (a) eggs, (b) nymphal stages, and (c) adults.

Insects

Insect pests of turf may be grouped into three categories based on the type of damage caused: root-feeding insects, shoot-feeding insects, and burrowing insects. The first two pose a direct threat to the turfgrass community, while the last category includes insects that are a nuisance because of their disruptive effect on the turf's surface.

Root-Feeding Insects

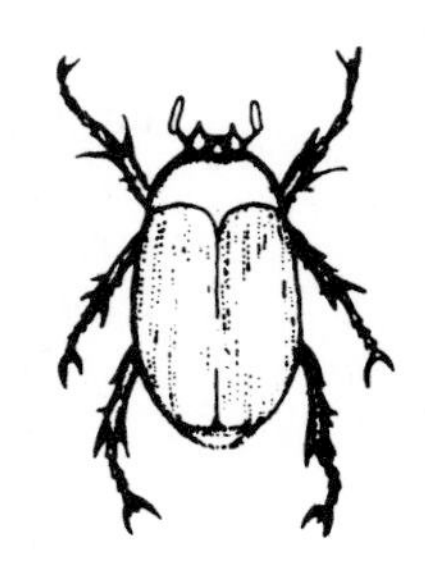

Grubs are the larvae of several species of beetles. These include May beetles, northern and southern masked chafers, European chafers, Japanese beetles, and the black turfgrass ataenius. Except for the rice-grain-size ataenius, mature larvae measure 1/2 to 1 inch long and have a characteristic organization of hairs on the terminal segment. All have one-year life cycles except May beetles, which typically complete their life cycles in three years. The adults normally do not feed on turfgrasses, but the crescent-shaped larvae chew voraciously on turfgrass roots, especially during late summer and fall. In freezing climates, the larvae tunnel deeper in the soil, where they remain for the winter. In spring, they resume feeding for a while, pupate, and finally emerge as adult beetles. One exception is the black turfgrass ataenius, which overwinters as the adult.

Turfgrass damage results from the loss of much of the root system, especially where feeding is intense and weather conditions favor desiccation of the plants. Affected turfs can be easily lifted from the soil to expose the larvae. In addition to insecticide treatment, grub-infested turfs should be irrigated frequently to reduce further desiccation and promote the growth of new adventitious roots.

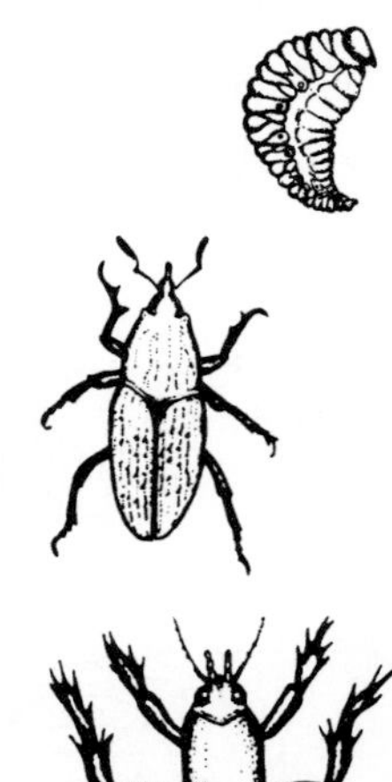

Billbugs feed on both roots and stems of turfgrasses. The overwintering adult lays its eggs after burrowing into the turfgrass stems. Upon hatching, the larvae chew on stems and later on roots. Damaged plants turn brown and are easily pulled up during the summer feeding period. A sawdustlike material (frass) is evident around feeding sites and serves as an important diagnostic feature. In the fall, the larvae usually pupate, and adult billbugs may be observed on driveways and sidewalks. The Phoenix and hunting billbugs are pests of bermudagrass and zoysiagrass, respectively; the bluegrass billbug feeds primarily on Kentucky bluegrass.

Mole Crickets feed on turfgrass roots as they burrow through the soil. Severe damage can result as plants are uprooted and the soil dries out, especially in newly seeded turfs. Mole crickets are most active at night, when they come to the surface to feed. The adults measure about 1.5 inches long and are covered with fine hairs that give them a velvety appearance. In late spring the females lay eggs, which hatch into nymphs during summer. The new generation overwinters and completes its life cycle the following spring.

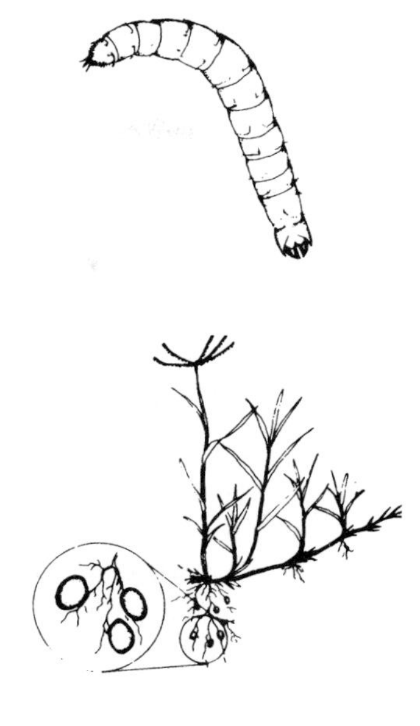

Wireworms are cylindrical, hard, wirelike larvae measuring about 1 inch long that chew on grass roots. They live from two to six years in the soil, depending on species. The adults, called click beetles, live for about one year and lay their eggs on grass roots. Damage is often unnoticed unless large populations of the insect develop; then irregular areas of brown grass may appear.

Ground Pearls are scale insects that feed on the roots of several warm-season turfgrasses, especially centipedegrass. These tiny insects, measuring up to 1/8 inch across, have needlelike mouthparts that enable them to suck juices from turfgrass roots. Mature females deposit their eggs in the soil. The newly hatched nymphs attach themselves to roots and form a protective pearllike shell or cyst around their bodies. Affected turfs appear chlorotic, then brown, in irregular patches. Insecticides are largely ineffective in controlling this insect.

Shoot-Feeding Insects

Sod Webworms are the larvae of lawn moths. They feed by chewing leaves near the base of the sheath. Intense feeding can result in substantial losses of turf, especially during warm weather, when the grass is under stress. At rest the adults fold their wings around their bodies. In flight the female moths drop their eggs at random over the turf. Once the eggs hatch the larvae live in the thatch during the day and feed at night. Full-grown larvae measure about 3/4 inch long and are usually tan colored and spotted. Usually, two generations occur in the course of a growing season. Overwintering occurs as larvae. Diagnostic features include chewed-off leaf sheaths near the surface of the ground at the periphery of damaged areas and the accumulation of green pellets of excrement (frass) found within the thatch of damaged turf.

Army Worms are caterpillars measuring approximately 1.5 inches at maturity. They have distinctive stripes along the sides of their bodies. Like sod webworms, they feed at night on grass leaves. From one to six generations may occur in the course of a growing season. The larvae overwinter in subtropical and tropical climates; poleward migration of adults can result in insect activity in temperate areas.

Cutworms are the larvae of night-flying moths. They measure 1.5 to 2 inches long. Except on greens, where they chew grass shoots down to the surface, they are considered a minor problem. From one to four generations may occur during the growing season, depending on species and climate. The pupae or eggs usually overwinter in the soil.

Annual Bluegrass Weevils or turfgrass weevils are related to beetles (order Coleoptera) but belong to the family Curculionidae. They are distinguished by their long slender snouts and measure less than 1/5 inch long. In the northeastern United States, both larvae and adults of the weevils feed preferentially on the stems and leaves of annual bluegrass, but the adults cause relatively little damage compared to the larvae. The adults overwinter below the soil surface and emerge in spring to mate and lay eggs between leaf sheaths of annual bluegrass. Upon hatching, the legless larvae begin feeding and cause extensive damage as grass stems are hollowed out or severed at their bases. Damage is initially observed by late spring, and complete loss of annual bluegrass stands can occur on severely infested sites.

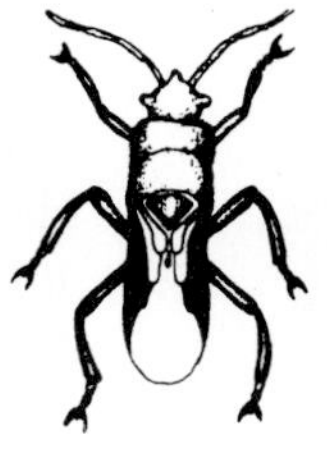

Chinchbugs are insects that feed by inserting their mouthparts in grass shoots and sucking out fluids. With the simultaneous injection of salivary fluids into the plant, affected turfgrasses become chlorotic and eventually die under hot dry conditions. The two principal chinchbug pests of turf are the hairy chinchbug, which feeds on cool-season turfgrasses, and the southern chinchbug of warm-season turfgrasses, especially St. Augustinegrass.

Adult chinchbugs measure 1/5 inch long and are black with white folded wings. The young nymphs are about 1/20 inch long, wingless, and red colored with a white stripe across their backs. Female adults insert their eggs in turfgrass leaf sheaths. Hatching of the eggs into nymphs is highly temperature dependent, requiring from ten to thirty days. Typically, there are five nymphal instars between hatching and adulthood, with each instar requiring from five to fourteen days. In the course of a growing season one to five generations may develop, depending on climate. Chinchbugs usually overwinter as adults; however, in tropical climates and warm subtropical climates feeding may proceed throughout the winter months.

In wet years damage from chinchbugs is usually much less than in dry years. This is due in part to a fungal disease of the insect that is most infectious under moist conditions. Infested turfs are most likely to show damage in sunlit areas; shaded areas often stay green, while the surrounding turf may become completely discolored or dead. Close examination of the turf, especially after brushing across the shoots with the fingers, reveals the tiny chinchbugs where feeding is taking place. Another diagnostic technique is to insert an open-ended cylinder (a coffee can with both ends cut off) into the turf, fill it with water, and add pyrethrin or household detergent; if present, chinchbugs will float to the surface in five to ten minutes.

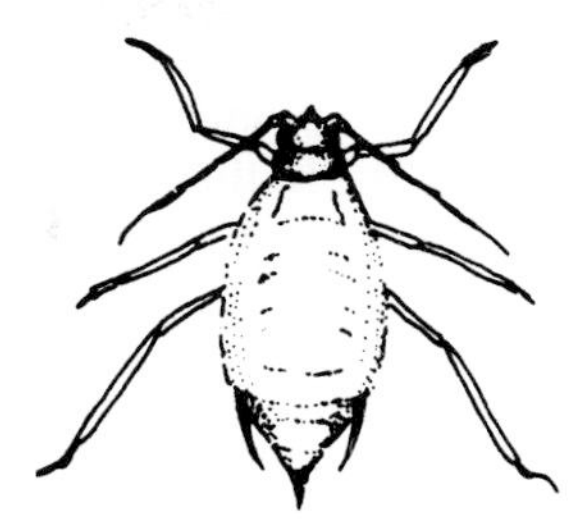

Aphids have not been considered serious pests of turf until recently. In the early 1970s the greenbug (*Schizaphis graminum* **Rondani.**) began causing serious damage to Kentucky bluegrass in midwestern United States. Greenbugs are light green soft-bodied insects measuring 1/16 inch long. They feed by inserting piercing mouthparts into turfgrass leaf blades and sucking out juices. Simultaneously, salivary fluids are injected into the plant, causing damage. Unlike most insects, adults give birth directly to young aphids. These mature rapidly and are capable of reproducing without mating. On infested leaves numerous greenbugs can be observed feeding side by side in crowded colonies. Turfgrass damage is most likely to occur in shaded areas at the base of trees and shrubs; however, open turfs can be invaded. Initially, an infested turfgrass appears yellow, then orange, and finally brown, indicating death of the plants. Damage typically occurs during the warm summer months.

Frit Flies are tiny black insects measuring about 1/16 inch as adults. Although the adult fly does not feed on turfgrasses, the larvae (maggots), measuring 1/8 inch long, tunnel into stems near the base of the plant, causing death during hot, dry weather. Overwintering occurs as larvae in grass stems. In spring the larvae pupate and adult flies soon emerge. After mating, female adults lay eggs on turfgrass leaves. The larvae emerge and begin burrowing into stem tissue. Several generations can occur during the course of a growing season. Bentgrasses and bluegrasses (especially annual bluegrass) are the primary hosts. The yellow to brown larvae are difficult to see because of their position within the shoot; however, the presence of adult frit flies can be determined by placing a piece of white cloth or paper onto an infested turf. As they are attracted to white, numerous black adults may be observed in a few minutes.

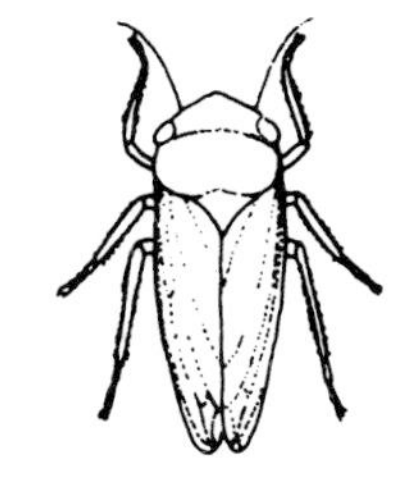

Leafhoppers are small wedge-shaped insects measuring up to 1/5 inch long. Adults fly or jump short distances when disturbed. Young nymphs resemble adults but are smaller and wingless. Both adults and nymphs suck juices from grass shoots, causing discoloration and retarded growth. Young turfgrass seedlings may be so severely damaged that replanting becomes necessary. The female inserts her eggs into plant tissues, hatching takes place in a few days, and the young nymphs initiate feeding on the same or adjacent plants. Several generations can occur during the growing season, but the largest populations occur during the summer months.

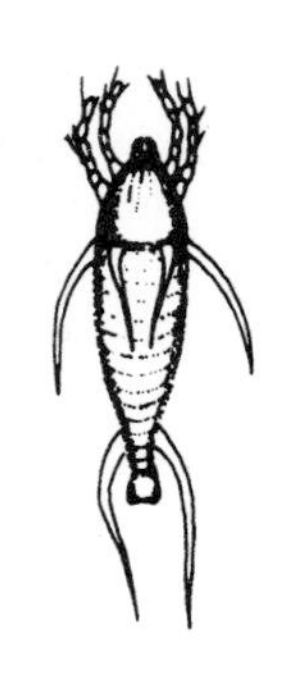

Mites are close relatives of insects; they have four pairs of legs and are extremely small. Several species feed on turfgrasses, causing a blotching or spotting of the leaves. Continued feeding may cause chlorosis and, eventually, death of leaves and other plant organs. The bermudagrass mite (*Aceria neocynodonis* **Keifer**) measures only 1/100 inch long and causes infested plants to grow abnormally, with shortened internodes and tufted leaves. Female mites deposit their eggs beneath the turfgrass leaf sheaths in spring. Several generations can develop during the course of the growing season. Microscopic examination is required to identify mites as causal organisms in affected turfgrasses.

Scale Insects are small inconspicuous insects measuring 1/16 inch across. They are mobile only in the early nymphal stages; then they settle on a specific location of the leaf or stem. They usually cover themselves with a shell-like protective shield while feeding with their needlelike mouthparts. Stems of infested plants have a whitish or moldy appearance; closer examination with a hand lens or microscope reveals the clam-shaped shields of the insects. The bermudagrass scale infests bermudagrass, particularly in shaded locations. The rhodesgrass scale (actually a mealybug) infests both bermudagrass and St. Augustinegrass; affected plants often wither and die. Ground pearls are also scale insects, but are discussed under root-feeding insects.

Burrowing Insects

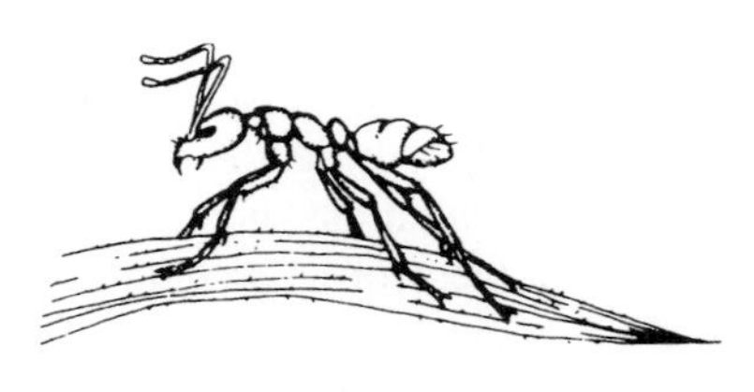

Ants are social insects that live in colonies in underground nests. Several species nest in turf; when their populations are large they can excavate considerable quantities of soil, resulting in large mounds at the surface. This not only disrupts the surface uniformity of a turf; it can also result in desiccation and burial of turfgrass plants surrounding nesting sites. On newly seeded sites ants may remove grass seeds and carry them into their nests.

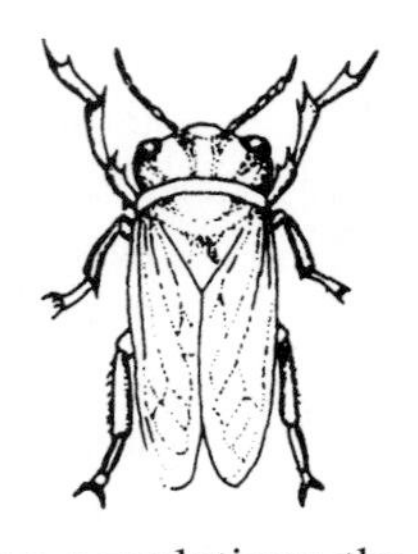

Periodical Cicadas are large insects measuring up to 2 inches long, with unusual life cycles. Adults emerge every thirteen or seventeen years from their burrows in the soil, leaving small holes that may be especially numerous under trees. After mating, the females lay eggs in twigs of trees. When the eggs hatch the nymphs drop to the ground, burrow into the soil, and feed on roots of trees and shrubs. Damage to turf occurs thirteen or seventeen years later when the nymphs emerge from their burrows. With large populations the number of holes that develop may be so numerous that the affected turf resembles a giant sieve. Control of the cicada nymphs is not possible because of their deep location in the soil; however, insecticides are sometimes applied to adult populations to reduce tree damage from egg laying.

Cicada Killers are insect predators of the periodical cicada. They are large wasps measuring 1.5 inches long. The wasps dig into a turf where cicada nymphs are located, leaving mounds of soil around each hole. The female wasp captures and stings a cicada, places it in a constructed nest, then deposits an egg in the cicada's body. The egg hatches and the emerging larva begins feeding on the cicada. Control is not necessary unless the wasps are so numerous that soil mounding becomes objectionable. If disturbed, the wasps will sting.

Bees and Wasps include some species that burrow into turf to form their nests. Wasps provision their nest with flies, grasshoppers, or crickets, while bees supply honey or pollen. If the entrance holes are numerous, it may be advisable to apply insecticides for control.

Insecticides

Insecticides typically used for controlling insect pests of turf are listed in Table 7.5. All are organic phosphorous compounds except bendiocarb, carbaryl, and dicofol. The banning of chlordane and other chlorinated hydrocarbon materials has had a major impact on insect control methods as the persistence of currently available materials is relatively short term (several weeks to several months) compared with several years or longer control from the chlorinated hydrocarbons. Timing of applications with nonpersistent insecticides for observed or anticipated insect infestations is therefore critical. Familiarity with insect life cycles and recognition of early signs of damaging insect activity are essential for preventing serious insect-induced damage to turf. Also, the position of the insect within the turf-soil profile and its method of feeding are important factors for determining the manner in which an insecticide should be used. Root-feeding insects inhabit the soil and thus are well below the turf's surface. For an insecticide to effectively control soil-inhabiting insects, it must be moved downward to the position of insect activity. This would normally require at least 1/2 inch of irrigation water following application of the insecticide. The same procedure applies when controlling burrowing insects. Irrigation should be performed as soon as possible following application to reduce losses of the insecticide to photodecomposition, metabolic processes, or volatilization, especially if the insecticide adheres to the turfgrass foliage. Foliar adherence is more likely from spray applications than from granular formulations.

Shoot-feeding insects often ingest insecticide residues adhering to the turfgrass foliage. Where these insects are the primary targets of insecticide application the turf should not be irrigated following application. For example, in controlling infestations of

Table 7.5. Insects of Turf and Insecticides for Controlling Them

		CONTROL WITH INSECTICIDES																					
		Acephate	*Bendiocarb*	*Bifenthrin*	*Carbaryl*	*Chlorpyrifos*	*Cyfluthrin*	*Dicofal*	*Diazinon*	*Ethion*	*Ethoprop*	*Fluvalinate*	*Fonofos*	*Halafenozide*	*Imidacloprid*	*Isazofos*	*Isofenphos*	*Lambda-cyhalothrin*	*Lindane*	*Malathion*	*Permethrin*	*Pyrethrin*	*Triclorfon*
Annual bluegrass														X	X	X	X	X	X		X		
weevil	*Listronotus* **spp.**	X				X	X	X								X		X	X	X	X	X	
Ants	Formicidae family	X																					
Aphids	Aphididae family		X	X					X									X	X	X	X		
Army worms	Noctuidae family																						
Common	*Pseudaletia unipuncta* **Haworth**	X		X	X		X		X			X		X		X	X	X	X		X	X	X
Fall	*Spodoptera frugiperda* **J. E. Smith**	X		X	X		X		X			X		X		X	X	X	X		X	X	X
Bees and wasps	Hymenoptera order				X																		
Billbugs	Curculionidae family																						
Bluegrass	*Sphenophorus parvulus* **Gyllenhal**		X	X	X		X		X		X		X	X	X	X	X	X	X		X		X
Hunting	*Sphenophorus venatus vestita* **Chttn.**		X	X	X		X		X		X		X	X	X	X	X	X	X		X		X
Phoenix	*Sphenophorus phoeniciensis* **Chttn.**		X	X	X		X		X		X		X	X	X	X	X	X	X		X		X
Chinchbug	Lygaeidae family																						
Hairy	*Blissus hirtus* **Montandon**		X	X	X	X	X		X	X	X	X				X	X	X				X	X
Southern	*Blissus insularis* **Barber**		X	X	X	X	X		X	X	X	X				X	X	X				X	X

Cicada killer	*Sphecius speciosus* **Drury**				X																		
Cutworms	Noctuidea family	X	X	X	X	X		X			X		X		X	X	X	X	X			X	X	
Grubs	Scarabaeidae family																						
Black turfgrass ataenius	*Ataenius spretulus* **Haldeman**			X		X	X				X		X		X	X	X	X	X	X	X		X
European chafer	*Amphimallon majalis* **Razoumowski**			X		X	X				X		X		X	X	X	X	X	X	X		X
Japanese beetle	*Popillia japonica* **Newm.**		X		X	X				X		X		X	X	X	X	X	X	X		X	
May beetle June bug	*Phyliophaga* **spp.**			X		X	X				X		X		X	X	X	X	X	X	X		X
Northern masked chafer	*Cyclocephala boraelis* **Arrow**			X		X	X				X		X		X	X	X	X	X	X	X		X
Southern masked chafer, annual white grub	*cyclocephala immaculata* **Olivier**	X		X	X			X		X		X	X	X	X	X		X	X		X		
Leafhoppers	Cicadellidae family				X	X	X					X											
Mites	Acarina order										X			X				X					
Mole crickets	Gryllotalpidae family	X		X	X	X	X	X			X		X		X		X	X	X	X		X	X
Periodical cicadas	*Magicicada septendecim* **L.**					X																	
Scale insects	Coccidae family																						
Bermudagrass scale	*Odonaspis ruthae* **Kotinsky**						X																
Ground pearls	*Margarodes meridionalis* **Morr**.																				X		X
Rhodesgrass scale[a]	*Antonina graminis* **Maskell**										X												
Sod webworms	*Crambus* **spp.**	X	X	X	X	X	X	X			X	X	X	X		X		X	X	X	X	X	

[a]Actually a member of the mealybug (Pseudococcidae) family.

sod webworm, at least twenty-four hours between insecticide application and irrigation should be allowed to enhance control. The preferred time for insecticide application is late afternoon, since many shoot-feeding insects feed at night. Under droughty conditions the turf should be irrigated prior to application of the insecticide to ensure that sufficient moisture is available for the turfgrass over the following one to three days.

Reducing turfgrass damage from insects is often dependent upon practices other than the application of insecticides. A healthy, vigorously growing turfgrass community is less likely to be damaged by a given population of insects than one that is slow growing. For example, grub-induced damage results from an inability of turfgrass to absorb sufficient moisture to offset evapotranspiration losses. Where the grub population is too small to substantially reduce the existing root system or where root growth is proceeding at a rate that is sufficient to offset losses, observable damage may not develop. Also, recovery from limited damage is usually faster where moisture and fertilizer nutrients are not limiting or where the turfgrass is relatively disease free. The discovery of the role of *Neotyphodium* and other endophytes in reducing the susceptibility of several cool-season turfgrasses to various shoot-feeding insects has introduced a new alternative to pesticides for controlling these insects, at least for the tall and fine fescues and perennial ryegrasses in which these endophytes currently occur.

CONTROLLING LARGE-ANIMAL PESTS

Occasionally, a turf may be damaged by rodents, dogs, and other large animals, including people. Control measures for these problems are extremely limited; often, the turfgrass manager's only recourse is to repair major damage and promote recovery through sound cultural practices.

Rodents

Rodents include ground squirrels, meadow mice, and pocket gophers (Figure 7.6). These animals can cause extensive damage to turf by burrowing holes and runways. They may also leave large mounds of soil at the entrance of tunnels as they dig. Rodent control is possible through the use of traps or poison baits or by introducing poison gases into the tunnels.

Ground Squirrels (*Spermophilus* **spp.**) are long-tailed animals measuring about 10 inches long. Their burrows measure 3 to 5 inches in diameter and are usually very extensive.

Meadow Mice (*Microtus* **spp.**) are soft furry animals measuring about 5 inches long. They dig vertical burrows 1 to 2 inches in diameter and cut runways between burrows.

Pocket Gophers (*Thomomys* **spp.**) are furry animals measuring about 7 inches long. Their long slender claws enable them to dig deeply into a turf and mound large quantities of soil at the surface.

Figure 7.6. Rodents (ground squirrel, top; meadow mouse, middle; and pocket gopher, bottom) that damage turf by burrowing through the soil.

Figure 7.7. A mole that damages turf while in search of insects and earthworms in the soil.

Moles

Moles (*Scalopus* **spp.**) are soft furry animals measuring about 8 inches long (Figure 7.7). They have a long slender snout, no external eyes or ears, and short heavy claws on shovel-like front feet. They feed mainly on soil insects and earthworms. Their burrowing activity results in cone-shaped mounds and ridges at the turf's surface. The extensive severing of roots above their tunnels may result in desiccation of affected turf. Although moles can be controlled using the same measures as for rodents, elimination of their food supply through the use of insecticides may also be an effective control. Similarly, controlling insect populations may prevent damage from skunks, armadillos, and birds that tear up turf in search of food.

Dogs

Dogs (*Canis* **spp.**) can be especially troublesome in residential areas when allowed to urinate on lawns and other turfs. Typical damage appears as brown spots measuring several inches in diameter surrounded by dark green grass. The problem is most evident in early spring as the turfgrass breaks dormancy; however, it can occur throughout the growing season. Deposition of fecal matter also results in localized injury to the turf and is an annoyance. Concentrated traffic along fences and other barriers can cause considerable wear or loss of turf. Except for local leash laws that restrict the movement of dogs in a neighborhood, there is little that can be done to prevent "dog injury." However, a sound cultural program that sustains the turf in a healthy, vigorously growing condition can partially offset damage as well as promote recovery.

People

One of the most destructive large-animal pests of turf is people. Extensive damage is caused each year on golf courses, parks, and other sites by vandalism. Sometimes people unknowingly cause damage with campfires, snowmobiles, motorbikes, automobiles, and the like. On restricted sites, fences and locked accessways may be all that is needed to prevent willful or inadvertent damage. A series of signs indicating use and nonuse areas may also be helpful. Proper placement of sidewalks and shrubs can aid in traffic flow and, thus, minimize traffic-induced damage on school grounds and other heavily trafficked sites. Most importantly, children should be taught to respect property so that the beauty and functional quality of turfgrass sites can be enjoyed and appreciated.

QUESTIONS

1. Define what is meant by a turfgrass **pest** and provide examples.
2. List and explain the **components** of an integrated turfgrass pest management program.
3. What are the desirable **attributes** of a turf pesticide?
4. Illustrate and explain the **fate** of pesticides in turf.
5. List and characterize the pesticide **formulations** available for use in turfgrass management.
6. Provide a detailed explanation of sprayer and spreader **calibration**.
7. Define a **weed** in the context of a turfgrass community.
8. Differentiate among **annual, biennial,** and **perennial** weeds.
9. Explain how you would **chemically control** annual grasses, perennial grasses, and broadleaf weeds in turf.
10. Differentiate between **herbicide types** based on their timing of application, site of application, and selectivity.
11. Define a **disease** in the context of a turfgrass community.
12. Differentiate between **parasites** and **saprophytes** as causal organisms of turfgrass diseases.
13. What is meant by the **disease triangle** and what does it suggest for controlling turfgrass diseases?
14. Differentiate between the following **turfgrass diseases:** brown patch and summer patch, downy mildew and powdery mildew, gray leaf spot and *Helminthosporium* leaf spot, red thread and pink patch, rust and smut, and *Typhula* blight and *Fusarium* patch.
15. Differentiate between **systemic** and **contact** fungicides.
16. Differentiate between endoparasitic and ectoparasitic **nematodes,** and list examples of each type.
17. Differentiate between **insects** that undergo simple and complex metamorphoses, and list examples of each.
18. List and characterize at least four **shoot-feeding** and four **root-feeding** insects, and explain how they can be controlled.
19. List and characterize at least three **large animal pests** of turf.

CHAPTER 8

Propagation

Many problems encountered in the culture of established turfs are directly related to mistakes or omissions made during propagation. Severe weed infestations, disease, poor drainage, scalping, and intolerance of stress may reflect unfavorable site conditions or improper planting and postplanting procedures. The rate at which a new turf becomes established is also influenced by specific procedures used during propagation.

Propagation includes site preparation, turfgrass selection, planting procedures, and postplanting culture.

SITE PREPARATION

Prior to planting, a new turfgrass site should be prepared in such a way that inherent problems are corrected and potential problems are avoided. For example, in constructing new buildings, existing topsoil on areas surrounding the buildings is often overlain by subsoil from excavation operations. Thus, an inferior soil of low fertility and limited aeration capacity is often substituted for one that might be far more suitable for supporting turfgrass growth. In addition, buried debris and compaction from large-equipment operation may render the soil unfavorable for turf.

Site preparation may involve various clearing operations along with tillage, grading, soil modification, installation of drainage and irrigation systems, and fertilization.

Clearing Operations

Clearing may be defined broadly as any practice designed to eliminate or reduce obstructions to successful propagation. On wooded sites trees and shrubs are removed either completely or selectively. Rocks and other debris that might interfere with subsequent construction operations or create unfavorable conditions for turf are also removed. Weed species that might pose a problem during or after turfgrass propagation (especially perennial grasses and sedges) are controlled. Some of the materials removed from the site may be sold if they have economic value; other materials can be deposited in landfills or burned on site.

Woody Vegetation

Woody vegetation includes standing trees and shrubs, fallen logs, stumps, and buried roots and stems. Standing trees may have considerable market value for lumber; this should be determined before removal so that provisions can be made for proper harvesting and transportation. The remaining stumps should be bulldozed or otherwise removed so that belowground residues do not decay and leave undesirable depressions that destroy the uniformity of the grade. Growth of mushrooms and puffballs in established turfs can often be traced to tree stumps that were allowed to remain at planting.

Rocks and Boulders

Removal of rock outcrops is an essential clearing operation. These should be cut to a depth of at least 15 inches below the final grade and replaced with soil. Without a sufficient soil depth above large rocks and boulders, the soil tends to remain wet for prolonged periods following irrigation or precipitation. Then these areas become hard and dry due to inadequate replenishment of moisture from lower soil depths.

Small rocks and stones in the surface 4 inches of soil interfere with cultivation operations and may also promote weed invasion where turfgrass root growth is blocked. Usually, most stones can be raked up and removed prior to planting. If they are not too numerous, stones that surface following planting can be picked up by hand or raked once the turfgrass seedlings are sufficiently rooted. A stone-picking machine can be used prior to planting where the number of stones is too great for removal by manual methods.

Preplant Weed Control

The presence of some weedy perennial grasses and sedges on planting sites can result in serious weed problems in the new turf. Even where objectionable plants have ostensibly been removed by raking following tillage or with a sod cutter, surviving vegetative propagules (rhizomes, stolons, tubers, and the like) remaining in the soil can often lead to reinfestation. The most effective means for controlling these weeds are fumigants and nonselective systemic herbicides. The principal herbicide used for this purpose is glyphosate (see Chapter 7). It should be applied when the weeds are several inches tall, and at least three to seven days should be allowed for absorption and translocation of the herbicide to belowground organs before initiating tillage operations. A period of fallowing is usually helpful after herbicide application for controlling weed populations. Belowground plant organs are brought to the surface, where they are subject to desiccation.

Additional herbicide applications may be necessary where the volume of vegetative propagules is large. If additional applications of a herbicide are made, they should be delayed until new green growth is evident. Fallowing is best done during the summer months. Some weeds of perennial grasses may still infest the new turf, but they usually occur in localized patches. Spot treatment with glyphosate where these patches occur is usually adequate to prevent widespread infestations.

Fumigation is the process by which a highly volatile chemical is applied to the soil to control weed seeds, vegetative propagules, disease-causing organisms, nematodes, and other potentially troublesome organisms. Topdressing soil is sometimes fumigated to prevent the introduction of weed seeds where the soil is applied to an established turf. Fumigation is a laborious and expensive procedure, but it may be justified where a large reservoir of annual bluegrass seed or vegetative propagules of perennial weedy grasses and sedges exists in the soil. Prior to fumigation the soil should be deeply tilled so that the chemical vapors of the fumigant can be adequately distributed to all locations containing weed propagules and other target organisms. The soil should be moist but not wet; a dry soil may tightly adsorb the fumigant molecules and thus restrict their mobility in the soil. The temperature of the soil should be at least 60°F; otherwise the activity of the fumigant may be substantially reduced.

The principal fumigants used for turf soils are methyl bromide, chloropicrin, and metham. Methyl bromide is a highly toxic odorless gas that is released under a polyethylene tarp covering the treated site. It is usually mixed with a small amount of chloropicrin (tear gas) as a warning agent. Two methods exist for applying methyl bromide: large-area application by ground soil fumigation rigs with automatic tarp-laying equipment and manual application to small areas, and manual application. This involves propping the tarp approximately 1 foot above the soil surface by means of blocks or other devices placed 30 feet apart, sealing the edges of the tarp with soil or boards, and injecting the gas into the covered area by means of polyethylene tubing from the fumigant cans to evaporation pans located within the tarped area. After twenty-four to forty-eight hours the tarp is removed and the soil is allowed to aerate for an additional forty-eight hours before planting.

Metham is a sprayable material that should be soil-incorporated and/or irrigated in immediately following application. A period of at least three weeks is required between application and planting of treated sites for decomposition of the chemical in the soil.

Tillage

Tillage is any operation performed to prepare a soil for planting. On large sites it may involve a succession of operations, including plowing, disking, and harrowing. On small sites rototilling the soil once or twice may produce similar results (Figure 8.1). The soil is tilled to promote rapid infiltration and percolation of water, satisfactory retention of water for turfgrass growth, adequate aeration capacity, minimal resistance to root penetration, and surface stability against erosion and traffic. Except for sands, soil should be well granulated in response to tillage operations. Tilling when it is too wet results in hard clods

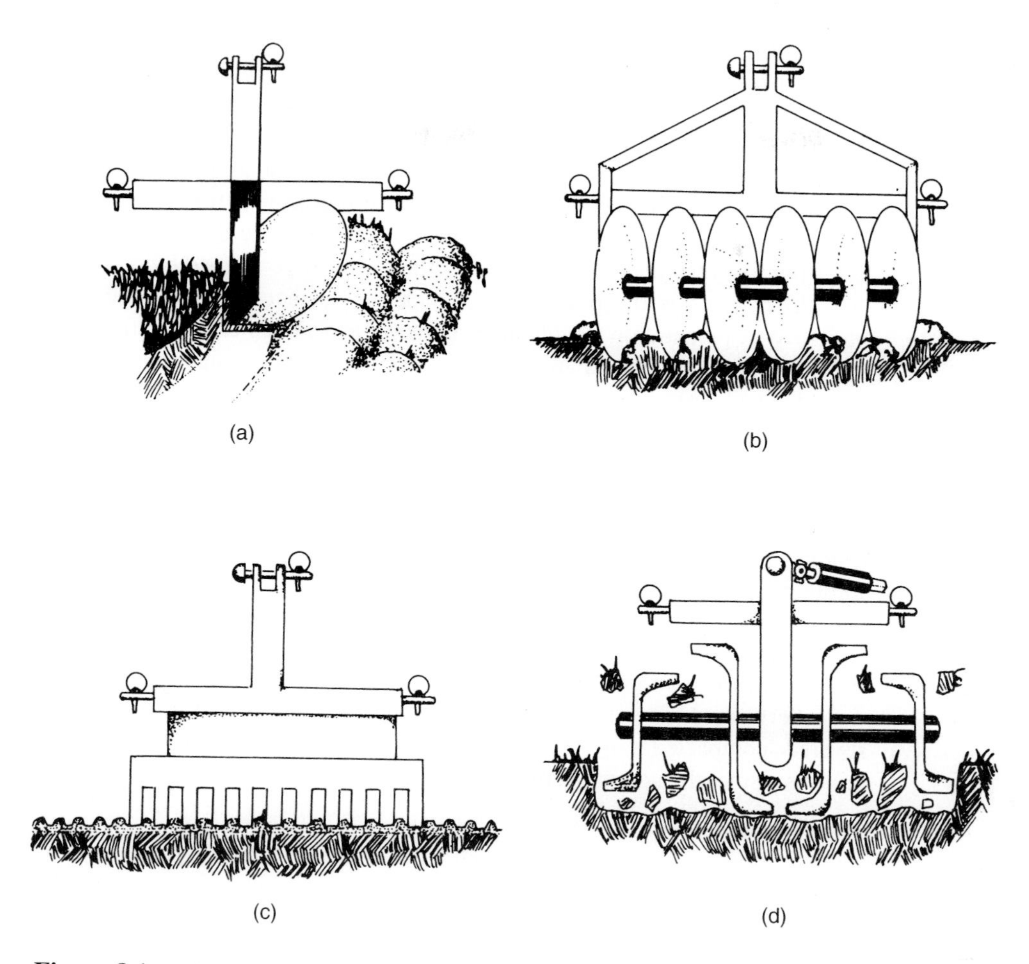

Figure 8.1. Tillage by (a) plowing, (b) disking, (c) harrowing, and (d) rototilling.

or puddling, depending on the amount of water present; when too dry, soil structure can be destroyed, resulting in a powdery and highly compactable surface. To test whether soil moisture is within a favorable range for tillage, squeeze a handful of soil tightly and then try to break it apart with the thumb. If the soil mass crumbles easily, tillage operations can proceed. A too-dry soil will be very hard to break up, while a too-wet soil will tend to smear under pressure.

Plowing is a procedure by which a furrow slice (plow depth) of soil is cut loose, granulated, and inverted. The resulting surface is nonuniform, with furrows and ridges occurring parallel to the direction of plowing. This is most pronounced in soils bound by a deep network of turfgrass roots. Where substantial amounts of plant residues (thatch, verdure) are turned under by plowing, sufficient time should be allowed for these residues to decompose. Fall plowing and overwintering of the overturned sod favor decomposition of organic residues, especially in climates with cold winters. Where a subsurface compacted layer or pan exists, this can be broken up by a chisel plow, providing the pan is not beyond the penetration depth of the plow.

Disking is performed on plowed or loose soil to break up clods and surface crusts, thereby improving soil granulation and surface uniformity. It may be performed shortly after plowing or at a later date after organic residues have decomposed. Where summer fallowing for weed control is practiced, this is usually performed by disking.

Harrowing is the practice by which the surface soil is further granulated and smoothed in preparation for planting. Furrows and ridges resulting from previous plowing and disking operations are smoothed out; for turfgrass plantings, additional smoothing of the soil surface is usually desired and will be discussed under fine grading.

Rototilling is a very intensive form of tillage employed primarily on relatively small sites such as golf tees and residential lawns. An adequately powered rototiller can cut through an existing turf and finely granulate the soil in one or two operations. However, prior removal of the sod is usually desired to eliminate the thatch and other organic residues near the surface. Otherwise, these may form objectionable clumps in or on the soil that interfere with subsequent planting operations unless a sufficient period is available to allow for their decomposition. Rototilling is an excellent method for incorporating organic amendments into the soil and for mixing different soil layers to yield a soil of relatively uniform composition.

Performed excessively, all tillage methods can destroy soil structure and result in a severely compacted soil. The intensity of tillage operations should be just adequate to provide a smooth and well-aerated planting bed suitable for turfgrass propagation.

Grading

Grading includes all operations for providing a smooth planting bed of desired contour. Since grading may involve relocating large quantities of soil, it may be advisable to remove and stockpile the topsoil before attempting any major changes in contour. The depth of the existing topsoil layer should be carefully determined prior to initiating construction operations. Then the topsoil layer can be scraped off and moved to a predesignated storage area. This is also advisable for areas adjacent to construction sites for new homes and other buildings where deep excavations are performed. Failure to do so may result in burial of the topsoil under excavation soil.

Grading operations are usually divided into two types: rough grading and fine grading.

Rough Grading

Rough grading refers to the contouring of a subgrade following removal of the topsoil. This usually involves cutting down elevated sections and filling in depressions. To ensure a grade free of surface irregularities, marked stakes should be set at intervals between fixed grade levels to establish the desired level throughout the area. Where the fill soil is loose and subject to settling it should be placed at a higher level than that desired in the final grade. Approximately 15% (1.5 to 2 inches per foot) settling should be anticipated with fine-textured soils, but less will occur with coarse-textured soils. Where large amounts of fill are required, it should be added in 1-foot increments and rolled to accelerate settling.

A favorable slope for surface drainage is an approximately 2% or 1/4-inch drop per foot of linear distance. On homesites the slope should always be away from the house in all directions to prevent water from seeping into basements and crawlspaces. Athletic fields should be crowned so that surface drainage proceeds away from the center of the field. Golf greens, tees, and fairways should slope in one or more directions toward the roughs.

Steep slopes should be avoided if at all possible because of the difficulties that may be encountered during and after propagation. The potential for soil erosion and loss of planting material increases in direct proportion to the angle of slope. Turfs established on steep slopes, especially those facing the midday sun, are more subject to heat and drought stresses during the summer months than relatively level turfs. Mowing is more difficult and scalping injury is more likely along steep slopes. Finally, fertilizers and pesticides are susceptible to runoff whenever excess water from precipitation or irrigation flows downslope; as the angle of slope increases so does the potential for runoff. In view of these problems, consideration should be given to constructing retaining walls to limit the angle of slope of the turf where major elevation changes cannot be avoided.

As the subgrade must conform to the final grade after topsoil is replaced, grade stakes should be marked to indicate the topsoil depth required. At least 6 inches (settled depth) of topsoil should be provided. Again, some settling should be anticipated, depending on soil texture. Where the subsoil is of substantially different texture and structure, it may be advisable to mix 2 inches of topsoil with the surface 2 inches of subsoil to effect a more gradual transition from topsoil to subsoil. This would have two direct benefits: alleviation of surface compaction of the subsoil and reduction of problems associated with an abrupt transition at the topsoil-subsoil interface.

Fine Grading

Fine grading serves to smooth the soil surface in preparation for planting (Figure 8.2). It is also effective for working in starter fertilizer. Hand raking is usually the preferred method for small areas where the operation of large equipment would be cumbersome. A heavy steel mat pulled with a rope tied to two corners is also suitable for producing a smooth surface. Large areas generally require specialized equipment such as soil blades, harrows, heavy steel mats, plank drags, or tiller rakes.

Prior to fine grading, sufficient time should be allowed for settling of the soil; otherwise, the operation of heavy equipment can result in deep depressions in the surface and, therefore, an uneven surface in the turf once it is established. Settling can be accelerated by intensive irrigation or rainfall. Rolling can also serve as a means of firming the soil surface. As settling of the soil may occur differentially, resulting in localized depressions and high spots, fine grading is often necessary to restore a smooth and uniform surface before planting operations are initiated.

Figure 8.2. Fine grading by hand raking (top) or through the operation of a tiller rake (bottom).

Fine grading should be delayed until just prior to planting; otherwise, the soil surface may crust over and thus require further conditioning. As with other grading and tillage operations, soil moisture should be within a suitable range before attempting to fine grade the site.

Soil Modification

Soil modification involves the incorporation of amendments to improve the physical and chemical properties of a soil. Insufficient moisture and nutrient retention, on one hand, or inadequate aeration, on the other, are conditions that can be corrected through soil modification. However, improper selection and use of amending materials may actually result in less favorable soil conditions than where no amendments were incorporated. One example is the improper use of sand for improving aeration capacity; often, a small quantity of sand added to a fine-textured soil produces a cementlike mixture that is unsatisfactory for supporting turfgrass growth. According to Madison, better results are obtained when a small amount of soil is mixed with sand than when a small amount of sand is mixed with soil. Individual sand grains do not provide improved aeration when added to a fine-textured soil; the sand grains simply displace pore spaces that would otherwise exist between adjacent silt or clay particles. This is analogous to the addition of a marble to a cup of flour; although the porosity of the flour is low, the porosity of the marble is zero. It is only when individual sand particles are in direct contact that aeration porosity increases. Obviously, this requires large quantities of sand relative to the amount of fine-textured soil particles in the mixture. Therefore, sand as an amendment should be avoided unless laboratory tests indicate that results with specific mixtures would be beneficial.

Numerous synthetic amendments, such as perlite and vermiculite, are used extensively in the production of container-grown plants. However, these are generally considered unsuitable for use in turf soils because of their inability to hold up under the compressing force of traffic. Vermiculite particles tend to flatten out, while perlite crushes easily. Other locally available materials are sometimes suggested as soil amendments, but these should be selected only when sufficient data are available to indicate that they will provide satisfactory results.

The most widely used organic amendment is peat moss. It can be purchased in large bales, is relatively lightweight, and can be incorporated into a soil without too much difficulty. Used in sufficient amounts, peat moss also represents a substantial investment. A 2-inch layer spread over the soil for subsequent incorporation requires more than 3 $yd^3/1000\ ft^2$. In fine-textured soils peat moss dilutes clay particles and spreads them apart. This improves aeration capacity while promoting soil aggregation. In coarse-textured soils peat moss increases water and nutrient retention while providing improved resiliency in an established turf.

Other organic amendments may yield results as favorable as those obtained with peat moss; however, their qualities should be carefully examined before a selection is made. Some highly decomposed organic media may contain appreciable amounts of dispersed clay or silt particles that can clog soil pores and thus reduce aeration capacity. Other organic amendments with high carbon-to-nitrogen (C:N) ratios, such as straw or undecomposed sawdust, can cause a nitrogen draft in the soil (see Chapter 4) that can adversely affect turfgrass growth. Local authorities should be consulted before selecting an organic amendment for modifying problem soils.

In constructing greens and some athletic field turfs, specially prepared media are frequently used to provide adequate tolerance to intensive traffic and play. In many instances much of the native soil is removed and replaced by an entirely different medium of carefully prescribed characteristics. As this is really soil substitution rather than soil modification, construction of specialized turfgrass sites will be discussed in Chapter 9.

Drainage and Irrigation Systems

Installation of drainage and irrigation systems can be scheduled after rough grading of a new site. The advantage of scheduling at this point in the construction process is that settling of soil in trenches is less of a problem than it might be following replacement of the topsoil. Drain tiles are normally placed at a depth of 18 to 36 inches below the turf's surface and lines are spaced 15 to 60 feet apart on sites where internal drainage critically affects turf quality. In semiarid climates placement depth may be up to 6 feet where salinization of the surface soil by upward movement of groundwater is a potential problem. If installed prior to the addition of topsoil, placement depth should be calculated to include the topsoil depth.

Drain tiles may be arranged in a herringbone or gridiron pattern (Figure 8.3) or simply placed in low areas where water collects from surface runoff. Suitable discharge facilities must be available with sufficient capacity for accommodating water under the most extreme conditions encountered in a particular area. Storm sewers, lakes, and streams frequently serve as receptacles for discharge water from drainage systems.

Traditionally, drain tiles constructed of clay or concrete have been used. However, in recent years perforated plastic pipe has received wide acceptance because of its light weight and ease of handling. Plastic pipe can be purchased in rigid sections or as flexible corrugated tubes in large rolls. It is usually desirable to place gravel around the drain tiles to prevent clogging by fine soil particles. At strategic locations the gravel may extend up the soil surface and thus function as a catch basin for surface runoff water. Catch basins may also be constructed using cement wall cylinders extending from the surface to the depth of the drain tiles. A perforated steel or plastic cover is placed atop the cylinder so that it is flush with the turf's surface.

Installation of irrigation systems presents some problems when scheduled prior to topsoil placement. Care must be exercised to ensure that damage to system components

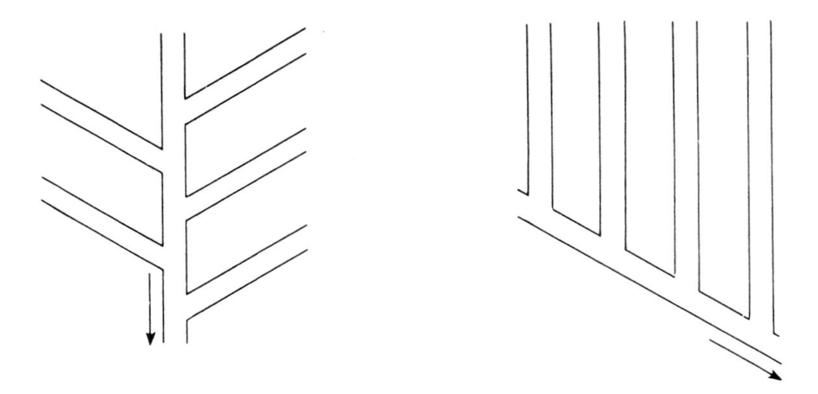

Figure 8.3. Two drain tile designs: herringbone (left) and gridiron (right).

does not result from grading operations. If the potential for damage is too great, installation of all or a portion of the system should be delayed until after the topsoil has been placed and graded to the final contour.

Fertilization and Liming

Because of nutrient deficiencies or unsuitable pH in the planting bed, basic fertilizer and lime may be required prior to planting. The term *basic fertilizer* refers primarily to phosphorous and potash but may also include secondary and minor nutrients. As preparation of planting beds presents a rare opportunity to incorporate fertilizer nutrients and lime into the turfgrass root zone, these materials should be tilled into the soil by disking, rototilling, or other means. Whenever possible, the amounts applied should be based on soil test results.

Nitrogen is generally applied just prior to final grading of the site and is referred to as a *starter* fertilizer. It usually is not tilled deeply into the soil because subsequent leaching may result in much of the nitrogen being lost before roots have developed sufficiently to absorb it. Where soluble nitrogen carriers are used, 1 to 1.5 lb N/1000 ft^2 should be applied prior to planting. If needed, this should be supplemented by a second application of 0.5 lb N approximately two weeks after seedling emergence. Care should be taken to prevent foliar burn, which could be lethal to young seedlings. The application should be made when the seedlings are dry, and irrigation should immediately follow. With slowly available nitrogen carriers the initial application rate can be two to three times that of the soluble carriers, and a second application several weeks later may not be necessary.

TURFGRASS SELECTION

The selection of turfgrasses for planting is one of the most important decisions made in turfgrass culture. The persistence and quality of a turf and the severity of weeds, diseases, insects, and other pest problems often reflect known characteristics of specific turfgrass genotypes.

With the initiation of formal investigations in the 1920s to find better-adapted bentgrasses for use on greens, naturally occurring ecotypes were selected, propagated, and observed in experimental plots. Some of these early selections were eventually adopted for commercial use, and many of them can still be found on older golf courses and other turfgrass sites.

The discovery and release of Merion Kentucky bluegrass had a dramatic influence on lawn turf culture during the 1950s and 1960s. Merion showed high resistance to *Helminthosporium* melting-out, a serious disease of Kentucky bluegrass during cool-weather periods. Since the severity of melting-out disease increased in response to closer mowing and higher rates of nitrogen fertilization, the introduction of the melting-out-resistant Merion allowed greater flexibility in mowing and fertilization practices. In recent years many new melting-out-resistant cultivars have been introduced that possess excellent resistance to other diseases, as well as improved tolerance to moderate shade, close mowing, traffic, and other adversities.

Superior cultivars of perennial ryegrass, creeping bentgrass, bermudagrass, zoysiagrass, and St. Augustinegrass are available, and efforts are underway to develop improved cultivars of other turfgrass species. The development and commercialization of many new turfgrasses has created some confusion—commonly expressed as *varietal dilemma.* The amount of testing required to clearly establish the environmental adaptation and cultural requirements of each new cultivar often exceeds the resources available at university experiment stations. Furthermore, conclusions drawn from results at one location may not apply to conditions existing at sites located several hundred miles away. Thus, complete information on the performance characteristics of newer turfgrass cultivars is usually not available until many years after a cultivar has been introduced.

One means by which turfgrass managers have attempted to compensate for the lack of detailed information on the performance characteristics of specific cultivars is through the use of blends. The thinking has been that, if we must guess which cultivar is best adapted to a specific set of conditions, several cultivars planted together offer better assurance of satisfactory results than does a single cultivar planted alone.

Blends and Mixtures

A blend is a combination of different cultivars of the same turfgrass species. In a mixture, two or more species are combined. Mixtures have been used for many decades; a traditional lawn-seed mixture for temperate climates has been Kentucky bluegrass, red fescue, and perennial ryegrass. Sites planted with this mixture often become predominantly Kentucky bluegrass in sunny locations, while shaded conditions favor red fescue. Perennial ryegrass, used primarily as a nursegrass to provide quick cover, usually disappears or is reduced to isolated patches after several years. Redtop and annual ryegrass have also been used as nursegrasses, but these species are so highly competitive in a seedling stand that they threaten the survival of more desirable turfgrasses. Isolated clumps of these grasses are highly visible and objectionable in a turf and often persist as weeds. Mixtures also containing bentgrass species often become predominantly bentgrass or have isolated patches of bentgrass scattered throughout the turf.

The trend in recent decades has been to limit the number of species in a mixture to avoid the variegated appearance that so often results where a multicomponent mixture is used. In many instances a single species is planted to achieve a uniform stand, usually of Kentucky bluegrass.

While single-species plantings may be relatively new to temperate climatic zones, they have been the rule for subtropical and tropical climates. Warm-season turfgrass species are generally incompatible in mixtures; they tend to segregate into isolated patches, resulting in a nonuniform turf. The selection of a turfgrass species for planting on a specific site must, therefore, be based on the range of environmental conditions encountered on the site. Bermudagrass forms an excellent turf on unshaded sites provided it receives adequate care, but it will not persist in shaded locations. St. Augustinegrass, which is more widely adapted to shaded and unshaded conditions, is often the preferred choice for residential lawns in warm subtropical and tropical climates.

A concern with any single-species turfgrass planting is its limited range of environmental adaptation. Light, moisture, temperature, fertility, soil aeration, and traffic can vary substantially from one location to another within a single turfgrass site. Wherever the adaptation limits of a particular turfgrass are exceeded, reduced turf quality is almost inevitable. Also, diseases are more likely to seriously reduce turf quality where a genetically homogeneous turfgrass community exists. Thus, gains realized from single-species plantings (for example, uniform appearance and composition) may be offset by losses from higher disease incidence and intolerance to certain environmental conditions. One contemporary approach to reconciling these concerns is through the use of blends.

Successful blending of different cultivars of a single species depends on compatibility among component cultivars, demonstrated differences in disease incidence or potential, and documented evidence of broad differences in environmental adaptation. Resulting primarily from the productive efforts of several turfgrass breeders, especially C.R. Funk at Rutgers University, numerous superior cultivars of Kentucky bluegrass and perennial ryegrass are available for use in blends. Through the use of properly composed blends of Kentucky bluegrass, many of the more important diseases of this species are no longer of primary concern. Even if a particular disease, such as stripe smut, does develop in one of the component cultivars, the other cultivars may assume its position and, therefore preserve the turfgrass community. Additional benefits of blending may include better shade adaptation, improved wear resistance, adaptation to mowing heights of 1 inch or even lower, and improved recuperative capacity following injury. However, where wide differences in appearance (color, texture, growth habit, and the like) exist among cultivars within a blend, a variegated or patchy turf may eventually develop. Reasonably compatible cultivars must be selected; otherwise, differences similar to those encountered with some mixtures can become evident. Since more information than is currently available is required to fully establish criteria for selecting compatible cultivars, there is still some guesswork involved in composing blends. Reasonable guidelines would probably include:

1. Select only those cultivars that have shown resistance to diseases of primary importance in a particular geographic area.
2. Ensure that all cultivars selected are fairly similar in their appearance and, if known, competitive ability.
3. Select at least one cultivar that is well adapted to any special conditions existing at the planting site (moderate shade, alkaline soils, and so on).
4. Choose at least three cultivars for the blend.

Seed Quality and Labeling

The principal factors influencing seed quality are purity and viability. Purity is the percentage of pure seed of an identified species or cultivar present in a particular seed lot. Where purity is less than 100%, the remaining percentage is composed of inert matter (chaff, dust, and the like), weed seeds, and other crop seeds. Viability is the percentage of seed that is alive and will germinate under standard laboratory conditions. The product of purity and viability is the percentage of pure live seed (percent P $\times$ percent V = percent PLS). For example, a seed lot with 90% purity and 75% viability will have 67.5% pure live seed. Dividing the cost per pound of seed by the percentage of PLS and then multiplying by 100 yields the actual cost per pound of pure live seed:

$$\$1.10 \div 67.5 \times 100 = \$1.63/\text{lb PLS}$$

Using these simple calculations it is possible to compare two or more packages of seed to determine which is the best buy.

The minimum acceptable purity and germination percentages for different turfgrass seeds are given in Table 8.1. In addition, the approximate number of seeds per gram is listed. Multiplying the percentage of PLS by the number of seeds per pound yields the number of pure live seed per pound. For example, Kenblue-type Kentucky bluegrass with 2,200,000 seeds per pound and a calculated PLS of 67.5% will actually contain 1,485,000 pure live seed per pound. If uniformly seeded at the rate of 2 lb/1000 ft^2, each square foot of the seedbed will have 2970 pure live seed of this turfgrass. This does not mean, however, that 2970 seedlings per square foot will emerge. An additional factor, called the *expected seedling mortality rate,* must also be taken into consideration. As indicated previously, seed viability is determined under standard laboratory conditions. In the field, seedling emergence and survival are often lower than that determined in the laboratory. Reasons for this include loss of viability during the storage interval between seed testing and planting, burial of some seeds at soil depths that are unfavorable for germination and seedling survival, and seedling mortality due to unfavorable temperatures or inadequate moisture. Once a seedling stand has emerged, further losses may result from disease, insects, erosion, traffic, or competition from weeds or other seedlings in the young plant community.

Seed Purity

Minimum acceptable purity of turfgrass seed varies with species. In Table 8.1, the range is from a low of 40% to a high of 95% on a *weight* basis. The purity of a seed lot is related to the method of production and the ease and care with which the seed is cleaned and processed.

Contamination of a seed lot with weed and other crop seeds may be a serious concern depending on the types and quantities of undesired seed in the lot. Although they may present a nuisance following germination, most broadleaf weed species can be easily controlled in the seedling stand with commercially available herbicides. Even annual grasses can be controlled selectively with preemergence or postemergence herbicides if necessary. Perennial grasses, however, present a special problem in that selective control with herbicides is often not possible. Thus, bentgrass, tall fescue, rough bluegrass, orchardgrass, and ryegrass (usually listed under "other crop" seeds) may be highly undesirable contaminants, even when they occur in very small quantities. Sod growers often pay premium prices to obtain seed free of annual bluegrass and creeping bentgrass where

Table 8.1. Quality Characteristics of Turfgrass Seeds and Seeding Information

Turfgrass	Approx. No. of Seeds Per Gram[a]	Seeding Rate, Lb/1000 Ft^2	% Minimum Purity (By Weight)	% Minimum Germination (By Numbers)
Bahiagrass	360	6–8	70	70
Bentgrass, colonial	18,000	0.5–2	95	85
creeping	14,000	0.5–1.5	95	85
redtop	11,000	0.5–2	90	85
velvet	24,000	0.5–1.5	90	85
Bermudagrass, common (unhulled)	3,900	1–1.5	95	80
Bluegrass, Canada	5,500	1–2	85	80
Kentucky	4,800	1–2	90	75
rough	5,600	1–2	90	80
Buffalograss (burs)	110	3–6	85	60
Carpetgrass	2,500	1.5–2.5	90	85
Centipedegrass	900	0.25–0.5	45	65
Fescue, meadow	500	4–8	95	85
red	1,200	3–5	95	80
sheep	1,200	3–5	90	80
tall	500	4–8	95	85
Gramagrass, blue	2,000	1–2	40	70
Ryegrass, annual	500	4–6	95	90
perennial	500	4–8	95	90
Timothy	2,500	1–2	95	90
Wheatgrass, fairway	700	3–5	85	80

[a]To convert to number of seeds per pound, multiply by 454. Considerable variability may exist among cultivars within a species, or even within a cultivar, due to growing conditions; some Kentucky bluegrass cultivars have only half as many seeds per gram as the figure given in the table.

these species result in reduced sod quality. In a Kentucky bluegrass seed lot 1% contamination with annual bluegrass may result in twenty to forty seeds of this species per square foot at planting. This may be sufficient to substantially reduce sod quality or render the sod unsuitable for some uses.

The amount of inert material in a seed lot may reflect the degree of care exercised in cleaning the seed after it has been harvested from the field. Through the use of properly sized screens and airflow mechanisms, a seed lot can be cleaned nearly free of most undesired materials. The greatest difficulty encountered in seed-cleaning operations is in attempting to remove weed seeds of the same or smaller size as the desired turfgrass seed. Annual bluegrass and rough bluegrass are extremely difficult to remove from Kentucky bluegrass as all are approximately the same size. Creeping bentgrass seed, although much smaller, may lodge between the floral bracts or behind the rachilla of Kentucky bluegrass and thus be difficult to remove or even detect. The best assurance of obtaining seed free of these and several other undesirable weeds is through careful weed monitoring and control in the seed production fields.

Seed Viability

Viable seeds are those that are capable of germinating and producing a normal seedling. Viability is therefore synonymous with germination capacity. Seed viability is usually highest at physiological maturity (several months following harvest) and declines with additional aging. The rate at which the decline in viability occurs depends on environmental conditions; dry cool storage conditions are most favorable for extending the period of seed viability. Reduced viability may be associated with mechanical injury during harvesting and processing operations, failure to remove poorly developed seeds during cleaning operations, unfavorable growing conditions, diseases, and insect-induced damage.

Different turfgrass species and cultivars vary substantially in the normal percentages of viable seed contained in a seed lot. Table 8.1 lists the minimum germination percentages by species. Unlike purity, which is reported on a weight basis, germination percentages are calculated from the numbers of seeds that germinate in a test sample.

Seed Labeling

The label accompanying a package of seed is an important guide for evaluating seed quality. It lists the amounts of each turfgrass species and, if known, cultivar contained in the package. The percentages given for purity, weed seeds, other crop seeds, and inert material must add up to 100. These values are determined by testing small samples (0.25 gram [g] for bentgrasses, 1 g for bluegrasses, 3 g for fine fescues, and 5 g for tall fescue and ryegrasses) taken from the seed lot. Usually, the specific weed and other crop seeds present are not listed except where required by state law. Noxious weeds are reported as numbers of seeds per pound of the seed lot based on analyses of samples weighing ten times the amount listed for each of the above species. Only those weeds listed as being noxious by state law are reported. However, noxious weeds in agricultural crops may not be troublesome in turfgrass communities, while "other crop" seeds such as bentgrass, timothy, orchardgrass, and tall fescue can cause serious problems. The user of the seed usually cannot distinguish between the weed and other crop seeds that are potentially troublesome and those that are not. It is only through a more thorough analysis of a representative sample from the seed lot that this can be determined. A system developed by Seed Technology at Marysville, Ohio, employs seed samples of 25 g or more per 10,000 lb lots; all weed and other crop seeds present are identified and placed into one of three categories: *no problem* (will not compete in turfgrass plantings or can be controlled easily), *controllable* weeds (can be controlled with selective herbicides), and *uncontrollable* weeds (cannot be controlled with selective herbicides). Until seed laws are changed to reflect the specific concerns in testing turfgrass seed lots, the individual recipient must either accept the risks of undesirable contaminants or have the seed tested independently.

Seed Certification

Seed certification is a program in which seed-production fields and seed lots are inspected to ensure genetic purity of the seed. The activities of various state agencies with responsibility for conducting the certification program are coordinated by the Association of Official Seed Certification Agencies (AOSCA).

The four seed classes established by AOSCA are breeder, foundation, registered, and certified. *Breeder seed* is the original source of all classes of certified seed. It is controlled by the sponsoring plant breeder or institution and provides the direct source of foundation seed. *Foundation seed* is produced in fields planted with breeder seed; its containers carry a white certification tag. *Registered seed* is produced in fields planted with foundation seed. Its production is primarily for increasing the supply of a particular cultivar. Containers of registered seed carry a purple certification tag. *Certified seed* is the seed available to the consumer; it is produced in fields planted with either registered or foundation seed. Containers of certified seed carry a blue certification tag. The color-coded seed certification tags accompany the seed label; while they do not provide a guarantee that the seed is of highest quality, they do ensure that the cultivar listed on the label is true-to-type.

Vegetative Planting Materials

Vegetative planting materials include sections of turfgrass communities, individual plants, and portions of plants (excluding seeds) with adequate reproductive capacity to produce a turf. They are used in place of seed for propagating those turfgrass genotypes that do not produce viable seed or that cannot be reproduced true-to-type from seed. They are also used for developing a new turf on sites where rapid establishment is desired. One common propagation method, called sodding, can be used for all turfgrasses. Other vegetative propagation methods include plugging, sprigging, and stolonizing. Some cultivars of creeping bentgrass, St. Augustinegrass, hybrid bermudagrasses, and zoysiagrass are routinely propagated by one or several of these methods.

Sod

Good-quality sod is uniform, pest free, sufficiently strong to hold together during handling, and capable of rooting within one to two weeks following planting. It should be cut as thin as possible. Rarely is more than 1/2 inch of soil containing the roots and other belowground organs necessary. It should have little or no thatch development. Some Kentucky bluegrass sod growers wash the sod free of soil to reduce weight, hasten rooting, and eliminate potentially troublesome soil layers where soil textures at production and transplant sites differ substantially. Sometimes thick-cut sod is used for emergencies when it is necessary to plant just prior to heavy play or traffic. Where this is practiced it is usually desirable to replace the sod as soon as possible to ensure satisfactory establishment and to eliminate potentially troublesome textural layers in the turf profile. Some types of sod are produced with plastic netting to enhance sod strength, especially when bunch-type turfgrasses (e.g., tall fescue, perennial ryegrass) are used.

Sod strips have traditionally measured 2 to 6 feet in length and 1 to 1.5 feet in width; however, large rolls of sod can be obtained for efficiently planting large sites. Sod strips are usually transported flat, folded, or rolled. Sod should be installed as soon after harvest as possible, preferably within twenty-four to forty-eight hours, to avoid damage from sod heating or desiccation, especially with cool-season turfgrasses. Sod heating is an increase in temperature resulting from respiration within the sod. It can cause deterioration or death of the turfgrass and is favored by high temperatures, long leaves, high tissue-nitrogen content, seedheads, disease, and poor air movement throughout the stacked sod. Vacuum cooling has been effective as an artificial means for minimizing sod heating, but high costs have drastically limited its practical use. Sod desiccation is the drying out and associated injury that occur in sod that has been exposed to prolonged dry periods. It is favored by high temperatures, low moisture in the soil or atmosphere, and high winds. A stack of sod may show desiccation injury in portions of exposed sod strips and heat damage in the unexposed sod. Both conditions can greatly influence sod appearance, growth, and survival.

Plugs

Plugs are small cylindrical- or block-shaped sections extracted from a field turf or cut from strips of sod. The amount of soil carried with plugs varies from 1/2 inch to several inches thick. Heating and desiccation may be just as much or more of a concern with turfgrass plugs as they are with sod, especially if the plugs are small. The turfgrass in plugs should be healthy, weed free, and vigorously growing at the time it is cut. As with all vegetative planting materials, plugs should contain only those cultivars desired in the new turf. Usually, only stoloniferous turfgrasses are propagated with plugs because of the requirement for vigorous lateral growth into areas between the plugs. However, in recent years strongly rhizomatous cultivars of Kentucky bluegrass have been successfully propagated by plugging.

Sprigs and Stolons

Sprigs and stolons are individual plants or sections of plants containing several nodes from which new plants may develop. Both may be bare stolons, but they usually have some roots and leaves attached. The basic difference between sprigs and stolons is the method of planting rather than the planting material itself. Turfgrasses most commonly planted by these materials include creeping and velvet bentgrasses, bermudagrass (other than common), zoysiagrasses, and St. Augustinegrass.

Usually, sod fields that are designated for stolon and sprig production are mowed at normal heights during the period of seed head development to prevent contamination by off-types resulting from sexually produced seed. Then mowing operations are halted for several months to promote the development of large stolons. When growth has been sufficient, the "sod" is harvested with as little soil as possible removed and the vegetative material is shredded or chopped to yield stolons and sprigs.

Stolons and sprigs should be planted as soon as possible after production to minimize damage from heating and desiccation. If temporary storage is necessary, the material should be kept in a cool moist environment.

PLANTING PROCEDURES

The two principal methods for propagating turfgrass are seeding and vegetative planting. The specific choice of propagation method depends on cost, time constraints, availability of genetically pure planting materials, and growth characteristics of the turfgrass. Planting seed involves the least expense and labor, but is often the slowest method for developing a new turf. Vegetative planting methods include sodding, plugging, sprigging, and stolonizing. Of these, sodding is the most expensive method, but yields the fastest results. With some turfgrass species, such as creeping bentgrass, any of these methods can be employed successfully. With other species some propagation methods may not be feasible because genetically pure or viable seed is unavailable, or the species lacks sufficient spreading ability to produce a turf from space-planted plugs or sprigs.

Seeding

Most cool-season turfgrasses can be propagated by seed; the only exceptions are several older cultivars of Kentucky bluegrass and creeping bentgrass that do not reproduce true-to-type from seed. Of the warm-season turfgrasses, centipedegrass, bahiagrass, carpetgrass, buffalograss, and common bermudagrass are commonly seeded (Figure 8.4). Seed of St. Augustinegrass and some zoysiagrasses is available, but these grasses, along with hybrid bermudagrasses, are usually propagated with vegetative planting materials.

Timing of Seeding

In theory turfgrasses can be seeded at any time of the year, even during winter when the soil is frozen. In actual practice, the potential for failure is often high when seeding is performed under conditions that do not favor rapid germination and vigorous seedling growth. Generally, the optimum time for seeding cool-season turfgrasses is late summer, while warm-season turfgrasses should be seeded in late spring to early summer for best results. This recommendation is based on two primary considerations: prevailing temperatures at the time of seeding and likely temperatures during the two-to three-month period following seeding.

In late summer soil temperatures are warm and highly favorable for seed germination. Cool-season turfgrasses germinate quickly at this time, and subsequent seedling

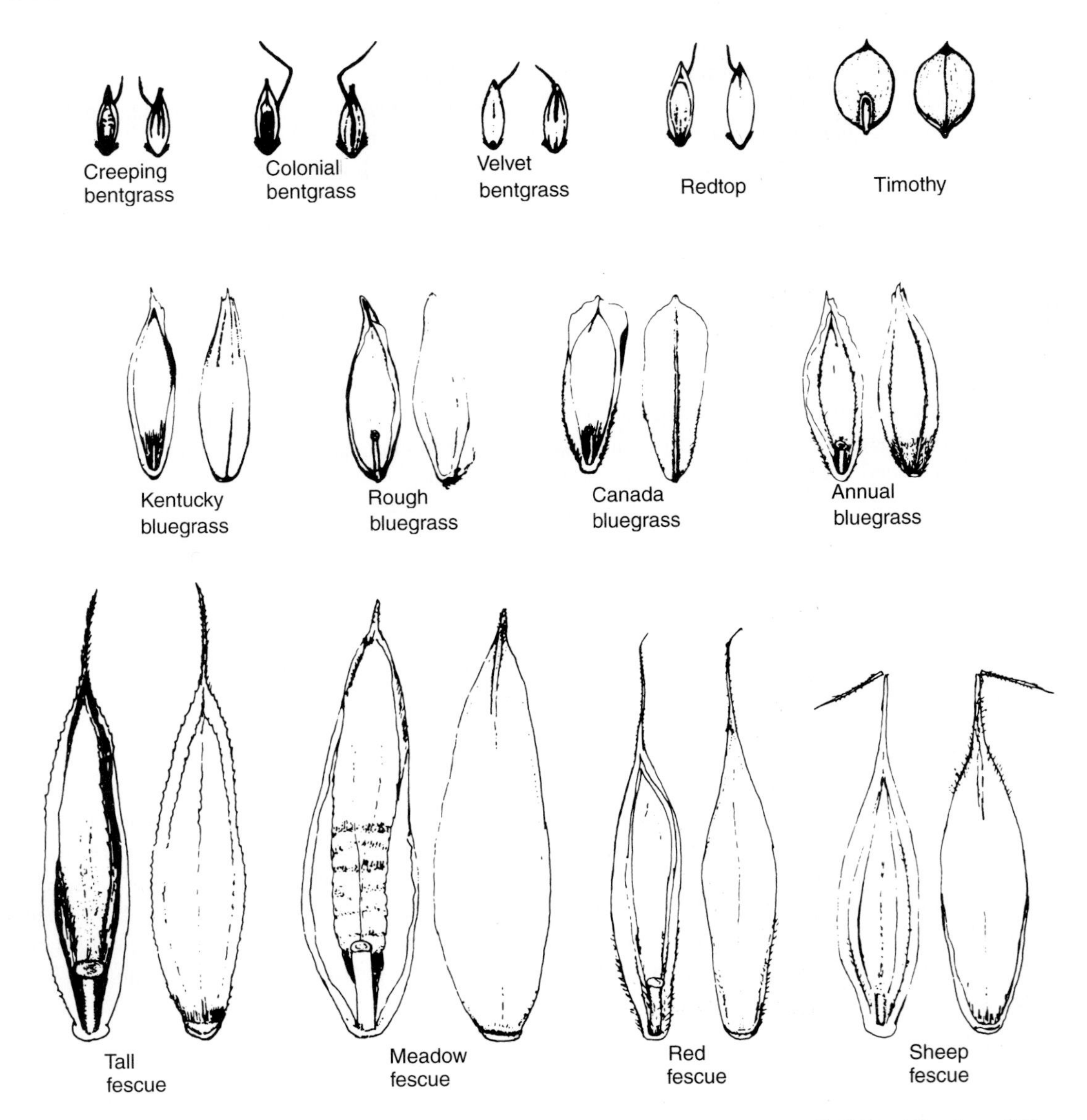

Figure 8.4. Turfgrass seeds illustrated by relative size. (Drawn from A.F. Musil, "Identification of Crop and Weed Seeds," *Agricultural Handbook No. 219,* Washington, DC: United States Department of Agriculture, 1963, p. 217.)

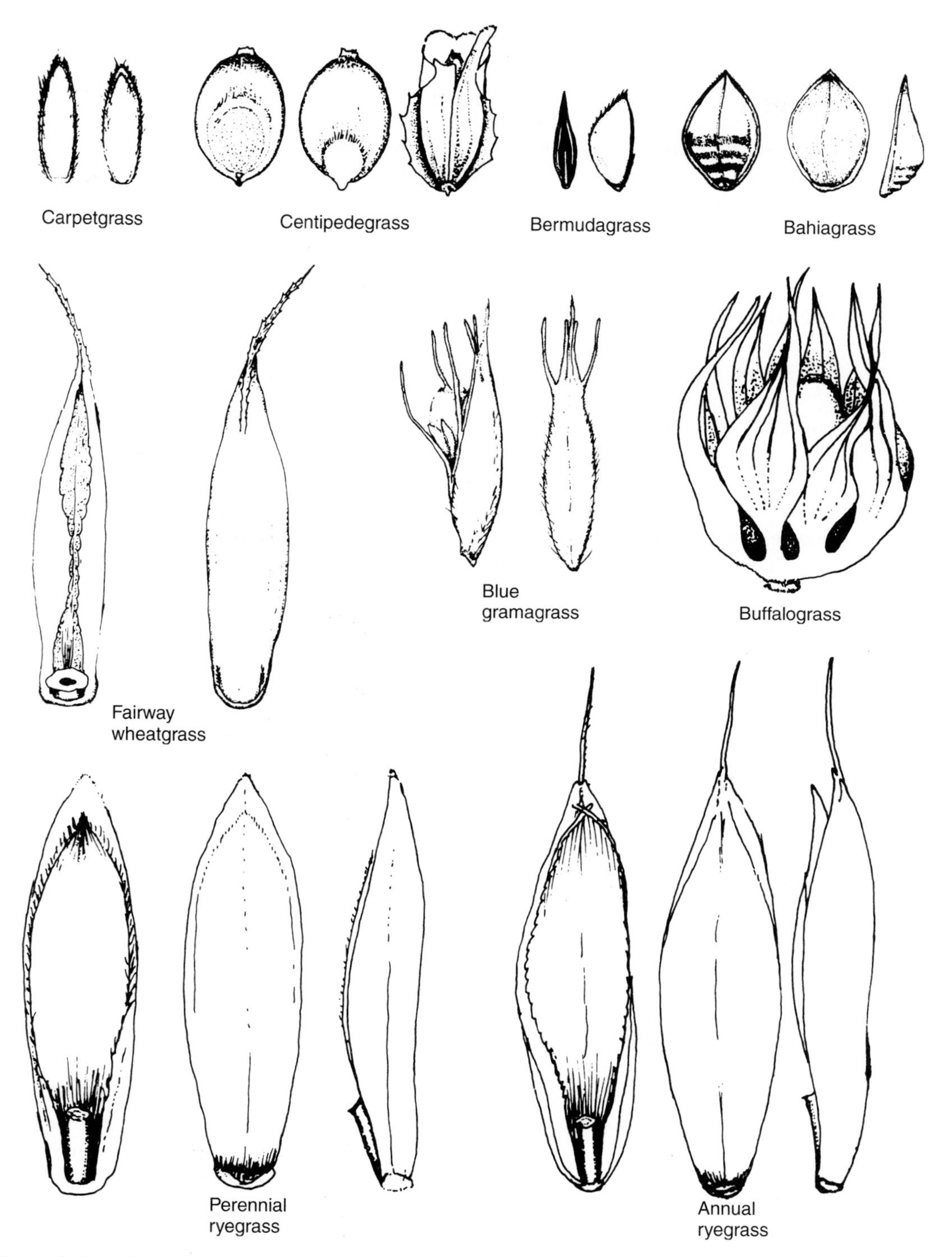

Figure 8.4. *(Continued)*

growth is vigorous as long as moisture, soil fertility, and light are not limiting. Weed-seed germination may also be favored, but cooler fall temperatures and frost conditions limit the growth and survival of many weed species (summer annuals) that are likely to be a problem. Seeding cool-season turfgrasses earlier in the summer increases the likelihood that the seedling plants will not survive heat and drought stresses and that summer-annual weeds will severely compete with the turfgrasses. If the seeding operation is delayed until well into the fall, temperatures may not be as favorable for germination and growth, and the seedling turf will overwinter as poorly developed, isolated plants. Frost heaving and subsequent desiccation can severely reduce the turfgrass stand. Ideally, the new turf should be sufficiently developed prior to winter so that the roots and creeping stems of individual plants are intermingled to provide some stability against frost heaving and soil erosion. Also, inadequately developed seedling turfgrasses often do not possess sufficient traffic tolerance to survive even moderate traffic from people, pets, and equipment.

Early spring to midspring seedings of cool-season turfgrasses may yield an acceptable turfgrass cover prior to midsummer stress, but because of cool soil temperatures, early development of the new turf is usually slower than that observed with late-summer seedings. Also, weed problems can be especially severe unless adequate measures are undertaken to control annual grasses and broadleaf weeds. There are some conditions, however, when spring seedings may be preferred. Attempts to propagate turfgrasses on tree-shaded sites may result in poor stands or even complete failure because of inadequate sunlight. Seeding prior to leaf development of deciduous trees is advantageous because of the higher light intensities available at this time. Of course, the turfgrasses selected for planting on tree-shaded sites must be adapted to low-light conditions or they will not persist.

Optimum growing temperatures for warm-season turfgrasses are substantially higher than those considered most favorable for cool-season turfgrasses. Therefore, while spring and fall weather may be highly favorable for the growth of cool-season turfgrasses, warm-season turfgrasses may grow very slowly until warmer weather develops in late spring or summer—the preferred time for planting. A sufficient period of warm weather should follow seeding in order to promote rapid turf establishment. Therefore, late-summer seedings, while favorable for germination, do not usually allow sufficient time for complete establishment. Summer annual weeds may emerge in new plantings, but warm-season turfgrasses are highly competitive during the summer months and, when necessary, selective herbicides can be used for control. A more important concern in propagating warm-season turfgrasses is the emergence of winter annual weeds, especially annual bluegrass, during cool fall weather. The turf should be well established by the end of summer to resist invasion by these weeds.

Pregermination of seed is sometimes practiced where it is important to achieve rapid cover to stabilize the soil or promote quick turf recovery from damage. This involves incubation of seed in a moist warm environment to promote coleoptile and root emergence, then carefully planting the pregerminated seed to minimize injury to the delicate seedling structures. Seed priming is a newer type of seed treatment that has shown promise for promoting rapid germination. This is a commercial process that involves the stimulation of biochemical activity within the seed, but not to the point at which seedling structures begin to emerge. Research at Penn State has shown that priming is most advantageous with slow-germinating species, such as Kentucky bluegrass, especially under cool suboptimal temperatures in spring and fall.

Dormant (winter) seeding is sometimes practiced with cool-season turfgrasses in temperate climates where construction and seedbed preparation have not been completed until late fall. Success depends on winter temperatures remaining low enough to inhibit germination, as germinating seed is highly susceptible to winter injury. A common difficulty experienced with dormant seeding is the washing of seed from planted sites by surface water movement. Since the seeds must remain on the surface for a prolonged period, it is usually necessary to stabilize them with suitable mulching materials.

Seeding Rate

A general guide for seeding is to apply a sufficient number of pure live seed to develop from 1000 to 2000 seedlings per square foot. Based on this guide, a Kentucky bluegrass cultivar with 2,000,000 seeds per pound and 72% pure live seed (90% purity, 80% germination) should be seeded at the rate of 0.7 to 1.4 lb/1000 ft^2. This calculation assumes that all pure live seed will produce a seedling; however, seedling mortality can amount to 50% or more depending on seed quality and environmental conditions following planting. Thus, 2 or more pounds of seed per 1000 ft^2 might be required for an acceptable stand of seedlings.

Other factors influencing seeding rate include seedling vigor and growth habit of the turfgrass species planted, desired rate of turf establishment, seed costs, anticipated weed competition, disease potential, and cultural intensity of the established turf.

Each turfgrass species differs in its growth characteristics. Rhizomatous and stoloniferous turfgrasses, once sufficiently developed, are capable of spreading well beyond the parent plant; therefore, relatively low seeding rates can result in acceptable turfs much faster than where bunch-type (noncreeping) turfgrasses are planted. Kentucky bluegrass sod growers often seed at rates well below those normally recommended because of the strong rhizomatous growth of this species. Since their objective is to produce harvestable sod as efficiently as possible rather than achieve a dense stand of seedlings, low seeding rates (30 to 40 pounds per acre) result in stronger rhizome and root growth due to less competition among seedlings.

Madison studied the influence of seeding rate on the population density of Kentucky bluegrass seedlings during an eight-month period following planting. Shortly after planting, seedling density closely paralleled seeding rate; plots seeded at 1 lb/1000 ft^2 had approximately 1500 seedlings/ft^2, while those seeded at 8 lb/1000 ft^2 had eight times as many seedlings (12,000/ft^2). During succeeding months, however, seedling populations converged so that at the end of eight months all plots had similar shoot densities, averaging 3500/ft^2. These results show that, at least with Kentucky bluegrass, the shoot density of a mature turfgrass community is ultimately determined by the carrying capacity of the environment and not by seeding rate. Sparse seedling stands eventually increase, and dense stands eventually reduce to a common level that can be sustained under prevailing environmental conditions. The concept of an "adequate" seedling stand, then, is based largely on two short-term objectives: optimum number of shoots for establishment and competition against weeds. Since many turfs are established for athletic or recreational purposes, it is implicit that the establishment period be as short as possible. Thus, a seedling density roughly equivalent to the shoot density of the turf at maturity is often desired even though much lower seedling densities may eventually provide the same result. Where viable weed seeds are present weed emergence tends to decrease as turfgrass seedling density increases. An adequate seedling stand, therefore, is also one that is reasonably competitive with germinating weeds.

Where extremely high seeding rates are used, the resulting seedling stand may be so dense that individual plants fail to develop properly. Also, the incidence of disease is often higher in very dense seedling stands. Presumably, this is due to increased disease susceptibility of seedling plants under intense competitive stress and to the existence of a favorable microenvironment for disease development in dense stands.

Larger seed size is frequently correlated with greater seedling vigor. Large-seeded cultivars of Kentucky bluegrass tend to be more vigorous as seedlings than smaller-seeded cultivars. Even though large-seeded cultivars have fewer seeds per pound than others, there is less likelihood of seedling mortality in field plantings, and comparable seeding rates may generate more surviving seedlings than are generated from smaller-seeded cultivars.

Cost is an important factor influencing seeding rate. Centipedegrass is commonly seeded at 0.25 to 0.5 lb/1000 ft^2, which is much lower than the rate required to yield 1000 to 2000 seedling plants/ft^2. The high cost of seed of this species often precludes seeding at higher rates.

The seeding rates provided in Table 8.1 are employed where only one species is planted. In seeding mixtures of two or more species the rates for each turfgrass should be reduced to favor development of the primary species in the mixture. For example, a 1:1 mixture of Kentucky bluegrass and fine-leaf fescues might be selected for sites where sunlight intensity varies due to differential shading from trees, buildings, and other structures. Since the primary species in this mixture is usually Kentucky bluegrass, seeding rates should be based on the recommended range for this turfgrass (1 to 2 lb/1000 ft^2). An alternative approach to ensuring that an adapted turfgrass is planted within different microenvironments is to plant seeds of different species in separate operations; Kentucky bluegrass would be seeded over the general area, while the fine-leaf fescues would be seeded only in shaded areas.

If a nursegrass such as perennial ryegrass is added to a seed mixture to promote rapid cover, its percentage should not be so high that it crowds out other slower-developing plants and thus becomes the dominant turfgrass. An old rule of thumb was to use no more than 15 to 20% perennial ryegrass in a seed mixture. However, with the introduction of many new improved cultivars of turf-type perennial ryegrasses, this rule may no longer be valid. For example, an aggressive cultivar of perennial ryegrass may become the dominant component of a turfgrass community where it comprises as little as 5 or 10% of a seed mixture with a nonaggressive cultivar of Kentucky bluegrass. Conversely, a 1:1 mixture of a nonaggressive perennial ryegrass and an aggressive Kentucky bluegrass may result in a predominantly Kentucky bluegrass turf. More information is needed on the competitive relationships within mixed turfgrass communities as influenced by new or different cultivars.

Finally, the cultural intensity planned for sustaining a new turf can dramatically influence seeding rate. A lawn turf of tall fescue should be planted with at least 6 to 8 pounds of seed per 1000 ft^2; however, a utility turf of this species along roadsides requires only 2 pounds or less per 1000 ft^2 because of the higher mowing heights employed. On golf courses where greens and fairways are seeded with the same turfgrass, the seeding rate is often higher on the greens because of closer mowing and higher shoot densities at desired maturity.

Seeding Practices

Turfgrass seeding practices are designed to uniformly distribute relatively large numbers of seed over a planting site and to incorporate them into the top 1/4 inch of soil. Deeper seeding or failure to incorporate the seed into the soil often results in reduced seedling stands. At excessive soil depths food reserves within the endosperm may be inadequate to supply the seedlings' nutritional requirements prior to the initiation of photosynthesis and nutrient absorption from the soil. At the soil surface seeds may wash away in surface runoff water if not sufficiently incorporated.

Proper placement of seeds requires a loosened surface soil for incorporation. Following seeding, the seedbed should be rolled to ensure firm contact between seeds and soil particles. Where rolling is not practiced, mulching materials should be applied to reduce moisture loss and to provide some stability against erosion.

Seeding can be performed by any method that results in uniform distribution of the seed over the planting site. Many turfs have been successfully established by hand seeding; however, this requires considerable skill and is usually not practical on large sites.

Fertilizer spreaders calibrated to deliver seeds at a desired rate can be used for small sites. Drop-type spreaders are preferred over rotary spreaders because distribution is less influenced by wind and differential seed size.

Cultipacker seeders (Figure 8.5) are widely used for seeding large sites; they are tractor-mounted units that not only distribute seed uniformly but also firm the seedbed after planting. The roller component of a cultipacker seeder is ridged to ensure proper seed placement. Although seed germination may occur uniformly across the soil surface, it is often largely confined to the parallel valleys created by the roller immediately after seeding.

Hydroseeding is a method by which seed is applied in a stream of water. A hydroseeder is essentially a large-capacity sprayer with a large single-nozzle delivery system. The advantages of hydroseeding are that fertilizers, mulches, and other materials can be applied with the seed in a single operation and the unit does not travel over the planting site during seeding. Hydroseeding is a popular method for turfgrass propagation on steeply sloped sites where other methods would be impractical. Since seed is deposited on the surface without any incorporation or rolling, mulching materials are usually required for satisfactory results.

Figure 8.5. A cultipacker seeder, showing the ridged roller that places seed at the proper planting depth.

Vegetative Planting Methods

Vegetative planting methods for propagating turfgrasses are sodding, plugging, sprigging, and stolonizing. Except for sodding, these methods are only practical for turfgrasses that have strong stoloniferous or rhizomatous growth habits. The principal advantage of vegetative propagation over seeding may simply be more rapid establishment. With some turfgrasses, however, seeding is not a practical alternative because viable seed cannot be produced or the genotype cannot be propagated true-to-type from seed. Regardless of the propagation method selected, the planting bed should be carefully prepared; vigorous turfgrass growth from both seed and vegetative propagules requires a well-aerated medium with abundant moisture and mineral nutrients.

Sodding

While sodding is the most expensive method of turfgrass propagation, it can result in an "instant turf" at virtually any time of the year. Although several weeks or months may be required before a newly sodded turf can withstand traffic or play, sodding does provide

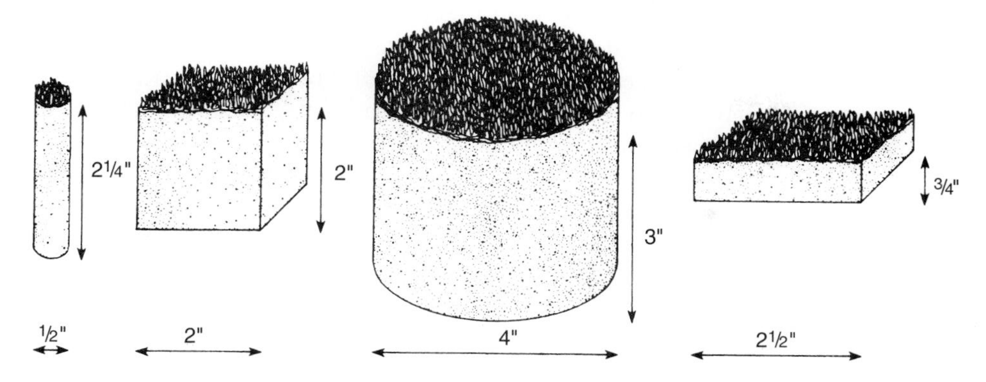

Figure 8.6. Various types of turf plugs used for vegetative propagation.

the appearance of an established turf. For this reason it is often used on poorly prepared sites where other propagation methods might not yield satisfactory results. However, unless the planting bed is properly prepared for sodding the resulting turf may be plagued by numerous problems during succeeding growing seasons.

Ideally, a sodbed should be moist, but not wet, at the time of sodding. If it is excessively dry, especially at high temperatures, sod rooting may be impaired regardless of subsequent irrigation practices. Sod strips should be as thin as possible to favor rapid rooting. Care should be exercised in handling the sod to avoid tearing or excessive stretching. Furthermore, the sod should be laid as soon as possible after being delivered to the planting site to avoid deterioration from desiccation and excessive sod heating. The likelihood of these problems occurring increases with increasing atmospheric temperatures.

Sod can be laid manually or by automatic unrolling from a horizontal bar at the rear of a tractor. Sod strips should be adjusted so that the ends of adjacent strips are staggered to minimize cracks from shrinkage. Individual strips should be firmly fitted against adjacent strips and lightly tamped to provide uniform contact with the soil. When laid on sloping terrain, each strip should be secured in place with a stake until sufficient rooting has occurred to stabilize the sod.

Desiccation of newly laid sod can be reduced by working screened soil into the seams between sod strips and along exposed edges. If at all possible, weed-free soil should be obtained for this purpose to minimize weed problems.

Plugging

Plugging is the planting of turf sections varying in size from small cores extracted during core cultivation to large plugs extracted with a cup cutter or similar device (Figure 8.6). The most common plugging method involves the use of square or circular plugs measuring approximately 2 inches across and 2 inches deep. These are inserted into a planting bed, usually at a 12- to 16-inch spacing, so that the tops of the plugs are flush with the soil surface. This plugging method is most commonly used for propagating zoysiagrass, but it can be used for other stoloniferous or strongly rhizomatous turfgrasses as well. In addition to propagating turfgrasses in bare soil, plugging is used to introduce a new species into an existing turf. For example, a Kentucky bluegrass lawn can be converted to bermudagrass or zoysiagrass by planting plugs; the process of conversion is usually slow, but it can be accelerated by adjusting cultural practices to favor the introduced turfgrass.

A second plugging method employs sections cut from sod strips. Cutting can be done by hand or mechanically. Mechanical pluggers employ a rotating drum with square knives for cutting plugs from sod. Sod strips are fed into a chute enclosing the drum, and the cut plugs are then deposited into furrows in the soil surface dug with one or more triangular knives. The furrows are filled in through the action of a V-shaped section of steel positioned between adjacent furrows. Finally, the planting bed is leveled and firmed by a roller located at the rear of the unit. Mechanical plugging by this and other equipment is an efficient planting method, especially for those turfgrasses that cannot be propagated by seed.

Large plugs measuring from 4 to 8 inches in diameter are used primarily for repairing damaged turfs such as greens, tees, and athletic fields. Since they must be extracted by hand using a cup cutter, this is a very laborious propagation method. These plugs are typically several inches deep and, therefore, can be inserted into recreational turfs during periods of heavy use.

A final plugging method employs small turf cores extracted during core cultivation of greens composed of stoloniferous turfgrasses (bermudagrass and creeping bentgrass). The cores are subsequently broadcast onto a planting bed and the area is rolled to provide a smooth surface. As the small turfgrass plants in the cores are highly susceptible to desiccation, the planting bed must be kept continually moist until sufficient rooting has taken place. This procedure is occasionally used by golf course superintendents to establish a nursery green of the same turfgrass found on the playable greens.

Sprigging

Unlike plugs and most types of sod, sprigs are essentially free of adhering soil. Therefore, they are more prone to desiccation under hot dry conditions. Sprigging is primarily used for propagating stoloniferous warm-season turfgrasses, but can also be used for creeping bentgrass. Sprigs are usually planted 2 to 3 inches deep in furrows spaced 6 to 12 inches apart. Depending on the spacing between individual sprigs in rows and between rows, 1 to 4 bushels of sprigs are required per 1000 ft^2. Each sprig should have from two to four nodes and should be planted so that a portion of the sprig emerges above the soil surface after the furrows have been filled in. The planting bed should be rolled and irrigated as soon as possible following sprigging.

Mechanical planting of sprigs can be accomplished with the same equipment described previously under plugging. Handfuls of sprigs, rather than sod, are fed into the chute of the machine and automatically planted in furrows.

Instead of being planted in furrows, sprigs may be placed on the soil surface and plunged into the soil using a thin notched stick.

Stolonizing

Stolonizing is, essentially, broadcast sprigging. The vegetative planting material is uniformly deposited on a moist, but not wet, soil surface. This usually requires from 5 to 10 bushels per 1000 ft^2. The planting bed is then topdressed to partially cover the stolons or the area is lightly disked to partially insert the stolons into the soil. Topdressing is usually the preferred method. The area should be rolled and irrigated as soon as possible after stolonizing. To reduce desiccation of the planting material, stolonizing should be done in 3- to 4-foot strips, followed immediately by topdressing and light irrigation.

Mulching

Mulching is the practice of applying foreign materials to a seedbed to reduce erosion and provide a more favorable microenvironment for germination and seedling development. On level planting beds where the capability for frequent irrigation exists, mulching may not be necessary during propagation. However, it is almost essential on sloping sites or where moisture availability largely depends on natural rainfall.

An effective mulch serves several functions. It stabilizes the soil and seed against erosion from wind and surface runoff water; it moderates temperature fluctuations at the soil surface and thus protects germinating seeds and seedlings from temperature-induced injury; it reduces evaporation of moisture from the soil surface and provides a more humid microenvironment in and above the soil; and it dissipates the energy of falling water droplets from rainfall and irrigation to reduce crust formation at the soil surface and thus favor higher infiltration rates. Not all mulches serve all these functions; some are more effective than others, and the selection of a particular material depends on the specific requirements of each site and the cost and local availability of materials.

Mulching Materials

Some mulches are produced specifically for use on seedbeds, while others are by-products of commercial manufacturing processes. For decades one of the most widely used mulches has been straw, primarily from wheat plantings. Applied at the rate of 1.5 to 2 tons per acre, it provides excellent results. If possible, weed-free straw should be used to reduce weed competition in the seedbed. As long as coverage is not too heavy (not more than 50% soil coverage), the straw need not be removed after seedling emergence.

Grassy hay is similar to straw in its use and effectiveness as a mulch. Early-season cuttings of hay are preferred, as they are less likely to contain substantial populations of weed seeds.

Loose wood mulches include wood cellulose fiber, wood chips, wood shavings (excelsior), sawdust, and shredded bark. Sawdust is the least desirable because of its capacity for drawing nitrogen from the soil. If sufficiently small in size, wood chips and shredded bark are helpful in reducing erosion, but they are not as effective as straw or hay in providing a favorable microenvironment for germination and seedling growth. Comparable results are usually obtained from slurries of wood cellulose fiber; the specific advantage of this material is that it can be applied in a stream of water with or without seed. Excelsior is available in loose form or in mats. When placed on the soil surface and wetted, the wood fibers expand to form a cover similar to straw. While comparable to straw in effectiveness, excelsior has the added advantage of being free of undesirable weed seeds.

Organic residues from some field crops have been used as mulches with varying degrees of success. These include bean, crushed corncobs, bagasse, sugar beet pulp, cocoa, peanut hulls, and tobacco stems. They are useful in reducing erosion potential, but rank poor to fair for enhancing germination and seedling development. The importance of these by-products is limited by local availability.

Synthetic mulches include fiberglass filaments, clear polyethylene covers, and elastrometric polymer emulsions. Fiberglass filaments applied with a compressed air gun, form a very persistent mulch which can interfere later with mowing operations. The use of fiberglass mulch has often been unsatisfactory. Clear polyethylene covers are sometimes used to accelerate germination during cool weather. When placed over a seedbed they cause a greenhouse effect; temperatures at the soil surface are increased, and relative humidity can increase to saturation. If the covers are allowed to remain in place too long, seedling diseases can quickly develop and destroy the stand. Elastrometric polymer emulsions are sprayable materials that can stabilize the seedbed against erosion; they are ineffective for enhancing germination or seedling development.

Jute netting can be placed over critical sites such as steep slopes or drainage ditches to stabilize a seedbed. Burlap strips are even more effective, but should be removed after seed germination to avoid excessive shading of the seedlings.

Application Methods

Straw and hay can be applied by hand to small sites. In windy areas they should be stabilized by crisscrossing the area with binder twine anchored down with stakes. Application to large sites is usually accomplished by using a mechanical mulch blower. This machine chops the material and blows it over the seedbed. An asphalt binder is usually sprayed on the mulch as it exits the blower, or after application, to stabilize the mulch on the seedbed. This same technique can be used for loose wood mulches and organic residues.

Wood cellulose fiber and elastrometric polymer emulsions are applied in water to the seedbed. The wood cellulose fibers form a slurry with water in the spray tank and are often mixed with seed and fertilizers for simultaneous application. Normal application rates of the mulch are from 1 to 2 tons per acre. Elastrometric polymer emulsions are diluted 9:1 with water and sprayed through a handgun for uniform application to the seedbed. The proper rate of application can be determined by the color of the treated site. Overapplication should be avoided, as this could seal the soil surface and impede seedling emergence.

Polyethylene covers, jute netting, excelsior mats, and burlap are simply laid on the seedbed and anchored with stakes or staples.

Only those mulching materials that interfere with seedling development need be removed following germination. These obviously include polyethylene covers, which, if left too long, can result in heat damage or disease. Excessive amounts of straw and hay should be removed by light raking when seedlings have grown to a height of approximately 1 inch. Where at least 50% of the soil is visible these mulches can be left to decompose naturally.

POSTPLANTING CULTURE

Following planting and, if practical, mulching, the planting bed should be irrigated to thoroughly moisten the soil. If vegetative propagation methods other than sodding have been employed, it may be necessary to topdress prior to irrigation to protect the plant material from desiccation. Frequent light irrigations may be necessary to prevent desiccation and to promote normal development of seedlings or vegetative sprouts. Traffic should be withheld from the planting bed until a solid stand has developed; even then, traffic should be kept at a minimum until the new turf has become fully established.

As the new plants show signs of development, a modified cultural program should be initiated to ensure proper development of the turf. This may include such practices as mowing, fertilization, irrigation, topdressing, and pest management.

Mowing

Mowing should be initiated when the new shoots reach a height of as low as 3/4 inch, or much higher, depending on turfgrass species and the intensity of culture planned for the turf. The one-third rule discussed in Chapter 5 is just as applicable to immature stands as it is to established turf. New greens are usually mowed at 1/2 inch until complete coverage has been reached; then the mowing height is gradually lowered as the turfgrass community develops toward maturity. This applies to both seeded and vegetatively propagated greens. On higher-cut turfs initial mowing is usually at the same height as that practiced at maturity.

Turfgrass seedlings are easily uprooted by mowing with a dull or improperly adjusted mower, especially when the soil is wet. The soil should be allowed to dry and become firm prior to mowing. All cutting edges should be sharp and mowers should be precisely adjusted to avoid tearing and bruising of delicate plant tissues. If possible, heavy mowing equipment should not be used on seedling turf for the first several mowings to avoid excessive turfgrass wear and soil compaction. If it is necessary to use heavy mowing equipment, the area should be sufficiently dry and firm to support the weight without damage.

Fertilization

Frequent light applications of fertilizer are helpful, and in some cases essential, for proper development of a new turf from seedlings, plugs, sprigs, and stolons. Light rates are used to ensure that nitrogen and other nutrients are in sufficient supply, but not so abundant that direct injury to the plants or restricted growth of roots and lateral shoots results. An appropriate rate for a seedling stand is about 0.5 lb of soluble nitrogen, or 1 lb of slowly available nitrogen, per 1000 ft^2. Higher rates may be used for vegetatively planted turfgrasses. In a seedling stand the first postplanting fertilizer application should be made, if needed, just prior to the first mowing. This may be important for replacing nutrients that may have been leached out of the shallow soil layer containing the young roots.

The frequency of subsequent fertilizer applications depends on soil texture and turfgrass growth. Soluble nutrients are more likely to be leached in coarse-textured soils than in fine-textured soils; therefore, new plantings in coarse-textured soils require more frequent applications of fertilizer. Leaching losses can be partially offset by using slowly available nitrogen carriers. The need for additional fertilizer is usually indicated by the color and growth rate of the plants. A light green or chlorotic appearance coupled with slow growth signals that inadequate nitrogen and other nutrients are limiting growth. As it is difficult to determine from a single observation whether nutrient deficiencies exist, the condition of the new plants should be monitored through frequent inspections. In this way definite trends in coloration and growth can be identified.

Although the nutrient most likely to be deficient is nitrogen, deficiencies of any of the essential nutrients can limit the growth of newly planted turfgrasses. Proper application of basic fertilizer materials during site preparation should ensure that most nutrients are in sufficient supply, except perhaps in very sandy soils. Where necessary, the occasional use of a complete fertilizer containing minor nutrients should be adequate to prevent or offset serious nutrient deficiencies.

Irrigation

Just after planting, irrigation is the most important cultural practice favoring seed germination or sprouting of vegetative propagules whenever rainfall is inadequate. Failure to provide sufficient moisture is a major cause of unsuccessful turf establishment. Later, as the new turf becomes partially developed, irrigation should be less frequent but more intensive (Figure 8.7). As individual plants develop, their roots occupy a greater volume of soil and their shoots become more hardy. Therefore, the soil surface need not be kept continually moist as long as the underlying soil within the root zone retains sufficient quantities of plant-available water. Also, as the new turf further develops, other cultural practices besides irrigation (mowing, fertilization, pesticide application) assume greater importance. The soil surface, therefore, must be sufficiently dry at times to support the weight of mowers, fertilizer spreaders, and other equipment.

Accompanying the reduction in irrigation frequency is improved soil aeration. As water evaporates and drains, air is pulled into the soil. Developing and mature plants require adequate concentrations of oxygen in the root zone for respiration (see Chapter 2).

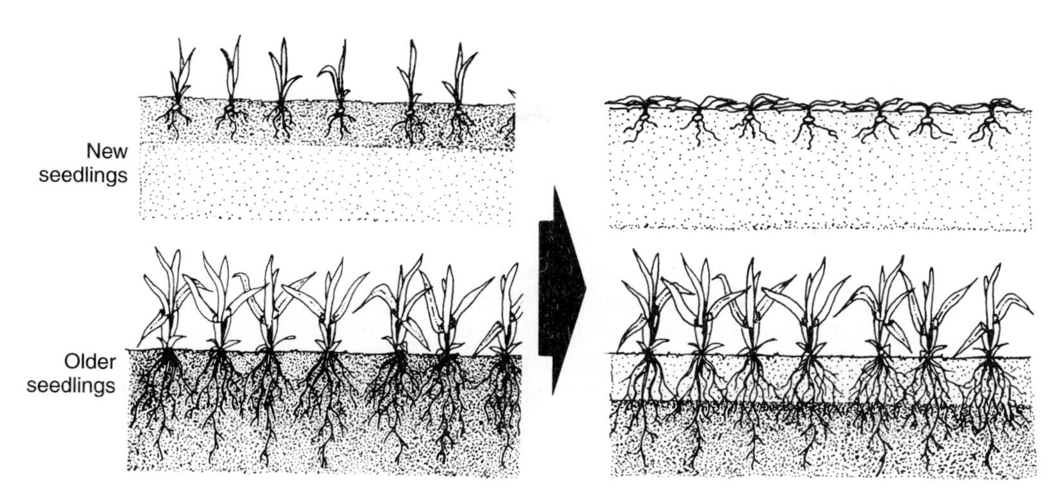

Figure 8.7. Different irrigation requirements of new and older turfgrass seedlings. New seedlings, with their roots in a shallow soil layer, require frequent irrigation for survival. Older seedlings, with their deeper, better-developed roots, require less frequent but more intensive irrigation to sustain growth.

Topdressing

Not all new turfs require topdressing. This practice is primarily designed to favor the development of turfs composed of stoloniferous turfgrasses that will be sustained under close mowing. Topdressing favors the growth of roots and aerial shoots from nodes along the stolons. It is also important for developing a smooth level surface on greens and tees for optimum playability.

The topdressing soil should be the same texture as the soil supporting turfgrass growth. If it is not, a surface layer could develop to impede the movement of air, water, and nutrients in the root zone.

Where surface irregularities exist due to differential settling of the soil, they can adversely affect the playability and mowing quality of a turf. Successive topdressings can be effective in filling in depressions. Care should be exercised to avoid excessive coverage of plant tissues with topdressing soil, as this could exclude enough light to be injurious.

Pest Management

Usually, weeds are the most troublesome pests in newly planted turfs. As discussed earlier under planting procedures, the timing of seeding operations is extremely important in reducing weed problems. Killing frosts in the fall in all but tropical climates effectively remove most annual weeds from the turfgrass community. Purchasing weed-free seed, vegetative planting materials, and mulches is important for preplant weed control. If necessary, fumigation of the planting soil and topdressing soil can eliminate most potential weed invaders. Summer fallowing followed by applications of nonselective systemic herbicides (principally glyphosate) after weed emergence can also reduce competition from weeds. Finally, an application of the preemergence herbicide siduron immediately after seeding cool-season turfgrasses can effectively control most summer annual grasses and some broadleaf weed species. However, even where these measures have been employed, some weeds may emerge and compete with developing turfgrass plants. Thus, postemergence applications of herbicides to the young plant community may be necessary during turf establishment.

Most herbicides are more toxic to young seedlings than to mature turfgrasses. Growth from vegetative planting materials, including sod, can be inhibited or reduced by some herbicides. Therefore, applications of most herbicides should be delayed until absolutely necessary to allow sufficient time for the new turfgrass community to develop. Postemergence application of 2,4-D, mecoprop, and dicamba should not be made on normally tolerant turfgrasses before the first mowing. Because seedling broadleaf weeds are more sensitive than mature weeds to these herbicides, half-normal rates (0.5 lb 2,4-D + 0.13 lb dicamba per acre for Kentucky bluegrass) may be quite effective, and pose less of a hazard to the turfgrass. Applications of organic arsenicals for controlling crabgrass and other summer annuals should be delayed even longer (after the second mowing) and, again, half-normal rates should be used. In newly sodded turfs preemergence herbicides may be necessary to prevent crabgrass emergence from cracks between adjacent sod strips during spring and summer. However, the application should be delayed for three to four weeks following planting to avoid inhibition of rooting. Where isolated plants of undesirable perennial grasses have emerged, they should be spot treated with glyphosate as soon as possible. Replanting of dead areas may be necessary if they measure more than 4 to 6 inches in diameter.

Most seedling diseases can be controlled by avoiding excessive irrigation and high seedling densities from planting at excessive seeding rates. Under some conditions it may be advisable to use fungicide-treated seed. For overseeding warm-season turfgrasses with cool-season species in subtropical climates, the seed is commonly treated with metalaxyl (Apron) for controlling *Pythium* blight. Postemergence applications are also made on a preventive or curative basis when conditions conducive to disease development exist.

Most insects usually do not cause problems in newly planted turf. One exception, however, is the mole cricket. Where this insect is active it can cause severe damage by uprooting seedling plants and drying out the soil through its burrowing activity. Control can be achieved with chlorpyrifos.

RENOVATION

A severely deteriorated turf may not require the kind of intensive reconstruction discussed previously in this chapter if the cause of deterioration is associated with undesirable shifts in the composition of the turfgrass community or unfavorable conditions in the surface medium supporting turfgrass growth. As long as the grade is satisfactory and the soil beneath the surface 2 inches is favorably structured, a less intensive form of turf establishment, called renovation, may provide satisfactory results. *Renovation* may be defined as turf improvement that involves partial or total replanting without complete tillage of the soil.

Conditions under which renovation might be considered include the following:

1. Turfgrass community composed entirely or predominantly of weeds that can be controlled with selective herbicides (summer annual grasses and broadleaf species)
2. Turfgrass community composed largely of perennial weedy grasses
3. Turf that has been extensively damaged by insects, disease-causing agents, or other pests
4. Poor-quality turf with excessive thatch, surface layers of different-textured soils, or severe compaction in the surface 1 to 2 inches of soil
5. Combinations of the above

Prior to initiating a renovation program some consideration should be given to the identifiable causes of turf deterioration as well as to appropriate corrective actions. Perhaps a weed-infested seed mixture was used initially to establish the turf, or proper weed- and other pest-control measures were not employed, or pesticides were misused. Poorly adapted turfgrasses which subsequently failed to survive may have been planted. Some primary or supplementary cultural operations may have been performed improperly or may not have been performed when needed. In such instances corrective action to prevent reoccurrence following renovation simply involves pursuing a turfgrass cultural program based on well-established principles and practices. Other causes of deterioration may require some major corrective action prior to renovation. Poor surface and subsurface drainage are common examples. Extensive tree-root development is another. If these problems are not satisfactorily resolved, attempts to renovate the turf may fail, or the renovated turf may simply deteriorate once unfavorable conditions recur.

The sequence of operations employed in renovating a deteriorated turf is a modification of the sequence discussed earlier in this chapter; it includes site preparation, turfgrass selection, planting procedures, and postplanting culture.

Site Preparation

Site preparation includes any practice designed to prepare the deteriorated turf for planting. An evaluation of the site should be made to determine the type and magnitude of weed populations, the amount of thatch present, and the physical condition of the underlying soil.

If substantial weed populations are present, a decision must be made to use either selective or nonselective herbicides. Usually, the determining factor in this decision is the distribution of perennial weedy grasses in the turf. Since most weed populations containing only annual grasses and broadleaf weeds can usually be selectively controlled with herbicides, desirable turfgrass populations may be worth salvaging if they are sufficiently distributed in the turf. The presence of large populations of weedy perennial grasses, however, usually dictates that the entire area be treated with a nonselective herbicide. This may also be advisable where the desirable turfgrass populations are insufficient to warrant any attempts to salvage them.

Once the turf has been treated with herbicides, a waiting period of several days or weeks is necessary to allow for sufficient absorption and translocation of the herbicide by target plants and for the dissipation of toxic herbicide residues in the soil. Glyphosate and the organic arsenicals leave essentially no toxic residues in most soils, but they may require up to two or three weeks to control target weeds. Paraquat, a nonselective contact herbicide, is the fastest-acting herbicide for general vegetation control, but it is not as effective against perennial weedy grasses as glyphosate. Although paraquat is inactivated once in contact with clay-containing soils, its toxic residues can persist in thatch and prevent seedling emergence unless the thatch is removed or intermixed with soil. The principal herbicides for broadleaf weed-control, 2,4-D, mecoprop, and dicamba, may persist in the surface soil for several weeks, and they usually require two or three weeks for control.

After all or a portion of the plant community has been killed and toxic herbicide residues have dissipated, the site is ready for cultivation. Where a substantial thatch layer exists it should be reduced by deep vertical mowing or, in extreme cases, with a sod cutter, and the debris should be hauled away. Where little or no thatch exists several vertical mowings or core cultivations are usually required to prepare a planting bed. Severe compaction of the surface soil or the presence of soil layers differing in texture in the surface 2 inches are conditions that require intensive core cultivation. After the cores have partially dried they should be broken up and matted in.

A complete fertilizer and, if necessary, lime, should be applied prior to working the soil. If possible, application rates should be based on soil test results. One pound of soluble nitrogen per 1000 ft^2 or higher rates of slowly available nitrogen should be adequate to promote favorable growth of the seedlings.

Turfgrass Selection

While it is possible to develop a new turfgrass community by vegetative propagation methods, most renovated sites are seeded once a suitable planting bed has been prepared. Turfgrass cultivars that are well adapted to natural environmental and cultural conditions should be selected for planting. Local authorities should be consulted for specific recommendations.

Planting Procedures

There are two basic methods for seeding a renovation site: broadcast and disk seeding. The broadcast method employs standard seeding rates. After application, the seed should be lightly raked or matted in and the area rolled. A disk seeder, illustrated in Chapter 6, applies seed directly into grooves created by a vertical mower at the front of the unit. Although lower quantities of seed are used compared with the broadcast method, results are often favorable. Matting and rolling are usually not necessary following disk seeding.

Postplanting Culture

Irrigation and other cultural practices following renovation are essentially the same as for tilled and graded sites. Where a portion of the previous plant community has been salvaged, however, mowing should be initiated as soon as growth from the older turfgrasses reaches appropriate heights.

TEMPORARY WINTER TURFS

In subtropical climates some warm-season turfgrass communities are customarily overseeded during the fall season with cool-season turfgrasses. This results in improved appearance and playability of the turf during the period in which the warm-season turfgrasses are dormant. In addition, the cool-season plant cover reduces injury from traffic. This practice is most common on greens and tees, but may be employed on fairways and lawns. Most overseeding is conducted on bermudagrass turfs, but St. Augustinegrass, bahiagrass, and centipedegrass lawns may be overseeded as well.

Site Preparation

In earlier decades extensive preparations were made just prior to overseeding to ensure successful establishment of the cool-season turfgrass community. These included scalping, vertical mowing, core cultivation, and topdressing for controlling thatch, reducing competition from the warm-season turfgrass, and preparing a seedbed. In recent years, however, procedures for preparing greens for overseeding have been modified in several important ways. Thatch control is practiced throughout the growing season rather than just prior to overseeding; preemergence herbicide applications are carefully timed to control annual bluegrass without impeding the germination of overseeded turfgrasses; and fertilization practices are conducted to minimize the competition from the warm-season turfgrass while enhancing the development of overseeded species.

Thatch Control

A thatch layer adversely affects the growth and survival of overseeded cool-season turfgrasses. Seedlings that develop in thatch are more susceptible to injury from cold temperatures, traffic, and other stresses than are seedlings that develop in soil. Therefore, methods that reduce the thickness of the thatch and that modify the remaining thatch layer so that it is a more favorable growth medium are important for ensuring optimum development of the cool-season turfgrass community.

At the time of overseeding, the warm-season turfgrass community should be uniform, weed free, and slow growing. Poor quality at this time usually results in a poor spring transition when the warm-season turfgrass resumes active growth. Frequent topdressing and light vertical mowings during the warm portion of the growing season are helpful in controlling thatch and sustaining a high level of turf quality. The final core cultivation for the year should be scheduled fifty to sixty days prior to overseeding. Coring just prior to overseeding often results in nonuniform development of the seedling stand, with tufts of growth at each coring hole.

Fertilization

Fertilizer applications should be withheld for two to four weeks prior to overseeding to minimize competitive growth from the warm-season turfgrass. However, the overseeding procedure should include fertilization so that sufficient nitrogen, phosphorus, and potassium are available to favor rapid development of the cool-season turfgrasses.

Weed Control

In closely mowed bermudagrass turf, annual bluegrass may germinate in the fall and severely compete with overseeded turfgrass species. The result is a substantial reduction in visual and functional turf quality. To control this problem, a preemergence herbicide is usually applied fifty to ninety days prior to overseeding. This is often done immediately following core cultivation and topdressing. Bensulide, benefin, and pronamide are the most commonly used herbicides for this purpose. Postemergence treatment of existing annual bluegrass populations with pronamide has resulted in excellent control. Where potentially toxic residues of these herbicides persist in the soil at overseeding time, activated charcoal should be applied to avoid inhibition of germination and seedling development of the overseeded turfgrasses. An increasingly popular alternative to these herbicides is fenarimol, a fungicide with preemergence activity against annual bluegrass. When applied once or twice prior to the emergence of annual bluegrass it has provided good to excellent control where perennial ryegrasses have been used for overseeding.

Turfgrass Selection

Prior to the 1960s most overseeding was done using annual ryegrass. Seeding rates as high as 60 lb/1000 ft^2 were used on greens to develop a dense, fine-textured, temporary turfgrass community suitable for putting. Annual ryegrass provided rapid cover, but its quality was subject to sharp declines in midwinter from frost injury in cooler subtropical climates. Due to its poor heat tolerance under close mowing, spring transition tended to occur abruptly, often before bermudagrass recovered sufficiently from winter dormancy.

During the 1960s mixtures of several cool-season turfgrasses were used for overseeding in place of or in addition to annual ryegrass. These mixtures included fine-leaf fescues, bentgrasses, rough bluegrass, Kentucky bluegrass, and perennial ryegrass. Planted alone, the fine-leaf fescues formed a fine-textured turf of good putting quality. Frost tolerance was superior to that of annual ryegrass, but performance during spring transition was erratic, with diseased areas sometimes appearing in the turf. The bentgrasses germinated rapidly, but turf establishment was slow due to low seedling vigor. The bentgrasses often persisted into early summer and thus retarded the development of bermudagrass. Germination and seedling development of Kentucky bluegrass were very slow; when finally established, a bumpy putting surface often resulted. Rough bluegrass formed an acceptable putting surface and exhibited good frost tolerance, but did not always transition well in the spring due to its sudden deterioration under warm temperatures. However, with the availability of improved cultivars of rough bluegrass, smoother spring transitions and better turf quality during the winter are now possible. Perennial ryegrass established an excellent putting surface fairly quickly but not as quickly as annual ryegrass. With the introduction of improved turf-type perennial ryegrass cultivars for overseeding, several highly desirable features of these grasses were observed, including better resistance to *Pythium* blight, excellent wear resistance, and good frost tolerance. Spring transition from perennial ryegrass to bermudagrass varied from poor to excellent depending on weather conditions; cool springs tended to favor the persistence of perennial ryegrass to the detriment of bermudagrass. Because of their fibrous leaves, some perennial ryegrass cultivars tend to fray at the tips, especially under warmer temperatures in spring, resulting in reduced putting quality.

Today bermudagrass greens are commonly overseeded with creeping bentgrasses, rough bluegrass, or turf-type perennial ryegrass alone or in mixtures with fine-leaf fescues. Seeding rates are much higher than those used for establishing permanent turfs in temperate climates. These average 10 lb/1000 ft^2 for rough bluegrass, 3 lb for creeping bentgrass, 35 lb for perennial ryegrass used alone, and 25 and 15 lb for perennial ryegrass and fine fescues, respectively, when used in combination. Some broader mixtures are still used, but they are not as popular as they were in earlier years. On lawn turfs of bermudagrass or other warm-season turfgrasses annual ryegrass continues as a popular choice for overseeding because of its relatively low cost, wide availability, and compatible texture. Seeding rates are typically between 5 and 10 lb/1000 ft^2.

Planting Procedures

Overseeding should be timed so that competition from the warm-season turfgrass community is minimal while environmental conditions are favorable for germination and seedling development of cool-season species. Delaying too long often results in considerably slower establishment due to lower temperatures. One recommendation has been to overseed twenty to thirty days prior to the average date of the first killing frost; however, the first frost date can vary widely from one year to another, with overseeding results varying accordingly. A more specific recommendation is to overseed when ambient midday temperatures drop to the low to mid-seventies (°F). In the southeastern United States, this usually occurs sometime between late September and late November, depending on location and year.

At the time selected for overseeding, the number of separate operations required for satisfactory results can vary from three to six or more. The simplest operational sequence involves application of the seed followed by matting and irrigation. Heavy steel mats may be ineffective for working the seed into the turf unless they are modified in some way to prevent flipping the seed into the air. Enclosing the mat within a burlap cover for this operation has given satisfactory results. Irrigation water is applied afterward to move the seed farther down into the turf and closer to the soil surface and also to initiate the germination process. Some golf course superintendents prefer to irrigate initially by hand rather than with sprinklers; the water droplets can thus be directed more forcefully to effectively work the seed into the turf.

This sequence of operations is often quite satisfactory for seeding into thatch-free turfs. Where a substantial thatch layer exists, however, it is usually necessary to precede the overseeding operation with vertical mowing in several directions. It may be desirable to topdress as well if the thatch layer measures 1/4 inch or more in thickness.

Topdressing is also practiced after overseeding to provide a more favorable medium for germination in a thatchy turf. Where thatch is not modified by the addition of soil, seed germination may occur, but seedling survival under traffic and cold temperatures may be reduced, resulting in fair to poor stands.

Postplanting Culture

The most serious concerns following overseeding are disease, competition from the warm-season turfgrass, and low-temperature injury to the seedlings. Fungicide-treated seed should be used for overseeding and a preventive fungicide spray program followed to minimize disease pressure, especially from *Pythium* fungi. Disease pressure is usually most severe during the first four weeks.

There is no guarantee that severe competition from the warm-season turfgrass can be avoided, especially when unseasonably warm temperatures occur. However, competition can be minimized by avoiding excessive rates of nitrogen fertilization. The fertilization program should be just adequate to sustain the cool-season turfgrass community.

The potential for low-temperature injury to the seedling turfgrasses is greatest during the first month following overseeding, especially in cooler subtropical climates. Where cold temperatures are anticipated, one possible safeguard is to raise the mowing height for greens from 3/16 to 5/16 inch for about the first three weeks or until the first evidence of tillering. Then the mowing height can be lowered gradually as midwinter is approached and the plants become more hardy.

QUESTIONS

1. List the four types of activities performed in **turfgrass propagation.**
2. What is **tillage** and why is it important in turfgrass propagation?
3. Differentiate between **rough** and **fine grading.**
4. What is meant by **soil modification** and what materials may be used as soil amendments?
5. Illustrate and explain two designs for installing **drainage tiles.**
6. Why is it desirable to incorporate **lime** and **fertilizer** into the soil?
7. Differentiate between turfgrass seed **blends** and **mixtures.**
8. What is meant by **pure live seed?**
9. Differentiate between seed **quality** and **viability.**
10. Explain the differences between **breeders**, **foundation**, **registered**, and **certified** seed.
11. List and describe several **vegetative planting materials** and explain how they are used in turfgrass propagation.
12. List and explain factors influencing the selection of a **seeding rate** for a particular turfgrass species.
13. Why is it desirable to **roll** a seedbed following planting?
14. What is the significance of **mulching** in turfgrass propagation?
15. Explain the important practices employed in **postplanting culture.**
16. Differentiate between turf **establishment** and **renovation.**

CHAPTER 9

Cultural Systems

A cultural program is an assemblage of all primary (Chapter 5) and supplementary (Chapter 6) cultural practices, as well as pest-management (Chapter 7) and propagation (Chapter 8) procedures, for establishing turf and sustaining it at a particular intensity of culture. Cultural programs have been developed for managing greens, tees, fairways, athletic fields, lawns, and utility turfs under nearly all climatic conditions. When a cultural program is designed to meet the specific requirements of a particular turf to achieve a desired level of quality, it is called a cultural system. This systems approach to turfgrass management requires the recognition of specific conditions in a turf and the adjustment of cultural practices for those conditions.

CULTURAL INTENSITY

The selection of specific cultural practices and the intensities with which they will be employed depend on the intended uses of the turf, the desired level of turf quality, the conditions existing at the turfgrass site, and the financial resources available for pursuing the program of culture.

The cultural variable of greatest importance, and the one that primarily determines the number and intensity of other cultural operations, is mowing height (Figure 9.1). Generally, as the cultural intensity increases mowing height is reduced and mowing frequency is increased. The range of mowing practices extends from a few mowings per year at 6 inches or higher for some utility turfs to daily mowing at 3/16 inch or lower for greens. As mowing height is lowered the cultural operations that are likely to be intensified or added

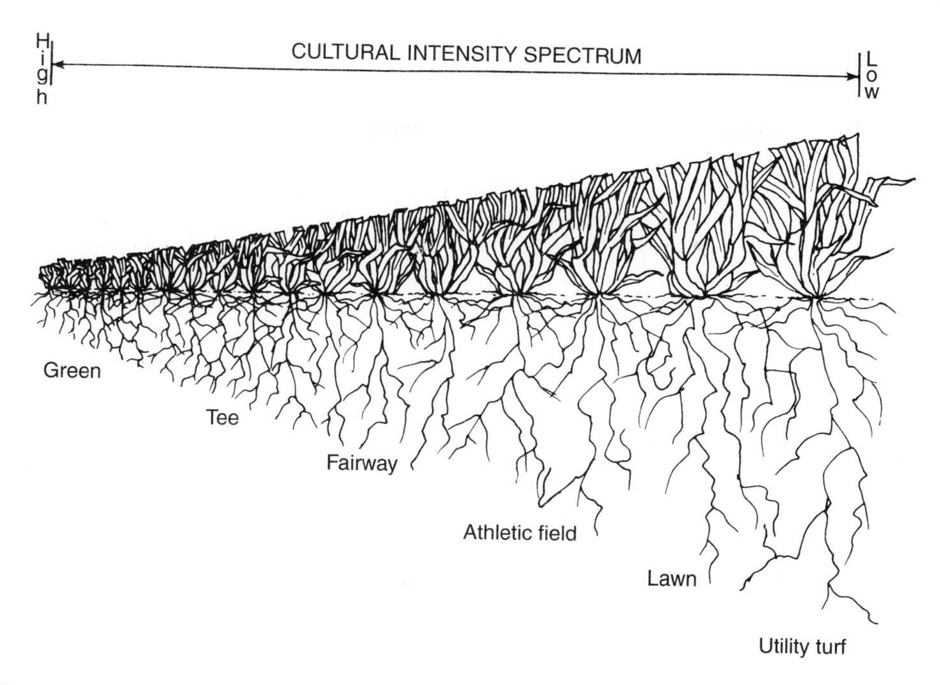

Figure 9.1. The turfgrass cultural intensity spectrum showing relative positions of greens, tees, fairways, athletic fields, lawns, and utility turfs.

include fertilization, irrigation, cultivation and other supplementary cultural practices, various pest-management procedures, specialized site preparation prior to planting, and careful selection of turfgrass genotypes.

Before development of a cultural system is attempted, the specific functions of the turf should be determined. Where the primary turf function is soil stabilization the required number and intensity of cultural operations are relatively low. An example is utility turf along roadsides; this may require little more than occasional mowing following successful establishment. As the esthetic appearance of a turf becomes increasingly important, the intensity of culture must increase in proportion to the desired level of turf quality. Many lawn turfs are mowed weekly, fertilized several times during the growing season, irrigated as often as necessary to maintain color and growth, cultivated annually to control thatch, and treated with pesticides as needed to control weeds, diseases, and insects. Finally, where a turf is used to support recreational activities the specific nature of these activities determines to a large extent how the cultural program should be composed. Sports turfs, including greens, tees, fairways, and athletic turfs, typically receive closer and more frequent mowing and are subjected to more intense traffic than most lawns. The number, intensity, and frequency of various cultural operations, therefore, are also usually higher.

GROWTH FACTORS

One of the principal purposes of turfgrass culture is to ensure that all the resources required to sustain turfgrass growth are in adequate supply. These include light, water, oxygen, mineral nutrients, and carbohydrates. Of these, moisture and mineral nutrients can be directly supplied through irrigation and fertilization practices. The available supply of other resources, however, may be influenced, but not directly controlled, by cultural practices.

The oxygen supply within the turfgrass root zone is influenced by soil physical conditions, cultivation, and irrigation practices. In fine-textured soils that are compacted or waterlogged oxygen may be so deficient that the absorption of water and nutrients by roots is impeded. Any cultural practice that improves drainage, reduces soil compaction, or otherwise ensures that a sufficient concentration of oxygen exists within the turfgrass root zone can, in effect, remove an important growth-limiting factor.

Usually, light intensity cannot be controlled except where obstacles to light penetration are reduced or eliminated. Under low-light conditions the capacity to sustain a turf is greatly influenced by mowing and fertilization practices. Excessively close mowing or excessively high rates of nitrogen fertilization can exhaust carbohydrate reserves so that photosynthetically produced carbohydrates are the limiting factor in turfgrass growth. Even under full sunlight, a depletion of carbohydrate reserves can result where these cultural practices are performed too intensively.

LEVELS OF SUPPLY

The intensity of each cultural operation affects not only turfgrass growth and survival but also the efficiency with which resources are utilized by the turfgrass community (Figure 9.2). To achieve a specific turfgrass response level under uniform conditions, a critical level of supply of a particular resource must be available. Within a concentration range above this level, called the adequate range, additional supplies of a resource may result in little or no further response. Above and below the adequate range the resource is present either in such short supply that it limits growth or in such excess that some other factor may limit growth. For example, soil moisture must be maintained at or above a critical level to sustain a desired rate of turfgrass growth. When moisture is substantially above this level, but within the adequate range, increases in turfgrass growth may be relatively small or not apparent. Where soil moisture is maintained above the adequate range (oversupply) the soil oxygen level may be in such short supply that it becomes the factor that limits turfgrass growth.

Depending on the intensity of culture and the particular turfgrass genotype employed within the cultural system, the adequate range can vary widely. A closely mowed dense shallow-rooted turfgrass community may have a very limited adequate range. Since the likelihood of encountering short-supply or oversupply conditions in this example is high, the cultural program must be meticulously controlled to keep all resources within the adequate range. In many lawn and utility turfs the adequate range is

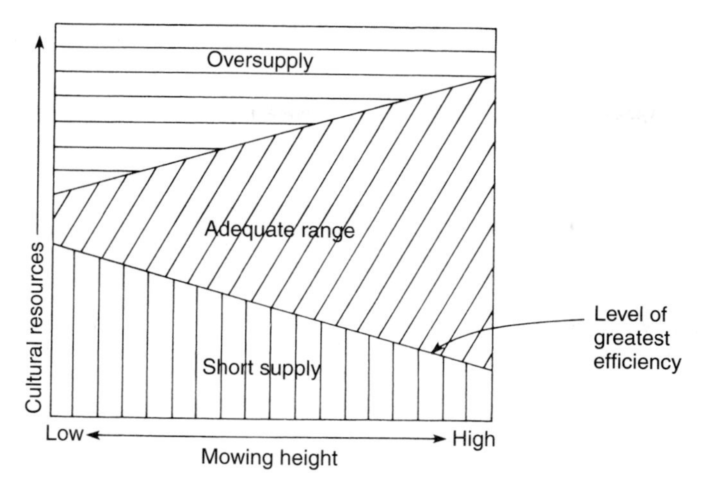

Figure 9.2. The dimensions of supply as influenced by the concentrations of cultural resources affecting turfgrass growth. The magnitude of the adequate range (the concentration range of a cultural resource that exists between short supply and oversupply) decreases as mowing height is reduced.

considerably wider and, therefore, the technical expertise required to manage these turfs satisfactorily is much less.

LIMITING FACTORS

The composition and sequence of limiting factors influencing turfgrass growth depend on the local environment and the morphological and physiological development of the turfgrass community to secure the resources that are available.

A possible sequence of limiting factors is provided in Figure 9.3. A shaded turfgrass community may be more limited in its growth by light than by any other single factor. In full sunlight, light is not likely to be limiting except under very high shoot densities or prolonged periods of cloudy weather. Temperature conditions, especially temperature extremes encountered during winter and summer, may so limit turfgrass growth and survival that improper selection of species can almost guarantee failure.

Plant-available moisture and mineral nutrient concentrations are usually the limiting factors of next importance; in most turfs they control turfgrass growth and esthetic appearance more than do any other environmental factors.

Disease-causing agents, insects, and other pests vary in their importance as factors influencing growth depending on the inherent susceptibility of the turfgrass genotype and the activity of these pests during the growing season. Some pest problems are so severe that a particular turfgrass cannot survive, or its survival and performance depend on a carefully conducted program of pesticide application.

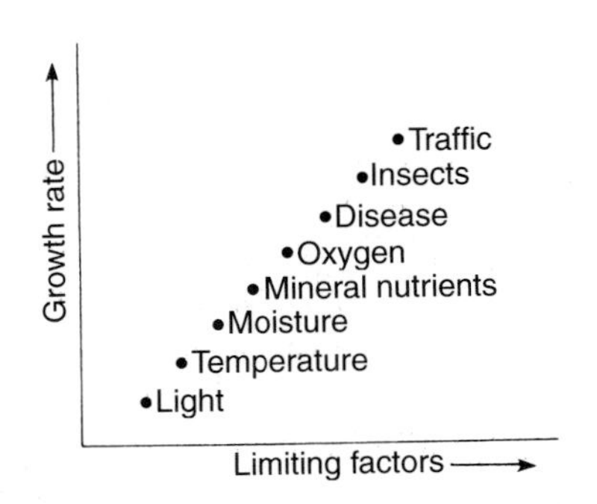

Figure 9.3. Sequence of factors that are likely to limit turfgrass growth.

Finally, traffic may limit turfgrass growth. Traffic-worn turfgrass either fails to survive or its vigor is reduced. As discussed previously, traffic-induced soil compaction can limit the oxygen supply within the turfgrass root zone. Also, highly compacted soils resist root growth and may adversely affect the growth of the entire plant.

The limiting-factor concept arose from Liebig's law of the minimum that stated, "If one necessary element is deficient and all others are adequate, growth will be limited by the one that is missing." This is a valuable concept in turfgrass management, as it can be used to develop a cultural system that is both efficient in its use of cultural resources and highly effective in promoting optimum turfgrass quality. The limiting-factor concept emphasizes the resources that are in short supply and thus eliminates trying one procedure after another to "force" turfgrass growth. It relies instead on the identification of the specific factor that is limiting growth and, subsequently, the adjustment of that factor to release the turfgrass community from one or more constraints imposed by the environment. With the successive removal of each limiting factor, the turfgrass community should automatically respond with additional increments of growth as long as no new limitations are inadvertently imposed.

CULTURAL INTENSITY SELECTION

Once the specific purposes for a turf have been determined, a cultural program can be designed to provide the desired level of turf quality. Then a turfgrass species or cultivar that is adapted to that cultural intensity must be selected for planting. Where the turf is already established we must determine whether the turfgrass populations within the plant community are adapted to the cultural intensity that has been selected. If not, different turfgrasses should be introduced into the existing turfgrass community or the turf should be completely reestablished.

The condition of the turfgrass site should be examined to determine if it is suitable for its intended use. A green established on native soil may not withstand the traffic intensity that accompanies normal use. Specialized construction techniques and

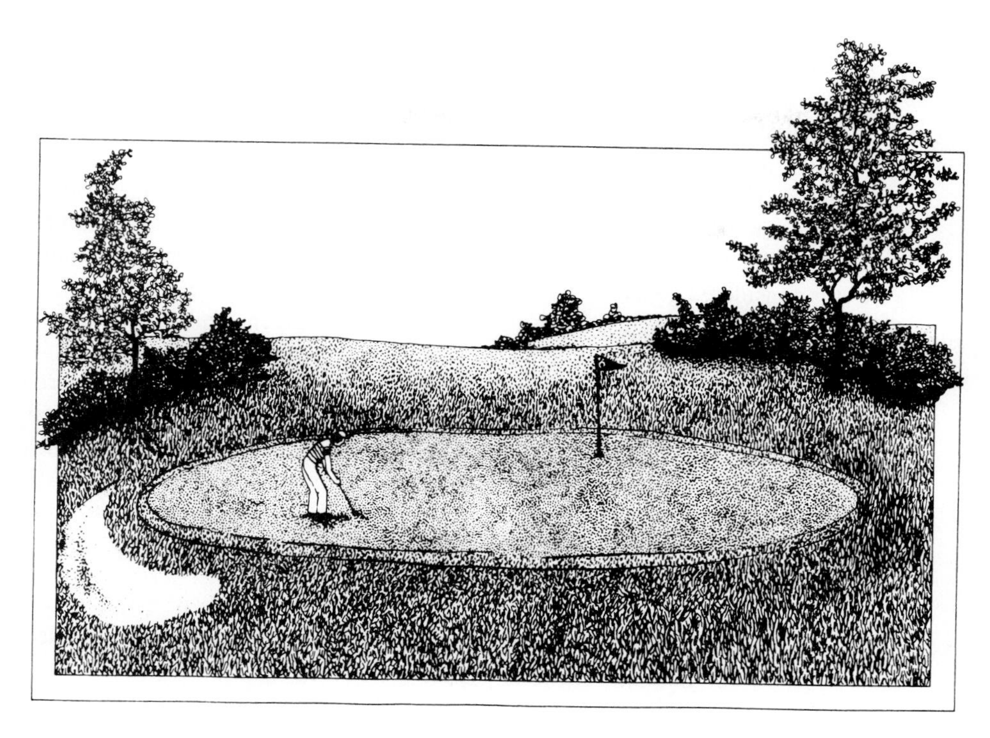

Figure 9.4. A putting green.

carefully prescribed soil mixtures may be necessary to sustain the green at desired levels of esthetic and functional quality.

Given an adapted turfgrass and suitable site conditions, a cultural program can be designed to sustain turf at any position along the cultural intensity spectrum illustrated in Figure 9.1.

Greens

Greens are very closely mowed turfs of high shoot density and uniformity that must provide smooth ball-roll characteristics (Figure 9.4). They should also be sufficiently resilient to cushion the impact of a ball as it lands on the surface. Specific types of greens are putting greens on golf courses, bowling greens, and croquet courts.

Turfgrass Selection

Turfgrasses used for establishing greens must be adapted to mowing at 3/16 inch. The turfgrasses best adapted to this intensity of culture are creeping bentgrass for temperate, subarctic, and some subtropical climates (seeded types: Penncross, Seaside, Emerald, and Penneagle; vegetatively propagated types: Toronto, Cohansey, and Pennpar) and hybrid bermudagrass (principally Tifgreen and Tifdwarf cultivars) for subtropical and tropical climates. Other turfgrasses that have been used for greens are common bermudagrass, velvet and colonial bentgrasses, and zoysiagrasses. Common bermudagrass forms an inferior turf compared to the hybrid bermudagrasses and is no longer widely used. Velvet bentgrass is a challenge to manage because of its thatch-proneness and high disease susceptibility. Colonial bentgrass is used in some areas with a temperate oceanic climate, but it is considered generally inferior to creeping bentgrass.

Annual bluegrass often invades established greens and may become the dominant turfgrass within the community because of its vigorous growth during cool-weather periods and its capacity to produce viable seed throughout much of the growing season. Finally, as discussed in Chapter 8 under Temporary Winter Turfs, several cool-season turfgrasses, including ryegrasses and fine fescues, are used for temporary cover on dormant bermudagrass greens in subtropical climates.

Propagation

Many greens are subjected to intense traffic during periods of active play. For par golf, thirty-six of seventy-two strokes are made on the green and another eighteen strokes are played to the green. Therefore, on an eighteen-hole golf course of one hundred or more acres, much of the play is concentrated on approximately two acres of greens turf, or about 2% or less of the total area of the golf course.

Propagation procedures for establishing greens usually include the use of specialized soil mixtures to provide an inherent tolerance to intense traffic. In the past large quantities of sand and some organic amendments were combined with native soil in an attempt to produce a coarse-textured mixture that would retain sufficient aeration porosity under traffic. Results varied depending on the constituents used in the mixture and their relative proportions. Given the variability of obtainable sources of sand, organic matter, and soil, consistent results were difficult to duplicate from one green to the next.

In 1960 the USGA Green Section first published its specifications for greens construction based on research conducted at Texas A & M University. These specifications called for a soil mixture that was predominantly sand, but included some organic matter and clay-containing soil. The exact proportions of these constituents depended on precise laboratory measurements of certain physical properties, including infiltration and percolation capacities, total and aeration porosities, bulk density, and water-retention capacity.

The most recent modification of these specifications calls for a root-zone mixture with the following physical properties: total porosity: 35–55%, air-filled (aeration) porosity: 15–30%, saturated (hydraulic) conductivity: 6–12 inches per hour for the normal range and 12–24 inches per hour for the accelerated range (where water quality is poor, cool-season turfgrasses are grown outside their range of adaptation, or dust storms or high rainfall events are common), and an organic-matter content of 1–5% (ideally, 2–4%). The specific components of this mixture should contain a minimum of 60%

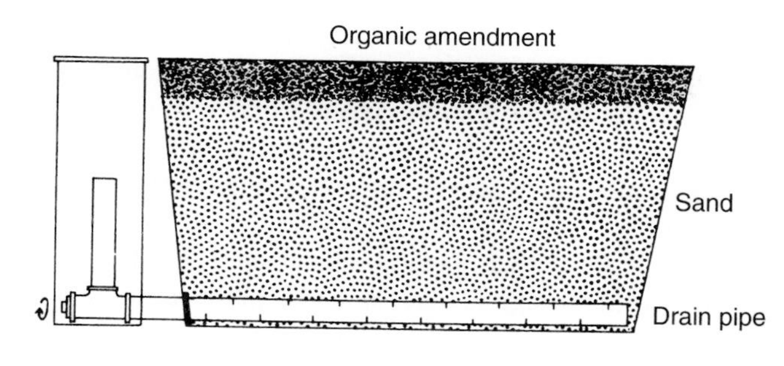

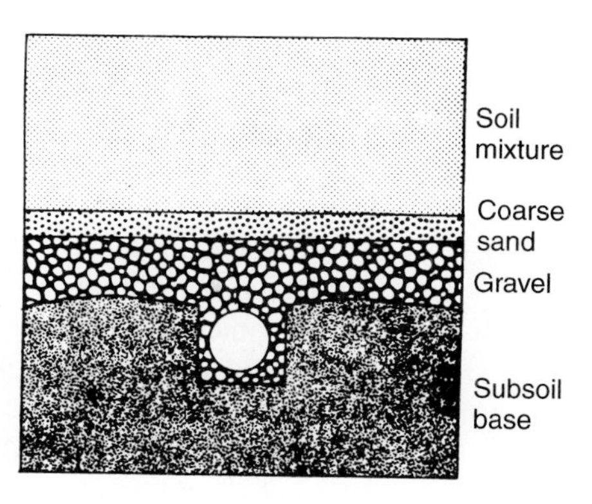

Figure 9.5. USGA Green Section design (right) and Purr-Wick (left) designs for greens construction. The USGA Green Section design uses a root-zone mix composed primarily of medium sand mixed with organic matter and clay-containing soil in proportions determined by laboratory tests. The Purr-Wick design uses a medium-fine sand with organic amendments tilled into the top few inches.

medium and coarse sands, not more than 20% fine sand, not more than 5% very fine sand and not more than 10% of a combination of very fine sand, silt, and clay. At the coarse end of the spectrum the mixture should not have more than 10% of a combination of very coarse sand and fine gravel; the fine gravel component should not exceed 3%. The depth of the root-zone mixture after settling should be 12 inches, with a tolerance of ± 0.5 inch. To meet these specifications, samples of sand, organic matter, and soil must be submitted to a soil physics laboratory for testing. The USGA Green Section supports a soil physics laboratory to provide these tests.

Construction of a USGA Green Section green requires excavation to a depth of at least 16 inches below the proposed final grade (Figure 9.5). Trenches are dug at least 6 inches wide and 8 inches deep into the subgrade for installation of drain lines surrounded by pea gravel or crushed stone. Additional pea gravel or crushed stone is applied over the entire subgrade to a depth of at least 4 inches. Gravel materials suspected of lacking mechanical stability to withstand common construction traffic should be tested using ASTM procedure C-131 (test value should not exceed 40), and materials of questionable weathering stability should be tested using ASTM procedure C-88 (test value should not exceed 12% loss by weight). Where an intermediate layer is used between the gravel blanket and the root-zone mixture, at least 65% of the gravel should fall within a range of 3/8 and 1/4 inch in diameter; where an intermediate layer is not used, specific recommendations are based on the particle size distribution of the root-zone mixture.

An intermediate layer is sometimes recommended to prevent migration of materials from the root-zone mixture into the gravel blanket. It should be 2 inches thick and composed of sand and gravel particles falling within a range of 1 to 4 millimeters in diameter. This layer need not be included, however, if properly sized gravel is used.

The significance of the gravel and intermediate layers is that they allow for rapid transmission of percolating water once it drains from the root-zone mixture. Because of the comparatively finer texture of the root-zone mixture, however, a perched water table occurs above the interface between the root-zone mixture and the underlying coarse-textured intermediate layer (see Chapter 4 under Soil Moisture). This is a desirable feature of the design in that it provides a moisture reservoir in what would otherwise be a very droughty medium.

Another contemporary design is the Purr-Wick method developed at Purdue University (Figure 9.5). The soil mixture for this design is pure sand except in the surface 1 to 2 inches, where organic amendments (usually peat moss) are incorporated into the sand. The depth of the sand varies from 14 to 22 inches depending on particle size; very fine sand should be placed to a depth of 18 to 22 inches, while a predominantly medium sand layer should be between 14 and 16 inches deep. Prior to sand placement, the excavated site is covered by two layers of 4-mil-thick plastic sheeting. Plastic drain pipes measuring 2 inches in diameter and with numerous slits on alternate sides are spaced 10 to 20 feet apart on the plastic sheeting. These are connected to a common pipe, which extends through the plastic sheeting into an adjacent service pit containing a drain plug and adjustable riser. Varying the orientation of the riser can maintain the water table within the sand at a desired level for prolonged periods, thus reducing the irrigation requirement of the green.

Both USGA Green Section and Purr-Wick green designs employ essentially compaction-resistant growth media of limited moisture retention capacities, while sustaining plant-available moisture reservoirs through the placement of barriers to continuous percolation. These unique designs overcome, at least partially, the primary objections to using predominantly sandy media to support turfgrass growth. While resisting the compacting effects of intense traffic, predominantly sandy media typically lack moisture- and nutrient-retention capacities and require frequent irrigation and fertilization. However, because of the induced retention of moisture and nutrients achieved through installation of barriers to continuous water percolation and leaching of nutrients, these greens can be managed without the kind of meticulous culture that would otherwise be required.

Primary Culture

The mowing height used for greens is typically 3/16 to 1/4 inch. Slightly higher mowing heights may be used under conditions of severe environmental stress as long as desired playability characteristics can be maintained. Closer mowing, down to 1/8 inch, is possible, but is usually practical only on those greens receiving relatively light traffic. Mowing frequency may be as low as three or four times per week and as high as seven times per week (daily), depending on mowing height. Generally, closer mowing is accompanied by more frequent mowing. Mowing direction should be varied as much as possible to reduce graininess. Brushes or combs attached to a greens mower can also reduce graininess by forcing decumbent shoots to "stand up" just prior to mowing.

Except under unusual circumstances, clippings should be removed with mowing. The extremely high shoot density of greens turf usually results in clippings remaining on the turf's surface, and thus interfering with playability, where clippings are returned.

Greens mowers are reel-type mowers with very short "clips of the reel" to reduce marcelling (see Chapter 5). Mowers must be precisely adjusted and backlapped often for acceptable cutting quality.

Care should be exercised to repair ball marks and control puffiness associated with thatch development to avoid scalping during mowing.

Fertilization should be performed only as necessary to sustain desired growth and density. Given the variability in soil mixtures, atmospheric conditions, turfgrass genotypes, and fertilizer carriers, it is difficult to prescribe the proper amount and application frequency of nitrogen and other fertilizer materials for greens. A reasonable guide for nitrogen is to apply an average of 3/4 to 1 1/2 lb (actual N)/1000 ft^2 per growing month at two- to six-week intervals, depending on the fertilizer carrier.

Phosphorus and potassium applications should be based on soil test results; phosphorus requirements are usually small in established greens, while the potassium requirement (as K_2O) is usually considered to be at least half that of the nitrogen requirement, or more where potassium is readily leached from the soil.

Iron should be applied as needed, usually in combination with fungicides, to maintain color. Other minor nutrients may be required from time to time, especially in greens constructed predominantly with sand. Based on soil pH measurements, lime or sulfur may be necessary to adjust pH to a favorable range of 6.0 to 6.5.

Irrigation water should be applied as often as necessary to sustain growth and prevent wilting. The specific amount and frequency of irrigation depend on many factors, as discussed in Chapter 5; however, the irrigation requirement for a green is usually greater than that of other less intensively cultured turfs. During especially hot weather, some midday syringing may be necessary to cool the turfgrass when wilting symptoms appear.

Supplementary Culture

The number of supplementary cultural practices is greater for greens than for any other turf types. Core cultivation may be performed from one to several times each growing season. Depending on soil conditions, the cores may be broken up and matted in or removed from the green. An unfavorable soil or a layering condition near the surface may necessitate removal, followed by topdressing.

Spiking may be performed at weekly intervals during the season of intense traffic. Coring and spiking are intended to reduce the effects of surface compaction and thus improve water infiltration and oxygen diffusion. As coring is the more disruptive practice, it should be scheduled for periods when temperatures favor rapid recovery of the turf. Spiking is less disruptive and can be scheduled almost any time turfgrass growth is active.

With the development of vertical-mowing units that can be mounted on greens mowers, shallow vertical mowing has become a popular practice for reducing graininess and thatch on greens. It may be performed at frequent intervals, perhaps weekly, during the growing season. Deep vertical mowing is rarely performed on greens because of its severely disruptive effect on the playing surface.

Topdressing is performed to control thatch, to smooth the playing surface, and, on some greens, to alter the characteristics of the soil medium that supports turfgrass growth. The rate and frequency of topdressing vary widely; in recent years the trend has been toward very frequent topdressing (every three to four weeks) at rates just adequate to fill the thatch. When practiced in conjunction with core cultivation, the topdressing rate should be increased to fill the core holes as well. Obviously, where the principal purpose of topdressing is thatch control its rate and frequency should be related to the rate at which thatch develops.

Pest Management

The most important pests of greens are disease-causing organisms. Fungicides are usually applied on a preventive schedule to control dollar spot, brown patch, and other diseases. In temperate and subarctic climates preventive fungicide applications are usually made for controlling snow-mold diseases during the overwintering period. During hot humid weather, *Pythium* blight may be controlled by preventive or early curative fungicide applications.

Insect pests such as cutworms and sod webworms should be controlled with insecticides as soon as the first symptoms of injury occur. Insecticides can be applied in conjunction with fungicides during periods of insect activity.

Injury from nematodes may be so severe in sandy soils in subtropical and tropical climates that nematicide applications are necessary to sustain the turf. As nematicides are highly toxic pesticides, they should be used with extreme care and only by trained personnel.

Finally, herbicides may be used to control various grasses and broadleaf weeds in greens. Because of the close mowing heights practiced on greens, many of the weeds that occur in lawns cannot survive in greens. However, white clover, chickweeds, and other broadleaf weeds can develop and seriously reduce playability. On bentgrass mecoprop is the principal herbicide used for broadleaf weed control, while for bermudagrass a wider selection of herbicides can be used safely (see Chapter 7).

Annual grasses such as crabgrass and goosegrass may be a problem in some greens, especially in warm climates. Preemergence herbicides can be used for control, but experience indicates that repeated use may result in reduced stress tolerance and increased disease incidence, especially in bentgrasses. Postemergence herbicides (organic arsenicals) have been used to control summer annual grasses in greens, but the potential for turfgrass injury is high, especially at high temperatures. Often, it is safer to remove these weeds by hand while attempting to improve turfgrass growing conditions in weed-infested greens than to rely on herbicides for control.

Generally, the most serious weed in greens is annual bluegrass. In bermudagrass greens several preemergence herbicides can be used to prevent emergence of annual bluegrass from seed. Also, postemergence applications of pronamide have been quite effective for removing existing populations. In bentgrass greens, however, the selection of safe and effective herbicides is more limited. Control of annual bluegrass depends more on sustaining the bentgrass in a healthy and competitive condition than in controlling annual bluegrass directly. Many greens in cool temperate and subarctic climates eventually become predominantly annual bluegrass; the cultural program must then be adjusted to sustain it as the primary turfgrass.

Other Practices

On putting greens the position of the cup is shifted several times per week to allow for recovery of the turf. The frequency with which the cup position should be changed depends on traffic intensity, turfgrass wear tolerance, and the compaction resistance of the soil. The contour of the green should be such that at least three to four major cupping areas exist. Because of the traffic intensity along approach lanes to the cup and between the cup and the next tee, the positions of the cup and tee markers should be coordinated to provide maximum distribution of traffic over the entire area. The placement of sand traps and other obstacles around the green can also influence traffic patterns and, therefore, turfgrass quality.

Tees

Tees and similar turfs, such as collars, aprons, or grass tennis courts, are usually mowed at a height of 1/2 to 1 inch. On some sites the cultural program may be similar to that practiced for greens, but it is usually less intensive and may be the same as that used for fairways (Figure 9.6).

Tees should be level and firm to provide for a firm stance in making tee shots. They should also be of sufficient size to allow for frequent movement of tee markers and recovery of worn turf.

Turfgrass Selection

The turfgrass species selected for establishing tees are often the same as those comprising the fairway turf. However, where high shoot density from close mowing is desired, bentgrass and bermudagrass are usually the preferred turfgrasses within their respective zones of climatic adaptation. In transition zones between temperate and subtropical climates zoysiagrass (*Z. japonica*) is often used because of its superior adaptation to the temperature extremes encountered in both winter and summer. It is also used as a substitute for bermudagrass on shaded sites.

A tee turfgrass must withstand severe traffic and be sufficiently vigorous to recover rapidly from divoting and other injury. Where adapted, bermudagrass is an excellent choice. Zoysiagrass also has excellent wear tolerance, but it lacks the recuperative capacity

Figure 9.6. Golf course tee.

of bermudagrass. In comparison, cool-season turfgrasses rank fair to poor depending on the specific species, cultivar, and growing conditions.

The recuperative capacity of creeping bentgrass is good, but its wear tolerance is at best only fair. The generally poor tolerance of Kentucky bluegrasses to close mowing usually limits their use to tees mowed above 1 inch; even then annual bluegrass and other weeds are likely to invade. However, with the introduction of new Kentucky bluegrass cultivars which appear to be better adapted to mowing as low as 3/4 inch, the use of this species as a tee turfgrass may increase.

Turf-type perennial ryegrasses are gaining favor in areas where they are adapted because of their superior wear tolerance and rapid establishment from seed. The ryegrasses are also used alone or in seed mixtures with other turfgrasses for seeding into divot scars. While they may not persist in competition with creeping bentgrass, they can provide temporary cover until the bentgrass has fully recovered.

Propagation

Tees may simply be cut out of fairways or constructed in the same elaborate fashion as greens. Usually, they are built on leveled deposits of topsoil amended with peat moss or other materials.

Tees are often partially surrounded by ornamental trees and shrubs. While this may be desirable for the protection of golfers and for providing shaded resting sites, the playable portion of the tee should not be heavily shaded. The wear resistance and recuperative capacity required of the turfgrasses are substantially reduced where trees provide too much shade.

Primary Culture

Mowing should be at a height that will permit a golf ball to stand clear of the turfgrass leaves without being teed too high. Only a few turfgrasses can be sustained indefinitely at the preferred height of 1/2 inch. Bermudagrass and zoysiagrass can be sustained at that height without great expense in their areas of climatic adaptation, while creeping bentgrass and annual bluegrass require greater care where they are used. Kentucky bluegrass and perennial ryegrass, which have lower cultural requirements, should be mowed above 3/4 inch.

Fertilization should be adequate to sustain vigorous recuperative growth but not so much that wear tolerance is reduced. This may require up to 1 lb or more of nitrogen per 1000 ft^2 per growing month, depending on the turfgrass. At least half as much potassium (as K_2O) should be supplied with the nitrogen; even larger amounts may be required where clippings are removed.

When irrigation is necessary to sustain vigorous turfgrass growth it should be timed so that the tee is dry and firm during play.

Supplementary Culture

Because of the intensive traffic to which tees are subjected, core cultivation is important for relieving soil compaction. Cultivations should be timed to avoid optimum conditions for the germination of weed species, especially annual bluegrass and summer annual grasses.

Topdressing may be necessary to control thatch where conditions favor its development. Under close mowing and severe traffic, however, thatch may not be a serious problem with some turfgrasses that are otherwise thatch prone. Reincorporation of soil cores following core cultivations may be all that is necessary to controll thatch on tees.

Vertical mowing is sometimes practiced on tees to control thatch. At some locations it is used more often to prepare the tee for supplemental seeding operations.

Pest Management

The serious disruptions of tee turf resulting from play often predispose the turf to weed invasion. While most broadleaf weeds can be controlled with selective herbicides, annual grasses may be particularly serious and difficult to control. Preemergence herbicide residues at the surface may be entirely removed or reduced with divoting, thus rendering the turf susceptible to invasion by annual grasses. In temperate and subarctic climates, invasion of cool-season turfgrass communities by annual bluegrass is extremely difficult to prevent, and its effective control may involve mechanical methods including resodding, plugging, or frequent applications of soil and seed to divot scars.

Depending on the turfgrass used on tees, regular applications of fungicides may be necessary for disease control. Disease control on tees is extremely important for resisting weed invasion and sustaining acceptable turfgrass cover. For the same reasons, damage from insects and other pests should be carefully monitored and, if necessary, controlled with appropriate pesticides.

Fairways

Fairways constitute the largest playable acreage on golf courses, ranging from 30 to 60 or more acres per 18 holes. The esthetic appearance of the course is more influenced by the condition of the fairways than any of the other turfs (Figure 9.7). Unlike those on tees, wood or iron shots from the fairways are played "on the grass." Therefore, the lie of the ball directly reflects the functional quality of the turf.

Figure 9.7. Golf course fairway.

Turfgrass Selection

The finest fairways in subtropical and tropical climates are of bermudagrass, especially improved cultivars such as Tifway. In transition zones between subtropical and temperate climates, zoysiagrass and some cold-hardy bermudagrasses form good fairway turfs. In temperate and subarctic climates, seeded bentgrasses can provide good to excellent fairways. With the introduction of improved turf-type cultivars, perennial ryegrasses have become very popular for establishing fairways in warm-temperate climates. And where it persists through the growing season, annual bluegrass can form an excellent fairway turf.

Kentucky bluegrass forms an attractive turf in temperate and subarctic climates, but usually must be mowed higher than other turfgrasses and irrigated less frequently to persist. With regular irrigation, especially under close mowing, it is likely to be invaded and eventually dominated by annual bluegrass or bentgrass. Some of the newer cultivars of Kentucky bluegrass have exhibited tolerance to mowing heights as low as 3/4 inch and may be used to establish fairways of excellent playing quality.

In subtropical dry summer and tropical highland climates kikuyugrass may invade and eventually dominate fairway turfgrass communities. Because of the difficulty in controlling this species, the cultural program is often adjusted to optimize its quality as a fairway turfgrass.

In semiarid regions buffalograss, fairway wheatgrass, and blue gramagrass are used for unirrigated fairways. With irrigation, the choice is usually bentgrass, zoysiagrass, bermudagrass, or possibly Kentucky bluegrass, depending on temperature conditions.

Any locally adapted turfgrass may be used for establishing fairway turf where playable quality is secondary to economic considerations. The best fairway turfgrasses are, however, usually the most expensive to sustain.

Propagation

Fairways are usually established with little or no modification of the native soil; rather, the topsoil is stockpiled prior to rough grading and subsequently replaced when final contours have been established.

Ideally, fairways should slope toward the rough to favor surface drainage. The surface should be as smooth as possible to avoid scalping of high spots and the higher effective mowing heights in small depressions.

Given the large acreage of fairways, the planting methods used are those that can be performed at the lowest cost. A cultipacker seeder is typically used for propagating seeded turfgrasses. Vegetative planting is usually performed by stolonizing, followed by disking in the stolons. Efficient methods for planting plugs may also be used.

Primary Culture

Fairways are mowed at heights between 1/2 and 1 1/4 inches, depending on the turfgrass and the playing quality desired. A range of 1/2 to 3/4 inch is preferred for best quality. Mowing frequency should be at least twice per week, while four times a week is often preferred for sustaining the highest possible density and the most uniform playing conditions.

The mowing pattern should be varied to avoid excessive tracking by mower wheels. With hydraulically operated mowing units, the outer unit can be lifted for every second mowing to reduce the differential soil compaction from wheels traveling the same lanes each time. Turns in front of the greens should be made carefully and varied, if possible, to minimize turfgrass wear.

Fertilization should be just enough to sustain high shoot density and an adequate growth rate. If clippings are not removed with mowing, the amounts of fertilizer nutrients required to sustain growth are substantially less than those required for greens and many tees. In temperate climates 2 to 3 lb N/1000 ft^2 per year are usually adequate. Where clippings are removed with mowing, additional nitrogen and other nutrients may be needed. More nutrients may also be required in subtropical and tropical climates, where the growing season is longer.

Irrigation is important not only as a means of providing a cultural resource for sustaining turfgrass growth, but also for providing uniform playing conditions. Dry hard fairways yield an advantage for the amateur golfer because of the additional distance the ball travels with each bounce. The professional must carefully place shots to achieve a tactical advantage over an opponent. Playing on a fairway that varies widely in its physical characteristics from one game to the next is not only troublesome, it is unfair.

Supplementary Culture

The size of fairways precludes many labor-intensive practices routinely performed on greens and tees. Mechanical reduction of thatch involving extraction and removal of organic debris can be done, but can often be avoided through the proper performance of other cultural operations.

Avoiding excessive nitrogen fertilization and carefully selecting pesticides that do not induce thatch formation can minimize the requirement for thatch control procedures. Core cultivation to mitigate the effects of soil compaction may, with the reincorporation of soil cores, effectively control the thatch that does develop.

Routine slicing of fairways during the growing season can improve infiltration in surface-compacted soils and thus aid in sustaining turfgrass quality.

Pest Management

Depending on the turfgrasses in the fairways, disease control may or may not require fungicide applications. Bentgrass and annual bluegrass fairways often require regular applications of fungicides, on a curative or preventive schedule, to sustain optimum playing conditions.

Weeds that develop in the fairways may require periodic applications of herbicides, including preemergence herbicides for annual grass control.

Treatment for insects and other problems may be necessary as problems develop. It is important to carefully monitor the condition of the fairways for early signs of white grubs and other insects so that control procedures can be implemented before serious losses occur.

Athletic Fields

While lawns, golf courses, and other turfs may not dominate the landscape in many parts of the world, athletic fields of one type or another can be found almost anywhere. In the United States, football and baseball fields can be found in nearly all residential communities. Over much of the rest of the world, soccer fields represent the most intensive use of turfgrasses. Other sports usually conducted on turfed surfaces include lacrosse, cricket, rugby, and polo. While the unique characteristics of each sport dictate the specific cultural requirements of the turf, certain features are desirable in all sports turfs—safety for the players and esthetic appearance for the spectators.

An athletic turf that is fairly representative of most is the football field (Figure 9.8). It must provide firm footing, adequate resiliency on impact, and resistance to tearing during play. In addition, it should provide a surface that is relatively free of dust, mud, and other unfavorable or unsafe conditions.

Figure 9.8. Football field.

Turfgrass Selection

Turfgrasses for athletic fields should be very wear resistant and vigorously growing to recover quickly from injury. Where adapted, bermudagrass is an excellent choice. In temperate and subarctic climates aggressive cultivars of Kentucky bluegrass have been used with favorable results. Also, turf-type perennial ryegrasses planted alone or in mixtures with Kentucky bluegrass have gained wide acceptance in many areas with temperate climates.

Tall fescue has been used in transition zones between temperate and subtropical climates, but it will not persist for long under the close mowing often demanded for athletic turfs. For low-budget facilities, however, it may be a satisfactory alternative to other turfgrasses.

Propagation

Like greens, athletic fields should be well drained and resistant to the compacting effects of severe traffic. The inability to sustain optimum playing conditions and acceptable appearance has resulted in the transition to artificial "turfs" at many college and professional football stadiums since the 1960s. Under some circumstances it is difficult, if not impossible, to sustain satisfactory athletic turf; however, failures have resulted more from improper construction and management than from an inherent incapacity of turfgrasses to survive the stresses of athletic activities. Other than in covered stadiums, turfgrasses can usually form acceptable athletic turfs provided a carefully designed cultural system is followed.

Many athletic turfs established on native fine-textured topsoil have been acceptable as long as provisions for internal and surface drainage have been made. The center of the field should be crowned at a height of 12 to 18 inches above the sidelines. A system of drain tiles installed throughout the field, and catch basins located along the sidelines are often effective in promoting satisfactory drainage.

Where the number of events held on athletic turfs collectively imposes more traffic stress than the turf can withstand, predominantly sandy soil mixtures and specialized design features similar to those used in constructing USGA Green Section and Purr-Wick greens may be necessary. One contemporary design, called the PAT (Prescription Athletic Turf) system, developed at Purdue University, is similar to the Purr-Wick green, except that the drain pipes are connected to a suction pump for accelerated internal drainage.

Planting procedures for football fields are described in Chapter 8; however, where sod is used, the soil carried with the sod should be nearly identical to that of the underlying soil in the field. Planting mineral- or muck-grown sod onto sand usually results in a perched water table within the sod that impedes water and air movement in the turf. Also, the sod tends to cleave along the interface between the two soil types under the stresses encountered during play.

Soil-less sod, from which the soil has been removed by mechanized washing, has shown promise for overcoming this problem. Limited research with soil-less sod has indicated that, once rooted, the new turf should be topdressed with sand to fully integrate the sod with the underlying sand medium. Otherwise, the organic material rests atop the sand and a serious thatch problem can develop.

Athletic turf that has been damaged from play should be repaired as soon as possible to maintain complete turfgrass cover. The two principal methods used are seeding and plugging. Turf-type perennial ryegrasses are commonly seeded on worn athletic turfs shortly after a game. Small damaged areas can be plugged with turf sections measuring 4 to 8 inches in diameter and at least 3 inches deep in the same fashion used for cupping golf greens. Selective sodding is sometimes recommended for repairing damaged athletic turf; however, at least several weeks of favorable growing conditions are necessary for sufficient rooting of the sod or it may be easily torn free during a subsequent game.

Primary Culture

Athletic turf should be mowed as closely as feasible to provide a dense fast surface for play while maintaining adequate wear tolerance and resiliency. Hybrid bermudagrass is typically mowed at 1/2 inch (common bermudagrass at 1 inch), while the mowing height for Kentucky bluegrass should be at least 1 1/4 inches or higher during the playing season. Higher mowing heights should be used when the field is not in play to favor root and rhizome development. Downward adjustments in mowing height should be made gradually over a period of several weeks prior to the playing season. Frequent mowing during the playing season promotes high shoot density while allowing slightly higher mowing heights to be used.

Fertilization practices prior to and during the playing season should be performed in accordance with the desired objectives of these practices. Prior to the actual playing season fertilization should be designed to achieve a well-developed turf with an extensive system of roots and lateral shoots. Aerial shoot growth should be just sufficient to sustain a dense turf.

During the playing season the fertilizer requirement, especially for nitrogen, is greater because of the need for rapid recuperative growth. Several weeks prior to the initiation of play nitrogen should be applied at the rate of 1 lb/1000 ft^2 from a quickly available source, or at 2 to 3 lb/1000 ft^2 from a slowly available source. Thereafter, light applications of quickly available nitrogen (1/4 to 1/2 lb/1000 ft^2) can be made as needed to sustain vigorous growth. Excessive application rates should be avoided, as they may result in a reduction in turfgrass wear tolerance. Potassium applications, preferably with potassium sulfate, should be made to promote adequate wear tolerance.

Irrigation practices prior to the playing season should be conducted to sustain a dense turf, while during the playing season an adjusted irrigation program should be followed.

The athletic field surface should be firm and dry during a game for optimum wear tolerance and playability. After a game the field should be irrigated to reduce desiccation injury and promote recovery. Subsequent irrigations may be necessary during droughty periods; however, irrigation water should not be applied for twenty-four to forty-eight hours prior to a game to allow sufficient time for drying.

Supplementary Culture

Since the football season is in the fall, most supplementary cultural practices should be performed in the spring. The field should not be in use at this time to allow for major corrective operations. Compaction from the previous year's play can render an athletic field nearly unsuitable for turfgrass growth. To correct this condition, the turf should be core cultivated six to eight times when the soil is moist, but not wet.

In temperate and subarctic climates core cultivation should begin in early spring to midspring. Then the field should be seeded with selected cultivars of Kentucky bluegrass and/or perennial ryegrass where turfgrass cover has been substantially reduced from play. On tall fescue fields this species should be used for seeding. A complete fertilizer and, if needed, lime, should be applied to provide adequate nutrients for the growth of seedlings and surviving turfgrass plants. Shallow vertical mowing can be used to break up the soil cores. Subsequently, the area should be matted to work the soil and other materials into the turf and produce a smooth surface. Irrigation water should be applied as needed to promote germination and subsequent growth.

In subtropical and tropical climates core cultivation should be performed after bermudagrass has initiated spring growth. Replanting is usually not necessary except where play and severe winter conditions have substantially reduced the turfgrass stand. If required, common bermudagrass seed or stolons of hybrid bermudagrass can be planted following extensive core cultivation. Subsequent fertilization, liming, and irrigation practices should be performed as necessary to restore the turf.

These operations should only be performed if conditions indicate the need to do so. Sand fields and others showing little or no injury from the previous season's play may not require any major corrective action. The condition of each field should be assessed as early as possible in the spring, and an appropriate cultural program implemented as needed.

During the actual playing season it may be necessary to cover the field with a tarp during the twenty-four- to forty-eight-hour period prior to a game when rainfall appears imminent to ensure that the field will be dry when the game starts. Even if rainfall occurs during the game, the damage from play will usually be less than when a game begins on a wet field. However, as soon as rainfall and cloudy conditions cease and the sun appears, the tarp must be removed to avoid serious turfgrass injury.

Immediately following a game a light rolling will press uprooted turfgrasses back into the turf and reduce further injury from desiccation, especially if followed by irrigation.

While bermudagrass is considered an almost ideal turfgrass for athletic fields, it may become dormant and turn brown during the football season as winter approaches. Excessive wear and pest-induced damage can also reduce the esthetic quality of the turf. Colorants may be used to mask the unfavorable appearance of an off-color athletic turf, but their effect is strictly cosmetic, and they do not improve the functional quality of a damaged turf.

Pest Management

The potential for weed invasion in athletic turf is high due to the damage resulting from play and intensive cultivation practices. Herbicides should be used judiciously to avoid injuring mature turfgrass plants or seedlings. Diseases, insects, and other pests should be controlled as soon as symptoms appear.

Artificial Turf

Artificial surfacing materials have gained some acceptance as substitutes for athletic turf. They offer several major advantages over the "real thing," including immunity from diseases, weeds, insects, and other turfgrass pests; freedom from the entire array of cultural practices necessary for establishing and sustaining turf; a fast playing surface when dry; adaptation to all types of environments, including severe shade; and complete accessibility for use whenever the ground is not frozen or snow covered.

The disadvantages of artificial turf are high initial cost; a usable life of as little as five years coupled with high replacement costs; costly damage from cigarettes, chewing gum, and other sources; substantial heat buildup at and above the surface during warm weather; high maintenance costs for cleaning and repairing the surface; and increased player injury from falling and running on the surface.

It is interesting to note that several major athletic facilities that constructed artificial turfs have since removed them and reestablished the "real thing."

Lawns

Lawns include most turfs planted in or around residences, schools, churches, military installations, government buildings, business sites, parks, and cemeteries. While we can cite examples of lawns sustained at all points along the cultural intensity spectrum illustrated in Figure 9.1, most lawns are designed to serve a principally decorative function (Figure 9.9).

The rapid development of a nationwide lawncare industry within the United States is having a major influence on contemporary lawncare practices. Many former do-it-yourselfers are now contracting to have professional organizations perform fertilization and pest management operations. The range of services offered by these organizations will probably expand to include cultivation and renovation as some lawns fail to respond satisfactorily to chemical treatments.

Figure 9.9. Residential lawn.

Turfgrass Selection

Virtually any turfgrass can be used for lawns. In most sections of the United States Kentucky bluegrass, tall fescue, perennial ryegrass, bermudagrass, St. Augustinegrass, and centipedegrass are commonly used within their respective areas of adaptation. These and other turfgrasses can be found in many other parts of the world but lawns tend to be smaller.

Propagation

All propagation methods are used for lawn establishment. Since the 1960s, however, sodding has become an increasingly popular method for planting lawns in the United States and Canada. This reflects not only the affluence of these societies, but also the rapid expansion of the sod-producing industry.

Primary Culture

Lawns composed of Kentucky bluegrass, alone or in mixtures with fine fescues and perennial ryegrasses, are typically mowed at heights of 1 1/2 to 2 inches and at a frequency of once per week. Where tall fescue, centipedegrass, bahiagrass, or St. Augustinegrass is planted, the mowing height may be the same or slightly higher. Bermudagrass and zoysiagrass lawns are usually mowed at heights between 1/2 and 1 inch.

Optimum fertilization rates vary from 2 to 6 or more lbs of nitrogen per 1000 ft^2 per year depending on turfgrass species, location, and other cultural practices.

Irrigation varies from none to approximately twice per week during droughty periods.

Supplementary Culture

The principal cultivation practice used on lawns is vertical mowing for thatch control. Some core cultivation is practiced on large sites, but not to the extent to which it is employed on sports turfs. The opportunities for improving many severely compacted or thatch-prone lawns by core cultivation suggest that this practice may be more widely used in the future.

Pest Management

Preemergence herbicides have become a standard treatment for preventing the emergence of crabgrass and other weeds in lawns. Postemergence herbicides are increasingly available in small packages for homeowners, for both liquid and granular application.

Insecticides are widely used for controlling white grubs, sod webworms, chinchbugs, and other lawn insects.

Fungicides, while not used extensively for lawn disease control, are being more widely used for lawns affected by *Helminthosporium* melting-out, summer patch, red thread, and other diseases.

Much of the increased use of pesticides on lawns in the United States is due to the growth of the commercial lawncare industry.

Utility Turfs

Utility turfs are established principally to stabilize soils along roadsides, airport runways, ski slopes, and other minimum-use areas (Figure 9.10). They represent the lowest intensity of culture used in turfgrass management.

Turfgrass Selection

Turfgrasses used for establishing utility turfs include mixtures of Kentucky bluegrass, fine fescues, and ryegrasses in temperate and subarctic climates; buffalograss, wheatgrasses, and blue gramagrass in semiarid climates; and common bermudagrass and bahiagrass in subtropical and tropical climates. In addition, tall fescue is widely used within a broad transition zone between temperate and subtropical climates.

Figure 9.10. Utility turf along roadside.

Propagation

The principal concern in the culture of utility turfs is in their establishment. Site preparation can usually be performed efficiently and economically over vast areas. Planting is usually by seed. Hydroseeding is widely used, especially on steeply sloping sites, because of the ease with which seed, fertilizer, and other materials can be applied in a single operation. Sprigs of some turfgrass species can also be applied with a hydroseeder. Some sodding of bahiagrass, bermudagrass, and Kentucky bluegrass is done on steep slopes where erosion potential is high. The sod strips must be staked until sufficient rooting has taken place for stability.

Mulching is very important for stabilizing the seedbed and promoting germination and seedling survival. Successful turf establishment usually depends on natural rainfall, as irrigation is either not possible or extremely limited.

Primary Culture

Except at planting, fertilization of utility turfs is minimal. One pound of nitrogen per 1000 ft^2 applied every two years is usually the most a utility turf will receive. On many sites little or no fertilizer is applied once the turf has been established.

Where mowing is practiced the mowing height is usually 3 to 6 inches. Mowing frequency averages between two and four times per year.

Utility turfs are almost never irrigated.

Supplementary Culture

Chemical growth retardants are used on some sites to reduce mowing requirements or control seed head formation. Currently available materials include maleic hydrazide (MH), chlorflurenol (Maintain CF-125), and mefluidide (Embark), amidichlor (Limit), imidazolinone (Event), metsulfuron (Escort), and chlorsulfuron (Telar). Results vary with the rate and timing of application and turfgrass species.

Pest Management

Herbicides may be used for controlling various broadleaf weeds in utility turfs. The most widely used material is 2,4-D; however, mecoprop or dicamba may be added to extend the spectrum of control. Care should be exercised to ensure that spray drift of the chemicals does not damage sensitive crops located nearby.

Other pesticides are rarely used except under special circumstances.

Other Practices

A utility "turf" need not be formed from turfgrasses. Some unaltered stands of native prairie vegetation can be found in the midwestern United States that provide an interesting study for the enthusiast. Attempts to reestablish native prairies should be encouraged or, at least, respected. Various ground covers, low-growing ornamental shrubs, and wildflowers are used in place of turfgrass species along roadsides to provide diversity and enhance the beauty of the landscape.

QUESTIONS

1. What is meant by the **cultural intensity spectrum** in turfgrass management?
2. What are the **growth factors** of importance in turfgrass management?
3. Explain the **limiting factor concept.**
4. Design cultural programs for **greens** in Chicago and Tampa.
5. In what important ways do cultural programs for **tees** and **fairways** differ from greens?
6. Design cultural programs for **football fields** in Detroit and Atlanta.
7. Design cultural programs for **lawns** in Toronto, Pittsburgh, Raleigh, St. Louis, Los Angeles, Austin, Birmingham, and Miami.
8. Design a cultural program for **roadside turfs** in Minneapolis, Lincoln, Tulsa, and Orlando.

References

App, B.A., and Kerr, S.H. "Harmful Insects." In *Turfgrass Science,* edited by A.A. Hanson and F.V. Juska. *Agronomy* 14:336–59. Madison, WI: American Society of Agronomy, 1969.

Bavor, L.D. *Soil Physics,* 3rd ed. New York: John Wiley & Sons, Inc., 1956, 489 pages.

Beard, J.B. *Turfgrass: Science and Culture.* Englewood Cliffs, NJ: Prentice-Hall, Inc., 1973, 658 pages.

Beard, J.B. *Turf Management for Golf Courses.* (For the United States Golf Association.) Minneapolis MN: Burgess Publishing Co., 1982, 642 pages.

—, and Rieke, P.E. "Producing Quality Sod." In *Turfgrass Science,* edited by A.A. Hanson and F.V. Juska. *Agronomy* 14:442–61. Madison, WI: American Society of Agronomy, 1969.

—, —, Turgeon, A.J., and Vargas, J.M., Jr. "Annual Bluegrass (*Poa annua* **L.**), Description, Adaptation, Culture, and Control." Research Report 352. East Lansing, MI: Michigan State University, Agricultural Experiment Station, 1978, 32 pages.

Benson, D.O. "Mowing Grass with a Reel-type Mower." Unpublished report. Minneapolis, MN: Toro Manufacturing Corp., 1962, 25 pages.

Britton, M.P. "Turfgrass Disease." In *Turfgrass Science,* edited by A.A. Hanson and F.V. Juska. *Agronomy* 14:288–335. Madison, WI: American Society of Agronomy, 1969.

Carrow, R.N and A.M. Petrovic. "Effects of Traffic on Turfgrass." In *Turfgrass,* edited by D.V. Waddington, R.N. Carrow, and R.C. Shearman. *Agronomy* 32:285-330. Madison, WI: American Society of Agronomy, 1992.

Collings, G.H. *Commercial Fertilizers,* 5th ed. New York: McGraw-Hill Book Co., 1955, 617 pages.

Couch, H.B. *Diseases of Turfgrasses,* 2nd ed. Huntington, NY: Robert E. Krieger Publishing Co., 1973, 348 pages.

Danneberger, K.T. *Turfgrass Ecology & Management.* Cleveland, OH: G.I.E, Inc., 1993, 201 pages.

Donahue, R.L., Miller, R.W., and Shickluna, J.C. *Soils, an Introduction to Soils and Plant Growth,* 4th ed. Englewood Cliffs, NJ: Prentice-Hall, Inc., 1977, 626 pages.

Duble, R.L. *Southern Turfgrasses: Their Management and Use.* College Station, TX: TexScape, Inc., 1989, 335 pages.

Engel, R.E., and Ilnicki, R. D. "Turf Weeds and Their Control." In *Turfgrass Science,* edited by A.A. Hanson and F.V. Juska. *Agronomy* 14:240–87. Madison, WI: American Society of Agronomy, 1969.

Etter, A.G. "How Kentucky Bluegrass Grows." *Annual of Missouri Botanical Gardens* 38:293–375, 1951.

Ferguson, M.H. "Putting Greens." In *Turfgrass Science,* edited by A.A. Hanson and F.V. Juska. *Agronomy* 14:562–83. Madison, WI: American Society of Agronomy, 1969.

Fick, G.W., and Luckow, M.A. "What We Need to Know about Scientific Names: An Example with White Clover." *Journal of Agronomic Education* 20:141–47, 1991.

Gould, F.W. *Grass Systematics.* New York: McGraw-Hill Book Co., 1968, 382 pages.

Hall, D.W., McCarthy, L.B., and Murphy, T.R. "Weed Taxonomy." In *Turf Weeds and Their Control,* edited by A.J. Turgeon. Madison, WI: American Society of Agronomy, 1994, 1–28.

Hanson, A.A., Juska, F.V., and Burton, G.W. "Species and Varieties." In *Turfgrass Science,* edited by A.A. Hanson and F.V. Juska. *Agronomy* 14:370–409. Madison, WI: American Society of Agronomy, 1969.

Harivandi, M.A., Butler, J.D., and Wu, L. "Salinity and Turfgrass Culture." In *Turfgrass,* edited by D.V. Waddington, R.N. Carrow, and R.C. Shearman. *Agronomy* 32:208–31. Madison, WI: American Society of Agronomy, 1992.

Harper, J.C., II. "Athletic Fields." In *Turfgrass Science,* edited by A.A. Hanson and F.V. Juska. *Agronomy* 14:542–61. Madison, WI: American Society of Agronomy, 1969.

Heald, C.M., and Perry, V.G. "Nematodes and Other Pests." In *Turfgrass Science,* edited by A.A. Hanson and F.V. Juska. *Agronomy* 14:360–69. Madison, WI: American Society of Agronomy, 1969.

Hottenstein, W.L. "Highway Roadsides." In *Turfgrass Science,* edited by A.A. Hanson and F.V. Juska. *Agronomy* 14:603–37. Madison, WI: American Society of Agronomy, 1969.

Hull, R.J. "Energy Relations and Carbohydrate Partitioning in Turfgrasses." In *Turfgrass,* edited by D.V. Waddington, R.N. Carrow, and R.C. Shearman. *Agronomy* 32:175–207. Madison, WI: American Society of Agronomy, 1992.

Langer, R.M.H. "How Grasses Grow." *Studies in Biology No. 34.* London: Edward Arnold Ltd., 1972, 60 pages.

Latting, J. "Differentiation in the Grass Inflorescence." In *The Biology and Utilization of Grasses,* edited by V.B. Youngner and C.M. McKell. New York: Academic Press, Inc., 1972, 366–99.

Little, V.A. *General and Applied Entomology,* 2nd ed. New York: Harper & Row Publishers, Inc., 1957, 543 pages.

Madison, J.H. *Principles of Turfgrass Culture.* New York: Van Nostrand Reinhold Co., 1971, 420 pages.

—. *Practical Turfgrass Management.* New York: Van Nostrand Reinhold Co., 1971, 466 pages.

Marsh, A.W. "Soil Water—Irrigation and Drainage." In *Turfgrass Science,* edited by A.A. Hanson and F.V. Juska. *Agronomy* 14:151–86. Madison, WI: American Society of Agronomy, 1969.

McCarthy, L.B., and Murphy, T.R. "Control of Turfgrass Weeds." In *Turf Weeds and Their Control,* edited by A.J. Turgeon. Madison, WI: American Society of Agronomy, 1994, 209–45.

Metcalf, C.L., Flint, W.P., and R.L. Metcalf. *Destructive and Useful Insects.* New York: McGraw-Hill Book Co., 1962, 1087 pages.

Moore, R., W.D. Clark, and K.R. Stern. *Botany.* Dubuque, IA: Wm. C. Brown Publishers, 1995, 824 pages.

Musser, H.B. *Turf Management.* (For the United States Golf Association.) New York: McGraw-Hill Book Co., 1962, 356 pages.

—, and Perkins, A.T. "Guide to Seedbed Preparation, Guide to Planting." In *Turfgrass Science,* edited by A.A. Hanson and F.V. Juska. *Agronomy* 14:462–90. Madison, WI: American Society of Agronomy, 1969.

Neimczyk, H.D. "Destructive Turf Insects." Wooster, OH: Harry D. Niemczyk, 1981, 48 pages.

Richards, L.A., ed. "Diagnosis and Improvement of Saline and Alkali Soils," *USDA Agricultural Handbook No. 60,* 1954, 160 pages.

Rudall, P. *Anatomy of Flowering Plants: An Introduction to Structure and Development,* 2nd ed. Cambridge, Great Britain: Cambridge University Press, 1992, 110 pages.

Sacks, R.M. "Inflorescence Induction and Initiation." In *The Biology and Utilization of Grasses,* edited by V.B. Youngner and C.M. McKell. New York: Academic Press, Inc., 1972, pp. 351–65.

Shurtleff, M.C., and R. Randell. *How to Control Lawn Diseases and Pests.* Kansas City, MO: Intertec Publishing Corp., 1974, 97 pages.

Smiley, R.M., Dernoeden, P.H., and Clarke, B.B. *Compendium of Turfgrass Diseases,* 2nd ed. St. Paul, MN: American Phytopathological Society, 1992, 98 pages.

Smith, D. "Carbohydrate Reserves of Grasses." In *The Biology and Utilization of Grasses,* edited by V.B. Youngner and C.M. McKell. New York: Academic Press, Inc., 1972, pp. 318–34.

Smith, J.D., Jackson, N., and Woolhouse, A.R. *Fungal Diseases of Amenity Turf Grasses,* 3rd ed. New York: E. & F.N. Spon, 1989, 401 pages.

Taylor, S.A. and Ashcroft, G.L. *Physical Edaphology.* San Francisco, CA: W.H. Freeman and Co., 1972, 533 pages.

Trewartha, G.T. *An Introduction to Climate.* New York: McGraw-Hill Book Co., 1968, 408 pages.

Troughton, A. "The Underground Organs of Herbage Grasses," *Bulletin No. 44.* Hurley, Berkshire, England: Commonwealth Bureau of Pastures and Field Crops, 1957, 163 pages.

Vargas, J.M. *Management of Turfgrass Diseases,* 2nd ed. Ann Arbor, MI: Lewis Publishers, 1994, 294 pages.

Waddington, D.V. "Soil and Soil-Related Problems." In *Turfgrass Science,* edited by A.A. Hanson and F.V. Juska. *Agronomy* 14:80–129. Madison, WI: American Society of Agronomy, 1969.

Walker, J.C. *Plant Pathology,* 2nd ed. New York: McGraw-Hill Book Co., 1969, 707 pages.

Watschke, T.L., Dernoeden, P.H., and Sheltar, D.J. *Managing Turfgrass Pests.* Ann Arbor, MI: Lewis Publishers, 1994, 361 pages.

Watschke, T.L., Prinster, M.G., and Breuninger, J.M. "Plant Growth Regulators and Turfgrass Management." In *Turfgrass,* edited by D.V. Waddington, R.N. Carrow, and R.C. Shearman. *Agronomy* 32:558–83. Madison, WI: American Society of Agronomy, 1992.

Wilson, C.G., and Latham, J.M., Jr. "Golf Fairways, Tees, and Roughs." In *Turfgrass Science,* edited by A.A. Hanson and F.V. Juska. *Agronomy* 14:584–602. Madison, WI: American Society of Agronomy, 1969.

Wise, L.N. *The Lawn Book.* State College, MS: W.R. Thompson, 1961, 250 pages.

Youngner, V.B. "Physiology of Growth and Development." In *Turfgrass Science,* edited by A.A. Hanson and F.V. Juska. *Agronomy* 14:187–216. Madison, WI: American Society of Agronomy, 1969.

Glossary*

Abaxial—Located on the side away from the axis (lower side of leaf blade).

Acid soil—Soil whose reaction is below pH 7.

Adaxial—Located on the side toward the axis (upper side of leaf blade).

Adventitious root—A root that arises from any organ other than primary or seminal roots.

Aeration, mechanical—See Cultivation.

Aerify—See Cultivation.

Alkaline soil—Soil whose reaction is above pH 7.

Amendment, physical—Any substance added to the soil for the purpose of altering physical conditions (sand, calcined clay, peat, and others).

Annual, summer—Plant that completes its life cycle from seed in one growing season.

Annual, winter—Plant that initiates growth in the fall, lives over winter, and dies after producing seed the following season.

Anther—The pollen-bearing part of a stamen.

Anthesis—The period during which the flower is open and functional.

Apical meristem—Terminal growing point.

Apomixis—Reproduction by the formation of viable embryos without the union of male and female gametes.

Artificial turf—A fabricated rug of fibers simulating a turf.

Athlete field—A grounds devoted to athletic and recreational activities; usually outdoors and turfed.

Auricle—Clawlike appendages occurring in pairs at the base of the leaf blade or at the apex of the leaf sheath.

Awn—Hairlike projection usually extending from the midnerve of grass florets.

Axil—Upper angle formed between a leaf (or spikelet) and the stem axis.

Axillary bud—Vegetative bud growth arising from the junction of leaf and stem.

Ball mark—A depression and/or tear in the surface of a turf, usually a putting green, made by the impact of a ball.

Ball roll—The distance a ball moves (a) after striking the ground upon termination of its aerial flight, (b) as the result of a putting stroke, or (c) as a result of hand-imparted motion, as in lawn bowling.

Bed knife—The fixed blade of a reel mower against which the rotating reel blades scissor a shearing cut. The bed knife is carried in the mower frame at a fixed distance from the reel axis and a fixed distance above the plane of travel.

Bench setting—The height at which the bed knife of a mower is set above a firm, level surface.

Biomass—Total quantity of living and dead organisms occurring within a defined volume.

Blade—The flattened portion of the leaf located above the sheath.

Blend—A combination of two or more cultivars of a single turfgrass species.

Bract—Leaflike structure subtending a flower or occurring as scales of a vegetative bud.

Brush—To move a brush against the surface of a turf in order to lift nonvertical stolons and/or leaves before mowing, with the goal of producing a uniform surface of erect leaves.

Bulliform cells—Large, thin-walled, highly vacuolated, transparent epidermal cells present in the intercostal zones of leaf blades.

Bunch-type growth—Plant development by intravaginal tillering at or near the soil surface without production of rhizomes or stolons.

Calcined clay—Clay minerals, such as montmorillonite and attapulgite, that have been fired at high temperatures to obtain absorbent, stable, granular particles; used as amendments in soil modification.

Carbohydrate—Compound of carbon, hydrogen, and oxygen, as in sugar, starch, and cellulose.

Cart path—A roadway constructed of macadam, fine gravel, quarry dust, wood products, or other suitable materials to facilitate golf cart travel with minimum turf injury.

Caryopsis—Dry, indehiscent, one-seeded fruit with a thin pericarp fused to the seed coat.

Castings, earthworm (wormcasts)—Soil and plant remains excreted by earthworms and deposited on the turf surface or in the burrow; form a relatively stable soil granule that can be objectionable on closely mowed turf.

Cleavage plane sod—A zone of potential separation at the interface between underlying soil and an upper soil layer adhering to transplanted sod. Separation at this cleavage plane is most commonly a problem when soils of different physical textures are placed one over the other.

Cleistogamy—Pollination and fertilization within closed florets.

Clippings—Leaves and, in some cases, stems cut off by mowing.

Cold water insoluble nitrogen (CWIN)—Fertilizer nitrogen not soluble in cold water (25°C).

Cold water soluble nitrogen (CWSN)—Fertilizer nitrogen soluble in cold water (25°C).

Coleoptile—Protective sheath of an embryonic shoot.

Coleorhiza—Embryonic sheath of a monocot that protects the emerging primary root.

Collar—Light-colored band at the junction of the blade and sheath on the abaxial side of the leaf.

Colorant—A paintlike material, usually a dye or pigment, applies to (a) brown warm-season turfgrasses that are in winter dormancy; or (b) brown cool-season turfgrasses that are in summer dormancy; or (c) turfs that have been discolored by environmental stress, turfgrass pests, or the abuses of people. Its purpose is to maintain a favorable green appearance of the turf.

Comb—To use a comb with metal teeth or flexible tines fastened immediately in front of a reel mower for the purpose of lifting stolons and procumbent shoots so they may be cut by the mower.

Cool-season turfgrass—Turfgrass species adapted to favorable growth during cool portions (60°–75°F) of the growing season; may become dormant or injured during hot weather. Includes species of the festucoid subfamily.

Coring—A method of turf cultivation by which soil cores are removed using hollow tines or spoons.

Creeping growth habit—Plant development by extravaginal stem growth at or near the soil surface with lateral spreading by rhizomes and/or stolons.

Crown—A highly compressed stem located at the base of a vegetative aerial shoot.

Culm—Flowering stem of a grass plant.

Cultipacker seeder—A mechanical seeder designed to place turfgrass seeds in a prepared seedbed at a shallow soil depth, followed by firming of the soil around the seed. It usually consists of a pull-type or rear-mounted unit having a seed box positioned between the larger front, ridged roller and an offset, smaller rear roller.

Cultivar—An assemblage of cultivated plants distinguished by any characters (morphological, physiological, cytological, and the like) that, when reproduced sexually or asexually, retain their distinguishing features.

Cultivation—Applied to turf, cultivation refers to working of the soil and/or thatch without destruction of the turf; for example, coring, slicing, spiking, or other means.

Cup cutter—A hollow cylinder with a sharpened lower edge used to cut the hole or cup in a green for putting or to replace small spots of damaged sod, including a major portion of the roots and associated soil.

Cutting height—On a mower, the distance between the plane of travel and the parallel plane of cut.

Denitrification—Biological reduction of nitrate or nitrite to gaseous N (N_2,N_2O).

Dethatch—To remove an excessive thatch accumulation, usually by a mechanical practice such as vertical mowing.

Dicot—Plant having two cotyledons in the seed, as in broadleaf species.

Disarticulation—Separation at the joints or nodes at maturity.

Divot—A small opening in the turf from which the sod has been removed as a result of being struck with a golf club or twisted by a cleated shoe.

Dormant seeding—Planting seed during late fall or early winter after temperatures are too low for seed germination to occur until the following spring.

Dormant sodding—Transplanting sod during late fall or early winter after temperatures are too low for shoot growth and rapid rooting.

Dormant turfgrass—Turfgrasses that have temporarily ceased shoot growth as a result of extended drought, heat, or cold stress, but which are capable of reinitiating shoot growth from buds on the crown meristem when environmental conditions are favorable.

Ecosystem, turfgrass—An integration of a turfgrass community and its surrounding environment.

Edaphic—Pertaining to the influence of soil and other media on plant growth.

Epiblast—Small flat of tissue on the side of a grass embryo opposite the scutellum. Found in festucoid and eragrostoid embryos; absent in panicoid embryos.

Epiphytotic—A sudden or abnormally destructive outbreak of plant disease.

Evapotranspiration—Total loss of moisture through the processes of evaporation and transpiration.

Extravaginal—Shoot growth that occurs through an enclosing leaf sheath; opposite of intravaginal.

Fertigation—The application of fertilizer through an irrigation system.

Fertilizer burn—See Foliar burn.

Field capacity—The amount of moisture remaining in the soil after gravitational moisture has drained.

Flail mower—A mower that cuts turf by high-speed impact of inverted T-blades rotating in a vertical cutting plane relative to the turf surface.

Floret—A grass flower enclosed by a lemma and palea.

Foliar burn—Injury to shoot tissue caused by dehydration due to contact with high concentrations of chemicals, for example, certain fertilizers and pesticides.

Footprinting, frost—Discolored foot-shaped areas of dead leaf tissue created by walking on live, frosted turfgrass leaves.

Footprinting, wilt—Temporary foot-shaped impressions left on a turf because the flaccid leaves of grass plants suffering incipient wilt or wilt have insufficient turgor to spring back after treading.

Glabrous—Without hairs.

Glumes—A pair of bracts usually present at the base of a spikelet.

Grade—To establish elevations and contours prior to planting.

Grain—The undesirable, procumbently oriented growth of grass leaves, shoots, and stolons on greens; a rolling ball tends to be deflected from a true course in the direction of orientation.

Grass—Any plant of the family poaceae.

Herbicide—A pesticide used for controlling weeds.

Hot water insoluble nitrogen (HWIN)—Fertilizer nitrogen not soluble in hot water (100°C). Used to determine activity index of ureaforms. See also Nitrogen activity index.

Humus—The organic fraction of soil in which decomposition is so far advanced that its original form cannot be distinguished.

Hybrid—Product of a cross between individuals of unlike genetic constitution.

Hydroplant—To plant propagules (for example, stolons) in a water mixture by pumping through a nozzle that sprays the mixture onto the plant bed. The water-propagule mixture may also contain addends such as fertilizer and certain mulches.

Hydroseed—To seed in a water mixture by pumping through a nozzle that sprays the mixture onto a seedbed. The water mixture may also contain addends such as fertilizer and certain mulches.

Inflorescence—The flowering portion of a shoot; includes the spikelets and the supporting axis or branch system.

Insolation—Incoming solar radiation received by the earth.

Intercalary meristem—Region of dividing cells located at the base of the internodes in young grass shoots.

Intercostal—Areas of leaf tissue located between vascular bundles.

Internode—Portion of the stem between two successive nodes.

Interseed—To seed between plugs, sod strips, rows, or sprigs, or into turf to improve a stand or to alter its composition.

Intravaginal—Shoot growth that occurs within an enclosing leaf sheath; opposite of extravaginal.

Irrigation, automatic—Hydraulic-electric control of irrigation in response to a transducer that senses plant irrigation need. The term is commonly used more loosely to refer to hydraulic-electric control of irrigation by a manually preset program that is time based.

Irrigation, manual—Irrigation using hand-set and hand-valved equipment.

Irrigation, semiautomatic—Irrigation accomplished by direct response of valves to a manually operated remote control switch.

Irrigation, subsurface—Application of water below the soil surface by injection or by manipulation of the water table.

Keel—Prominent ridge on the abaxial side of a leaf blade or floral bract.

Lamina—Leaf blade; flattened portion of a grass leaf.

Lap, mower (backlap)—To run the reel of a mower backwards while applying grinding compound dispersed in a water-surfactant mixture or in oil between the reel and bedknife. Lapping mates the reel and bedknife to a precise fit for quality mowing.

Lateral shoot—A shoot originating from a vegetative bud in the axil of a leaf or from the node of a stem, rhizome, or stolon.

Layering, soil—Undesirable stratification within the A horizon of a soil profile; can be due to construction design, topdressing with different textured materials, inadequate onsite mixing of soil constituents, or blowing and washing of sand to other soil constituents.

Leaf area index (LAI)—Ratio of leaf area (one side) to ground surface.

Lemma—Lowermost of the two bracts enclosing the flower of a grass floret.

Ligule—Membranous or hairy appendage on the adaxial side of a grass leaf at the junction of the leaf and blade.

Liquid fertilization—A method of fluid fertilization by which dissolved fertilizer is applied as a solution.

Localized dry spot—A dry spot of sod amid normal, moist turf that resists rewetting by normal irrigation and rainfall; associated with a number of factors, including thatch, fungal activity, shallow soil over buried material, or elevated sites in the terrain.

Lodicule—Small, scalelike structures, usually two to three in number, at the base of the stamens in grass flowers.

Low-temperature discoloration—The loss of chlorophyll and associated green color that occurs in turfgrasses under low-temperature stress.

Marcelling—A wavy or washboard pattern on the surface cutting plane of mowed turfgrass; usually results where the clip of the reel exceeds the mowing height.

Mat—A tightly intermingled layer, composed of living and partially decomposed stem and root material and soil from topdressing or other sources, that develops between the zone of green vegetation and the soil surface.

Matting—A practice commonly associated with topdressing of greens in which steel doormatting is dragged over the turf surface to work in topdressing and smooth the surface. Also used for breaking up and working in soil cores from cultivation.

Meristem—Undifferentiated tissue with cells capable of division at the tip of a stem or root (apical) or at the base of a leaf (intercalary).

Mesocotyl—Internode between the scutellar node and the coleoptile in the embryo and seedling of a grass plant.

Metabolism—Complex of physical and chemical processes involved in the maintenance of life.

Microorganizm—Minute living organisms such as bacteria, fungi, or protozoa.

Mixture—A combination of two or more species.

Mole—(1) The weight of a substance in grams that is equivalent to the molecular weight of the substance. (2) A large-animal pest whose burrowing activity results in ridges of dislodged turf and the deposition of soil mounds onto the turf.

Monocot—Plant having one cotyledon in the seed; grasses are an example.

Mowing frequency—The number of times a turfgrass community is mowed per week, month, or growing season. (The reciprocal of mowing frequency is mowing interval—the number of days, weeks, or the like between successive mowings.)

Mowing height—The distance above the ground surface at which the turfgrass is cut during mowing.

Mowing pattern—The patterns of back-and-forth travel while mowing turf. Patterns may be changed regularly to distribute wear and compaction, to avoid creating "grain," and to create visually esthetic effects, especially for spectator sports.

Mulch—Any nonliving material that forms a covering on the soil surface.

Mulch blower—A machine that uses forced air to distribute particles of mulch over newly seeded turf. Developed for highway use, the machine accepts baled straw or hay, shreds it, and blows it onto the seedbed.

Nerve—A simple vein or slender rib of a leaf or bract.

Nitrification—Formation of nitrates and nitrites from ammonia by soil microorganisms.

Nitrogen activity index (AI)—Applied to urea formaldehyde compounds and mixtures containing such compounds, the AI is the percentage of cold water insoluble nitrogen that is soluble in hot water.

$$\mathrm{AI} = \frac{\%\mathrm{CWIN} - \%\mathrm{HWIN}}{\%\mathrm{CWIN}} \times 100$$

Node—The joint of a stem; the region of attachment of leaves to a stem.

Nursegrass—See Temporary grass.

Nursery, turfgrass—A place where turfgrasses are vegetatively propagated for increase and planting as stolons or sprigs, or where sod is grown for later transplanting by sodding or plugging. Sometimes also used for experimentation.

Off-site mixing—The mixing of soil and amendments during root-zone modification at a place other than the site where they are to be used.

Overseed—To seed onto an existing turf, usually with temporary cool-season turfgrasses, to provide green, active grass growth during dormancy of the original turf, usually a warm-season turfgrass.

Ovule—An immature seed contained within the ovary.

Palea—Uppermost of the two bracts enclosing the flower of a grass floret.

Panicle—Type of inflorescence in which the spikelets are not directly attached to the main axis.

Pedicel—The stalk of a single grass floret.

Pegging sod—The use of pegs to hold sod in place on slopes and waterways until transplant rooting occurs.

Pericarp—The fruit wall developed from the ovary wall.

pH, soil—A numerical measure of the acidity or hydrogen ion activity of a soil. A pH of 7 indicates neutrality; above 7 is basic (alkaline), while below 7 is acidic.

Phloem—The principal food-conducting elements of vascular plants.

Photoperiod—Period of a plant's daily exposure to light.

Photosynthesis—Process by which carbohydrates are produced from carbon dioxide, water, and light energy in chlorophyll-containing plants.

Phytomere—Basic unit of structure of the grass shoot; composed of an internode, a leaf and portion of a node at the upper end, and a vegetative bud and portion of a node at the lower end.

Pistil—Female structures of a flower; usually consisting of an ovary and one or more stigmas and styles.

Plug—To propagate turfgrasses vegetatively by means of plugs or small pieces of sod. A method of establishing vegetatively propagated turfgrasses as well as repairing damaged areas.

Pole—To remove dew and exudations from a turf by switching a long (bamboo) pole in an arc while it is in contact with the turf surface. The pole is also used to break up clumps of clippings and earthworm casts. The practice is usually confined to greens.

Pregerminated seed—The partial germination of seed prior to planting, by placing in a moist, oxygenated environment at favorable temperatures.

Prophyll—First leaf (bladeless) of a lateral shoot.

Puffiness—Descriptive of the condition where stolons of grass form a loose tangle on the surface. The opposite of tightly knit turf.

Pure Live seed (PLS)—Percentage of the content of a seed lot that is pure and viable.

Raceme—Type of inflorescence in which the spikelets are borne on pedicels attached directly to the main axis.

Rachilla—The axis of a grass spikelet.

Rachis—The axis of a spike or raceme.

Radicle—The primary root of the grass embryo.

Recuperative capacity—The capacity of turfgrasses to recover from injury; usually through vegetative growth from axillary buds.

Reel mower—A mower that cuts turf by means of a rotating reel of helical blades that passes across a stationary blade (bed knife) fixed to the mower frame; this action gives a shearing type of cut.

Renovation, turf—Turf improvement involving replanting into existing live and/or dead vegetation.

Resiliency—The capacity of the turf to spring back when balls, shoes, or other objects strike the surface, thus providing a cushioning effect.

Rhizome—An underground, elongated stem (or shoot) with scale leaves and adventitious roots arising from the nodes.

Roller, water ballast—A hollow, cylindrical body, the weight of which can be varied by the amount of water added, used for leveling, smoothing, and firming soil.

Root pruning, trees—Judicious cutting of tree roots to reduce their competition with an associated turf.

Rotary mower—A mower that cuts turf by high-speed impact of a blade rotating in a cutting plane parallel and incident to the turf surface.

Scald—A condition that exists when a turfgrass collapses and turns brown under conditions of intense sunlight, high water temperatures, and standing water, usually of a relatively shallow depth.

Scalp—To remove an excessive quantity of functioning, green leaves at any one mowing; results in a stubbly, brown appearance caused by exposing crowns, stolens, dead leaves, or even bare soil.

Scum—The layer of algae on the soil surface of thinned turfs; drying can produce a somewhat impervious layer that can impair subsequent shoot emergence.

Scutellar node—The node of the embryo axis at which the scutellum is attached.

Scutellum—Embryonic tissue located between the endosperm and the main body of the embryo; considered to be the single cotyledon of the grass seed.

Seed—A ripened ovule.

Seed mat—A fabricated mat with seed (and fertilizer) applied to one side; the mat serves as the vehicle to (a) apply seed and fertilizer, (b) control erosion, and (c) provide a favorable microenvironment for seed germination and establishment.

Semiarid turfgrass—Turfgrasses adapted to growth in semiarid regions without irrigation. Includes buffalograss, bluegrama, and sideoats grama.

Seminal root—The primary root and all other roots arising from the embryonic tissue below the scutellar node.

Settling, soil—A lowering of the soil surface resulting from a decrease in the volume of a soil previously loosened by tillage; occurs naturally and can be accelerated mechanically by tamping, rolling, cultipacking, or watering.

Sheath—The tubular, basal portion of the leaf that encloses the stem.

Shoot density—The relative number of shoots per unit area.

Slicing—A method of turf cultivation in which rotating, flat tines slice intermittently through turf and soil.

Slit trench drain—A narrow trench (usually 5 to 10 cm wide) backfilled to the surface with a porous material such as sand, gravel, or crushed rock; used to intercept surface or lateral subsurface drainage water.

Slowly available fertilizer—Designates a rate of dissolution less than that obtained for completely water soluble fertilizers; may involve compounds that dissolve slowly, materials that must be microbially decomposed, or soluble compounds coated with substances highly impermeable to water. Used interchangeably with delayed release, controlled release, controlled availability, slow acting, and metered release fertilizers.

Sod—Plugs, squares, or strips of turfgrass with the adhering soil; can be used in vegetative planting.

Sod cutter—A device used to sever sods of turf from the ground; with a mechanical sod cutting machine, the length and thickness of the sod cut are subject to operator control.

Sod harvesting—Mechanical cutting of sod, usually at soil depths between 0.6 and 3.8 cm (1/4 and 1 1/2 inches), into pieces ranging from 929 cm^2 (1 ft^2) to 8361 cm^2 (1 yd^2).

Sod heating—Heat accumulation in tightly stacked sod; can reach lethal temperatures if the stacks are held for a sufficient length of time under the right environmental conditions.

Sod production—The culture of turf to a level of maturity and quality that allows harvesting and transplanting.

Sod rooting—The growth of roots originating from nodes in the sod into the underlying soil; knitting has occurred when rooting has secured the sod to the underlying soil.

Sod strength—The relative ability of a sod to remain in its original condition during harvesting, handling, and transplanting without tearing and with minimal stretching.

Sodding—Planting turf by means of sod.

Soil mix—A prepared mixture used as a growth medium for turfgrass.

Soil modification—Alteration of soil characteristics by soil amendment; commonly used to improve physical conditions.

Soil probe—A cylindrical soil sampling tool with a cutting edge at the lower end.

Soil screen—A perforated, woven, or welded mesh used to remove clods, coarse fragments, and trash from soil; may be stationary, oscillating, or in the case of cylindrical screens, rotating.

Soil shredder—A machine that crushes or pulverizes large soil aggregates and clods to facilitate uniform soil mixing and topdressing application.

Soil warming—The artificial heating of turf from below the surface, usually by electrical means, for the purposes of preventing soil freezing and maintaining a green turf.

Spike—Type of inflorescence in which the spikelets are directly sessile (without pedicel) to the main axis.

Spikelet—The basic unit of grass inflorescence, consisting of two glumes and one or more florets.

Spiking—A method of turf cultivation in which solid tines or flat, pointed blades penetrate the turf and soil surface.

Spoon, coring—A method of turf cultivation by which curved, hollow, spoonlike tines remove small soil cores and leave a hole or cavity in the sod.

Sprig—A stolon, rhizome, or tiller used to establish a turf by planting in furrows or small holes.

Sprigging—Vegetative planting by placing stolons, rhizomes, or tillers in furrows or small holes.

Spring green-up—The initial seasonal appearance of green shoots as spring temperature and moisture conditions become favorable for chlorophyll synthesis, thus breaking winter dormancy.

Spud—To remove individual weedy plants by means of a small spadelike tool that severs the root so the weed can be lifted from the turf.

Stamen—The male organ of the flower, consisting of a pollen-bearing anther on a filament.

Stand—The number of established individual shoots per unit area.

Stigma—The feathery portion of the pistil that receives the pollen for fertilization.

Stolon—An elongated stem (or shoot) that grows along the surface of the ground and from which leaves and adventitious roots develop at the nodes.

Stolon nursery—A field used for the production of stolons grown for propagation.

Stolonize—Vegetative planting by broadcasting stolons over a prepared soil and, in most cases, covering by topdressing or press rolling.

Stomates—Openings in the epidermis of leaves and stems that function in the exchange of gases between the atmosphere and the plant.

Strip sodding—The use of strips of sod spaced at intervals, usually across a slope. Depends on the grass spreading to form a complete cover. Sometimes the area between the strips is seeded.

Style—The contracted portion of the pistil between the ovary and the stigma.

Subgrade—The soil elevation established so that the topsoil, root zone mix, concrete, or other material placed on it will have the desired thickness and final grade or elevation.

Summer dormancy—The cessation of growth and subsequent death of shoots of perennial plants due to heat and/or moisture stress.

Syringe—To spray turf with small amounts of water to (a) dissipate accumulated energy in the leaves by evaporating free water rather than by transpiring plant water; (b) prevent or correct a leaf water deficit, particularly wilt; and (c) remove dew, frost, and exudates.

Temporary grasses—Grasses not expected to persist in a turf; they are used as a temporary cover.

Texture, leaf—Texture imparted to turf by leaf width, taper, and arrangement.

Thatch—A layer of undecomposed or partially decomposed organic residues situated above the soil surface and constituting the upper stratum of the medium that supports turfgrass growth.

Thatch control—The process of (a) preventing excessive thatch accumulation by cultural manipulation and/or (b) removing excess thatch from a turf by either mechanical or biological means.

Tiller—A lateral shoot, usually erect, that develops intravaginally from axillary buds.

Tip burn—A whitening of the leaf tip resulting from a lethal internal water stress caused by wind desiccation or salt.

Topdressing—A prepared soil mix added to the surface of a turf and worked in by brooming, matting, raking, and/or irrigating to smooth a green surface, firm a turf by working soil in among stolons and thatch-forming materials, enhance thatch decomposition, and cover stolons or sprigs during vegetative planting. Also the act of applying topdressing materials to turf.

Transitional climatic zone—The suboptimal zone between temperate and subtropical climates.

Turf—A covering of mowed vegetation, usually a turfgrass, growing intimately with an upper soil stratum of intermingled roots and stems.

Turfgrass—A species or cultivar of grass, usually of spreading habit, which is maintained as a mowed turf.

Turfgrass community—An aggregation of individual turfgrass plants that have mutual relationships with the environment as well as among the individual plants.

UF—See Urea formaldehyde.

Urea formaldehyde—A synthetic, slowly soluble nitrogen fertilizer consisting mainly of methylene urea polymers of different lengths and solubilities; formed by reacting urea and formaldehyde.

Variety—See Cultivar.

Vascular bundles—Strands of tissue composed principally of xylem and phloem elements.

Vegetative propagation—Asexual propagation using pieces of vegetation (sprigs or sod pieces).

Verdure—The layer of aboveground, green, living plant tissue remaining after mowing.

Vertical mower—A mechanical device whose vertically rotating blades cut into the face of a turf for the purpose of reducing thatch, grain, and surface compaction.

Warm-season turfgrass—Turfgrass species adapted to favorable growth during warm portions (80°–95°F) of the growing season; includes species of the eragrostoid and panicoid subfamilies.

Wear—The collective injurious effects of traffic on a turf.

Wet wilt—Wilting of turf in the presence of free soil water when evapotranspiration exceeds the ability of roots to take up water.

Wind burn—Death and browning, most commonly occurring on the uppermost leaves of semidormant grasses and caused by atmospheric desiccation.

Winter desiccation—The death of leaves or plants by drying during winter dormancy.

Winter discoloration—See Low-temperature discoloration.

Winter overseeding—Seeding cool-season turfgrasses over warm-season turfgrasses at or near their start of winter dormancy; practiced in subtropical climates to provide green, growing turf during the winter period when the warm-season species are brown and dormant.

Winter protection cover—A mulch placed over a turf to prevent winter desiccation, insulate against low-temperature stress, and stimulate early spring greenup.

Winterkill—Any injury to turfgrass plants that occurs during the winter period.

Xylem—The principal water-conducting elements in vascular plants.

*Turfgrass terms were adapted from the preliminary draft of a glossary of turfgrass terms prepared by the C-5 Turfgrass Terminology Committee of the American Society of Agronomy.

APPENDIX 1

Conversion Tables

Liquid Volume Measures

CUBIC CENTIMETER OR MILLILITER (CC OR ML)	FLUID OUNCE* (FL OZ)	LITER (L)	U.S. GALLON**	U.S. BARREL
1	0.03815	0.001		
29.573	1	0.029573	0.00781	
1000	33.8147	1	0.26418	
3785	128	3.785	1	
			31.5	1

* 1 fl oz = 2 tablespoons (Tbsp) = 6 teaspoons (tsp)
** 1 gallon = 4 quarts = 8 pints = 16 cups

Dry Volume Measures

CUBIC INCHES (IN^3)	LITER (L)	QUART (QT)	CUBIC FEET (FT^3)	BUSHEL* (BU)	CUBIC YARDS (YD^3)
1	0.01639				
	1	0.9081			
67.2	1.1	1	0.039	0.031	
1728	28.32	25.71	1	0.80354	0.037
	35.24	32	1.25	1	0.046
			27	21.7	1

* 1 bu = 4 pecks

Weights

GRAM (G)	OUNCE (OZ)	POUND (LB)	KILOGRAM (KG)
1	0.035274	0.0022046	0.001
28.3495	1	0.0625	0.02835
453.592	16	1	0.45359
1000	35.2740	2.20462	1

Linear Measures

MILLIMETERS (MM)	CENTIMETERS (CM)	INCHES (IN)	FEET (FT)	YARDS (YD)	METERS (M)
1	0.1	0.03937			0.001
10	1	0.3937	0.03281	0.01094	0.01
25.4	2.54	1	0.08333	0.02778	0.254
304.8	30.48	12	1	0.3333	0.3048
914.4	91.44	36	3	1	0.9144
1000	100	39.37	3.2808	1.09361	1

Square Measures

SQUARE CENTIMETERS (CM^2)	SQUARE INCHES (IN^2)	SQUARE FEET (FT^2)	SQUARE YARDS (YD^2)	SQUARE METERS (M^2)	ACRES	HECTARES (HA)
1	0.155	0.00108				
6.45163	1	0.00694				
	144	1	0.11111	0.0929		
	1296	9	1	0.83613		
	1550	10.7639	1.19599	1		
		43560	4840	4046.87	1	0.40469
				10000	2.471	1

Weights Per Unit Area

OZ/YD^2	$OZ/1000\ FT^2$	$LB/1000\ FT^2$	$KG/100\ M^2$	KG/HA	LB/ACRE
1					302.5
0.009	1	0.0625			2.72
0.144	16	1	0.488	48.8	43.56
		2.05	1	100	89
		0.0205	.01	1	0.89
	0.37			1.12	1

APPENDIX 2

Calibration

To apply a pesticide, fertilizer, or other material at the desired rate to turf, the applicator (sprayer, spreader) must be properly calibrated. Applicator calibration is fundamental to controlling the level of cultural resources supplied as part of a turfgrass cultural program.

SPRAYER CALIBRATION

Sprayers are used to apply materials in water. Therefore, there are two objectives in sprayer calibration: measuring and controlling the rate at which water is applied to a known area (that is, spray volume or SV), and determining the correct dilution (D) of the material in water. To determine SV, the sprayer should be operated over an area of known size, and the amount of water applied should be carefully measured. For small sprayers, a test area of 1000 ft^2 (20 by 50 ft) would be suitable. Large, tractor-mounted sprayers may require a larger test area for accurate calibration. The SV can be measured by either of two methods: determining the water loss from the spray tank following spraying, or measuring the amount of water discharged from the sprayer during operation.

To determine SV by the water-loss method, fill the spray tank with a known volume of water and, after spraying, measure the amount of water remaining in the tank. To reduce measurement error, water contained in the hose and boom assemblies of the sprayer should be drained following application and added to the remaining volume of water in the spray tank prior to final measurement. The difference between initial and final water volumes divided by the area of the test site yields SV. For example, to calibrate a 2-gallon-capacity compressed-air sprayer, the following equation can be used:

$$SV = \frac{\overset{\substack{\text{initial volume} \\ \downarrow}}{256 \text{ fl oz}} - \overset{\substack{\text{final volume} \\ \downarrow}}{192 \text{ fl oz}}}{\underset{\substack{\uparrow \\ \text{test area}}}{1000 \text{ ft}^2}} = \frac{64 \text{ fl oz}}{1000 \text{ ft}^2} \tag{a}$$

Normally, the SV of a sprayer is given as gallons per acre (gpa). To convert fl oz/1000 ft^2 to gpa, conversion values from Appendix 1 for fl oz/gallon (128) and ft^2/acre (43,560) must be inserted in an equation as follows:

$$\frac{64 \text{ fl oz}}{1000 \text{ ft}^2} \times \frac{43{,}560 \text{ ft}^2}{1 \text{ acre}} \times \frac{1 \text{ gallon}}{128 \text{ fl oz}} = \frac{(64)(43{,}560) \text{ gallons}}{(1000)(128) \text{ acres}} = 21.78 \text{ gpa} \tag{b}$$

The SV of the sprayer is 21.78 gpa (or 0.5 gallons/1000 ft^2 by removing 43.560 ft^2 acre from the equation).

For calibrating large-capacity sprayers, a simple variation of the above method would be to use a calibrated stake to measure changes in water volume within the spray tank. To calibrate the stake, place 10 gallons of water in the spray tank, insert the stake vertically into the water, and mark the position of the water line. Follow this procedure after each addition of water in 10-gallon increments until the spray tank is completely filled. The calibrated stake is then useful for measuring initial and final water volumes in the spray tank, and it can be substituted for the aforementioned procedure. A test area of sufficient size to receive 20 to 50 gallons of water should be used for calibrating large-capacity sprayers. Obviously, the larger the test site, the more accurate the measurement with the calibrated stake.

The second method of sprayer calibration involves measuring the amount of water applied by the sprayer during operation. This can be accomplished by fixing a collection container onto one nozzle of the sprayer prior to spraying a test area of known size. The amount of water collected during spraying times the number of nozzles, divided by the area of the test site, yields the SV. For example, a 10-nozzle boom spraying 2 gallons of water per nozzle over 21,780 ft^2 (1/2 acre) would apply a total of 20 gallons of water (2 gal $\times$ 10 nozzles = 20 gal total) per 1/2 acre, or 40 gpa. A check of nozzle uniformity should accompany the calibration procedure, as this calculation assumes that each nozzle is spraying at the same rate.

Once a sprayer has been calibrated for water, the proper dilution for the material to be applied can be calculated. For example, dilution of a 4 lb/gal formulation of 2,4-D to supply 1 lb/acre of the herbicide active ingredient (a.i.) would be calculated as follows:

$$\frac{1 \text{ lb a.i.}}{\text{acre}} \times \frac{1 \text{ gal}}{4 \text{ lb a.i.}} = \begin{array}{l}0.25 \text{ gal/acre or } 1 \text{ qt/acre,} \\ \text{or } 32 \text{ fl oz/acre, or } 946 \text{ ml/acre}\end{array} \qquad \text{(c)}$$

Using a sprayer calibrated to deliver 40 gpa, the dilution factor (D) is calculated as follows:

$$\frac{\dfrac{1 \text{ qt}}{\text{acre}}}{\dfrac{40 \text{ gal}}{\text{acre}}} = \frac{1 \text{ qt}}{40 \text{ gal}} = D \qquad \text{(d)}$$

Therefore, 1 quart of the herbicide formulation should be added to each 40 gallons of water in the spray tank. If a 100-gallon tank were filled to capacity, this would require 2.5 quarts of the fomulation per tank:

$$\frac{1 \text{ qt}}{40 \text{ gal}} \times \frac{100 \text{ gal}}{\text{tank}} = 2.5 \text{ qt/tank} \qquad \text{(e)}$$

If a 75% wettable powder formulation (75WP) of the herbicide were used in place of the 4 lb/gal formulation in equation (c), the following calculation would be performed:

$$\frac{1 \text{ lb a.i.}}{\text{acre}} \times \frac{100}{75} = 1.33 \text{ lb} \qquad \text{(f)}$$

Therefore, 1.33 lb/acre of the 75 WP would be required to supply 1 lb a.i./acre. The formulation factor of 100/75 is necessary to convert pounds of active ingredient per acre to pounds of the 75 WP formulation per acre. Recalculation of equations (d) and (e) for the 75 WP formulation would be performed as follows:

$$\frac{\dfrac{1.33 \text{ lb (75 WP)}}{\text{acre}}}{\dfrac{40 \text{ gal}}{\text{acre}}} = \frac{1.33 \text{ lb (75 WP)}}{40 \text{ gal}} \qquad \text{(g)}$$

$$\frac{1.33 \text{ lb (75 WP)}}{40 \text{ gal}} \times \frac{100 \text{ gal}}{\text{tank}} = 3.325 \text{ lb (75 WP)/tank} \qquad \text{(h)}$$

SPREADER CALIBRATION

Spreaders are used to apply dry materials (usually granular formulations) directly to the turf. The procedures for spreader calibration are essentially the same as for sprayers. Since water is not used as the carrier for the material to be applied, there is no calculation or dilution. For example, to apply a 5% granular formulation (5G) of a herbicide to supply 12 lb a.i./acre, use the following calculation:

$$\frac{12 \text{ lb a.i.}}{\text{acre}} \times \frac{100}{5} = \frac{240 \text{ lb (5G)}}{\text{acre}} \tag{i}$$

The formulation factor of 100/5 thus converts pounds (12) of active ingredient per acre to pounds (240) of the 5G formulation per acre. The spreader must then be calibrated to deliver this amount of material per acre. Using a test area of 1000 ft^2, the amount of material required is calculated as follows:

$$\frac{240 \text{ lb (5G)}}{\text{acre}} \times \frac{1 \text{ acre}}{43{,}560 \text{ ft}^2} \times \frac{1000 \text{ ft}^2}{\text{test area}} = \begin{array}{l}5.5 \text{ lb (5G)/test area,} \\ \text{or 88 oz/test area, or 2500 g/test area}\end{array} \tag{j}$$

Given a suitable scale for measuring ounces or grams, a known amount of the 5G formulation should be placed in the spreader prior to application to the test area. Following application, the amount of material remaining should be weighed to obtain the difference between initial weight (IW) and final (FW), as in the following example:

$$700 \text{ g (IW)} - 375 \text{ g (FW)} = 325 \text{ g difference} \tag{k}$$

Since the desired rate was 250 g, or 75 g less than that measured in the first test (325 g), smaller spreader settings should be tried until a difference of 250 g is obtained after covering the 1000-ft^2 test area.

The same calibration procedure can be used for fertilizer applications. For example, to apply 0.75 pound of nitrogen (N) per 1000 ft^2 using an 18-5-9 ($N:P_2O_5:K_2O$) fertilizer, the following calculation should be used:

$$\frac{0.75 \text{ lb N}}{1000 \text{ ft}^2} \times \frac{100}{18} = \frac{4.17 \text{ lb } (18 - 5 - 9)}{1000 \text{ ft}^2} \tag{l}$$

The formulation factor of 100/18 thus converts pounds of actual nitrogen per 1000 ft^2 to pounds of fertilizer (18-5-9) per 1000 ft^2. An appropriate spreader setting can then be determined by adjusting the setting after each application to a test area until a difference of 4.17 lb of fertilizer between IW and FW is obtained.

APPENDIX 3

Common and Chemical Names of Pesticides Used in Turf

Common Name	Chemical Name
Herbicides and PGRs	
amidochlor	*N*-[(acetylamino) methyl]-2-chloro-N-(2,6-diethyl-phenyl) acetamide
amitrole	3-amino-*s*-triazole
asulam	methyl[(4-aminophenyl)sulfonyl]carbamate
atrazine	2-choloro-4-(ethylamino)-6-(isopropylamino)-*s*-triazine
benefin	*N*-butyl-*N*-ethyl-α,α,α,-trifluro-2,6-dinitro-*p*-toluidine
bensulide	*O,O*-diisopropyl phosphorodithioate *S*-ester with *N*-(2-mercaptoethyl) benzenesulfonamide
bentazon	3-isopropyl-1*H*-2,1,3-benzothiadiazin-4(3*H*-one 2,2-dioxide)
bromoxynil	3,5-dibromo-4-hydroxybenzonitrile
chlorflurenol	methyl 2 chloro-9-hydroxyfluorene-9-carboxylic acid
chlorsulfuron	2-chloro-*N*-[(4-methoxy-6-methyl-1,3,5-triazin-2-yl) aminocarbonyl]-benzenesulfonamide
dalapon	2,2-dichloropropionic acid
DCPA	dimethyl tetrachloroterephthalate
dicamba	3,-6-dichloro-*o*-anisic acid
diclofop-methyl	(±)-2-[4-(2,4-dichlorophenoxy)phenoxy]propionic acid
dithiopyr	*S,S*-dimethyl 2-(difluoromethyl)-4-(2-methylpropyl)-6-(trifluoro-methyl)-3,5-pyridinedicarbothioate
DSMA	disodium methanearsonate
EPTC	*S*-ethyl dipropylthiocarbamate
ethofumisate	2-ethoxy-2,3-dihydro-3,3-dimethyl-5-benzofuranyl methane-sulphonate
fenoxaprop-ethyl	(±)-ethyl 2-4-((6-chloro-2-benzoxazolyloxy)-phenoxy)propanoate

Common Name	Chemical Name
Herbicides and PGRs Continued	
fluazifop-P-ethyl	Butyl ®-2-[4-[[5-(trifluoromethyl)-2-pyridinyl]oxy]phenoxy] propanoate
flurprimidol	α-(1-methylethyl)-α-[4-(trifluoro-methoxy) phenyl] 5-pyrimidine-methanol
glufosinate	2-amino-4-(hydroxymethylphosphinly)-monoammonium salt
glyphosate	*N*-(phosphonomethyl)glycine
halosulfuron	Methyl 5-[[(4,6-dimethoxy-2-pyrimidinyl)amino] carbonylaminosulfonyl]-3-chloro-1-methyl-H-pyrazole-4-carboxylate
imazaquin	2-[4,5-dihydro-4-methyl-4-(1-methylethyl)-5-oxo-1H-imidazol-2-yl]-3-quinoline carboxylic acid
isoxaben	*N*-[3-(1-ethyl-1-methylpropyl)-5-isoxazoly]-2,6-dimethoxy-benzamide
maleic hydrazide	1,2-dihydro-3,6-pyridazine-dione
MAMA	monoammonium methanearsonate
mecoprop	2-[4-chloro-*o*-tolyl)oxy]propionic acid
mefluidide	*N*-(2,4-dimethyl-5-[(trifluoromethyl]sulfonyl) amino]phenyl) acetamide
metham	sodium methyl dithiocarbamate
metolachlor	2-chloro-*N*-(2-ethyl-6-methylphenyl)-*N*-(2-methoxy-1-methylethyl)acetamide
metribuzin	4-amino-6-*tert*-butyl-3-(methylthio)-*as*-triazin-5(4*H*)-one
metsulfuron	2-[[[[(4-methoxy-6-methyl-1,3,5-triazin-2-yl) amino] carbonyl]-amino] sulfonyl]-benzoate
MH	1,2-dihydro-3,6-pyridazinedione
MSMA	monosodium methanearsonate
napropamide	2-(α-naphthoxyl)-*N,N*-diethylpropionamide
oryzalin	3,5-dinitro-N^4N^4-dipropylsulfanilamide
oxadiazon	2-*tert*-butyl-4-(2,4 dichloro-5-isopropoxyphenyl)-Δ^2-1,3,4-oxadiazolin-5-one
paclobutrazol	(2*RS*,3*RS*)-1-(4chlorophenyl)-4,4dimethyl-2-1,2,4-triazol-1-yl penta-n-3-ol
paraquat	1,1′-dimethyl-4,4′-bipyridinium ion
pendimethalin	*N*-(1-ethylpropyl)-3,4-dimethyl-2,6-dinitro-benzenamine
pronamide	3,5-dichloro(*N*-1,1-dimethyl-2-propynyl) benzamide
sethoxydim	2[1-ethoxyimino)butyl]-5-(ethyl-thio)propyl]-3-hydroxy-2-cyclohexen-1-one
siduron	1-(2-methylcyclohexyl)-3-phenylurea
simazine	2-chloro-4,6-bis(ethylamino)-*s*-triazine
triclopyr	3,5,6-trichloro-2-pyridinyloxyacetic acid
trinexapac-ethyl	4-(cyclopropyl-α-hydroxy-methylene)-3,5-dioxo-cyclohexane-carboxylic acid ethyl ester
2,4-D	(2,4-dichlorophenoxy)acetic acid

Common Name	Chemical Name
Fungicides	
anilazine	2,4-dichloro-6-*o*-anilino-*s*-traizine
asoxystrobin	methyl (E)-2-{2-[6-(2-cyanophenoxy)pyrimidin-4-yloxy]phenyl}-3-methoxyacrylate
benomyl	methyl 1-(butylcarbamoyl)-2-benzimidazole-carbamate
chloroneb	1,4-dichloro-2,5-dimethoxybenzene
chlorothalonil	2,4,5,6-tetrachioroisophthalonitrile
cyproconazole	2-(4-chlorophenyl)-3-cyclopropyl-1-(1-*H*-1,2,4-triazol-1-yl)butan-2-ol
etridiazol	5-ethoxy-3-trichloromethyl-1,2,4-thiadiazole
fenarimol	2-(2-chlorophenyl)-2-(4-chlorophenyl)-5-pyrimidinemethanol
flutolanil	α,α,α,-trifluoro-3′-isopropoxy-2-toluanilide
fosetyl-Al	aluminum *tris*(O-ethyl phosphonate)
iprodione	3-(3,5-dichlorophenyl)-*N*-(1-methylethyl)-2,4-dioxo-1-imidazolidine carboxamide
mancozeb	coordination product of zinc ion and maneb
maneb	manganous ethylenebisdithiocarbamate
metalaxyl	*N*-(2,6-dimethylphenyl)-*N*-(methoxyacetyl)-alanine methylester
myclobutanil	α-butyl-α-(4-chlorophenyl)-1*H*-1,2,4-triazole-1-propenenitride
oxadixyl	2-methoxy-*N*-(2-oxo-1,3-oxazolidin-3-yl)-acet-2′,6′-xylidine
PCNB	pentachloronitrobenzene
propamocarb	propyl(3-(dimethylamino)propyl)carbamate monohydrochloride
propiconazole	1-[{2-(2,4-dichlorophenyl)-4-propyl-1,3-dioxolan-2-yl}methyl]-1H-1,2,4-triazole
thiophenate-methyl	dimethyl 4,4′-*O*-phenylene bis (3-thioallophanate)
thiram	tetramethylthiuram
triadimefon	1-(4-chlorophenoxy)-3,3-dimethyl-1-(1H-1,2,4-triazol-1-yl)-2-butanone
vinclozolin	3-(3,5-dichlorophenyl)-5-ethanyl-5-methyl-2,4-oxazolidinedione
Insecticides	
acephate	*O,S*-dimethyl acetylphosphoramidothioate
azadirachtin	complex plant extract from the Neem tree
bendiocarb	2,2-dimethyl-1,3-benzodioxol-4-yl-methyl-carbamate
bifenthrin	2-methylbiphenyl-3-ylmethyl(Z)-(1RS)-cis-3-(2-chloro-3,3,3-trifluoroprop-1-enyl)-2,2-dimethylcyclopropanecarboxylate
carbaryl	1-naphthyl *N*-methylcarbamate
chlorpyrifos	*O,O*-diethyl *O*-(3,5,6-trichloro-2-pyridyl)phosphorothioate
cyfluthrin	cyano(4-fluoro-3-phenoxyphenyl)methyl 3-(2,2-dichloroethenyl)-2,2-dimethylcyclopropane-carboxylate
diazinon	O,O,-diethyl O-(2-isopropyl-6-methyl-4-pyrimidinyl)phosphorothioate
dicofol	1,1-bis(*p*-chlorophenyl)-2,2,2-trichloroethanol
ethion	O,O,O′,O′-tetraethyl S,S′-methylene bis(phosphorodithioate)

Common Name	Chemical Name
Insecticides. Continued	
ethoprop	*O*-ethyl S,S-dipropyl phosphorodithioate
fluvalinate	α*RS*,2*R*)-fluvalinate {(*RS*)-α-cyano-2-phenoxybenzyl (*R*)-2-{2-chloro-4-(trifluoromethyl)anilino)-3-methylbutanoate}
fonofos	*O*-ethyl *S*-phenyl ethylphosphonodithioate
halofenozide	4-chloro-2-benzoyl-2-(1,1-dimethylethyl)hydrazide
imidacloprid	1-[(6-chloro-30pyridinyl)methyl]-*N*-nitro-2-imidazolidinimine
insofenphos	1-methylethyl 2-((ethoxy((1-methylethyl)amino)phosphinothioyl)-oxy)benzoate
isazofos	*O*-(5-chloro-1-[methylethyl]-1H-1,2,4-triazol-3-yl)*O,O*-diethyl phosphorothioate
lambda-cyhalothrin	[1α(S*),3α(Z)]-(+)-cyano-(3-phenoxyphenyl)methyl-3-(2-chloro-3,3,3-trifluoro-1-propenyl)2,2-dimethylcyclopropanecarboxylate
lindane	gamma isomers of 1,2,3,4,5,6-hexachloro-cyclohexane
malathion	O,O-dimethyl-S-(1,2-dicarbethoxyethyl)phosphorodithioate
permethrin	(3-phenoxyphenyl)methyl(±)cis-trans(2-2-dichloroethenyl)-2,2-dimethylcyclopropanecarboxylate
pyrethrin	Esters of pyrethrolone and various natural organic acids
trichlorfon	dimethyl (2,2,2-trichloro-1-hydroxyethyl) phosphonate
Nematicides	
ethoprop	O-ethyl *S,*S-dipropyl phosphorodithioate
fenamiphos	ethyl-3-methyl-4-(methyl thio)phenyl (1-methylethyl) phosphoramidate
Fumigants	
chloropicrin	trichloronitromethane
dichloropropane	dichloropropane
metham	(see herbicides)
methyl bromide	methyl bromide

Index

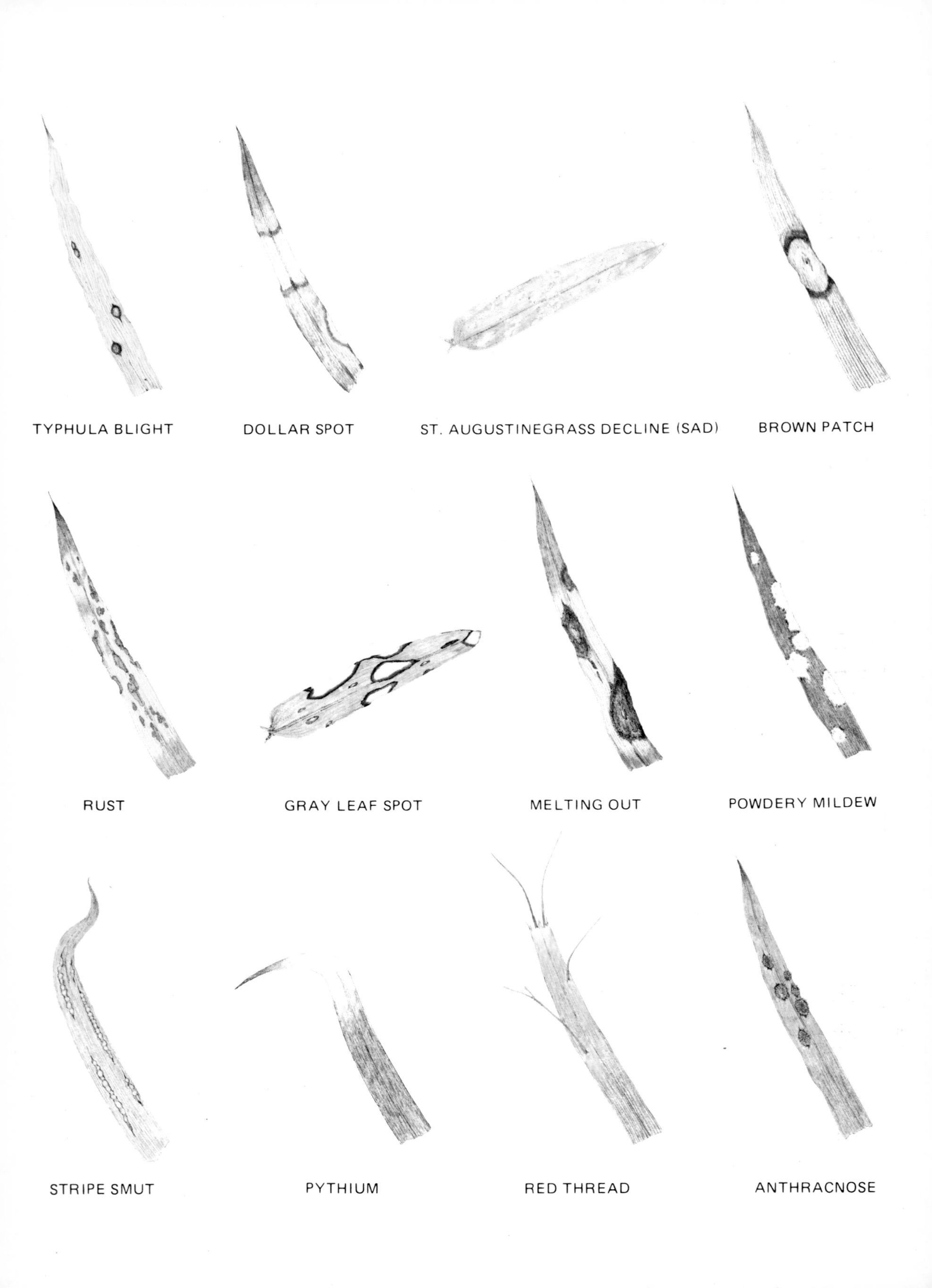
TYPHULA BLIGHT
DOLLAR SPOT
ST. AUGUSTINEGRASS DECLINE (SAD)
BROWN PATCH
RUST
GRAY LEAF SPOT
MELTING OUT
POWDERY MILDEW
STRIPE SMUT
PYTHIUM
RED THREAD
ANTHRACNOSE